CROSS-CULTURAL PERSPECTIVES ON THE (IM)POSSIBILITY OF GLOBAL BIOETHICS

Philosophy and Medicine

VOLUME 71

Founding Co-Editor
Stuart F. Spicker

Editor

H. Tristram Engelhardt, Jr., *Department of Philosophy, Rice University, and Baylor College of Medicine, Houston, Texas*

Associate Editor

Kevin Wm. Wildes, S.J., *Department of Philosophy and Kennedy Institute of Ethics, Georgetown University, Washington, D.C.*

ASIAN STUDIES IN BIOETHICS AND THE PHILOSOPHY OF MEDICINE 2

Series Editor

Ruiping Fan, *Department of Public & Social Administration, City University of Hong Kong, Hong Kong*

Editorial Advisory Board

Kazumasa Hoshino, *Kyoto Women's University, Kyoto, Japan*
Shui Chuen Lee, *National Central University, Chung-li, Taiwan*
Ping-cheung Lo, *Hong Kong Baptist University, Hong Kong*
Ren-Zong Qiu, *Chinese Academy of Social Sciences, Beijing, China*

The titles published in this series are listed at the end of this volume.

CROSS-CULTURAL PERSPECTIVES ON THE (IM)POSSIBILITY OF GLOBAL BIOETHICS

Edited by

JULIA TAO LAI PO-WAH

Associate Professor, Department of Public and Social Administration,
City University of Hong Kong
and
Director, Centre for Comparative Public Management and Social Policy
City University of Hong Kong

KLUWER ACADEMIC PUBLISHERS
DORDRECHT / BOSTON / LONDON

Library of Congress Cataloging-in-Publication Data is available.

ISBN 1-4020-0498-2

Published by Kluwer Academic Publishers,
P.O. Box 17, 3300 AA Dordrecht, The Netherlands

Sold and distributed in North, Central and South America
by Kluwer Academic Publishers,
101 Philip Drive, Norwell, MA 02061, U.S.A.

In all other countries, sold and distributed
by Kluwer Academic Publishers, Distribution Center,
P.O. Box 322, 3300 AH Dordrecht, The Netherlands

Printed on acid-free paper

Printed and bound in Great Britain by MPG Books Ltd., Bodmin, Cornwall.

TABLE OF CONTENTS

PART I / COMMUNITY AND CARE: LOST PERSPECTIVES?

PART II / THE (IM)POSSIBILITY OF GLOBAL BIOETHICS

49416028

PART III / MORAL DILEMMAS IN HEALTH CARE

PART IV / NEW STRATEGIES, NEW POSSIBILITIES

JULIA TAO LAI PO-WAH

PREFACE

The contributions to this volume grew out of papers presented at an international conference *Individual, Community & Society: Bioethics in the Third Millennium*, held in Hong Kong, Special Administrative Region of the People's Republic of China, between 25-28 May 1999. The conference was organized by the Centre for Comparative Public Management and Social Policy, and Ethics in Contemporary China Research Group, in the Faculty of Humanities and Social Sciences at the City University of Hong Kong.

The conference brought together scholars from east and west to investigate the challenges to caring and to traditional moral authorities that would confront bioethics in the third millennium. They explored the implications of moral loss and moral diversity in post-traditional and post-modern societies, and how these would shape the character of medical care and bioethics discourse in the new era. A proceedings volume under the same title of *Individual, Community & Society: Bioethics in the Third Millennium*, was published in May 1999 for the conference meeting.

The present volume is based upon papers selected from the conference collection. They have all been substantially revised on the basis of comments from commentators and reviewers. Together they explore one of the most searching questions in the field of global bioethics, namely: whether global bioethics is possible in the new century. The book assembles voices from China, including Hong Kong and Taiwan, Japan, Germany and the United States, and makes available deep reflections from diverse cultural and philosophical perspectives. The exploration took some of the papers into debating fundamental issues in moral philosophy at the normative and the meta-ethical level. They included issues about the universality of ethics, the meaning and justifiability of ethical claims, the nature of moral reasoning and the very idea of morality. There were also papers which focused on establishing a moral discourse to strike a balance between the challenge to universalize and generalize on the one hand, and the need to preserve entrenched particularistic moral conceptions and individual or group commitments on the other.

In many ways, the book is a rich, cross-cultural dialogue on global bioethics. The contributors have drawn on important insights from east and west, from both traditional and modern resources, to offer interpretations which are highly diverse, original, and inspiring. Together they have raised some fundamental questions about global bioethics. They serve to affirm the importance of nurturing the conditions under which moral diversity can flourish, since diversity is key to sustaining an on-going, fruitful global dialogue on global bioethics.

ACKNOWLEDGEMENTS

I wish to express my sincere appreciation to the Centre for Comparative Public Management and Social Policy, and Ethics in Contemporary China Research Group in the Faculty of Humanities and Social Sciences at the City University of Hong Kong for their valuable support and financial assistance in organizing the international conference *Individual, Community & Society: Bioethics in the Third Millennium* and in the publication of this book volume.

I am deeply indebted to Professor H. Tristram Engelhardt, Jr. for his insightful comments on the draft manuscript and to Dr. Ruiping Fan for his editorial assistance which has much facilitated the publication of this book. I am also grateful to Lisa Rasmussen and Ana Smith Iltis for their generous help with the editing work.

A note of special thanks is due to Dr. Yu Kam-por, Dr. Chan Ho-mun, Dr. Fung Ying-him and Mr. Francis Mok who were key members of the Conference Organizing Committee.

Their support and contribution have made the conference a great success, and the publication of this book, an exciting possibility.

Julia Tao Lai Po-Wah, Ph.D.
Associate Professor, Department of Public and Social Administration
City University of Hong Kong
and
Director, Centre for Comparative Public Management and Social Policy
City University of Hong Kong
Hong Kong

JULIA TAO LAI PO-WAH

GLOBAL BIOETHICS, GLOBAL DIALOGUE:
INTRODUCTION

This volume explores one of the central debates in the field of bioethics in
the new century. It analyses the important issue of the possibility of
global bioethics from multiple levels and perspectives, taking into
account the context of breathtaking technological and social changes, and
the challenge of moral pluralism, within which complex bioethical issues
demand solutions. The challenge requires us to explore some fundamental
questions in moral philosophy both at the normative and the meta-ethical
level. It also leads us to debate some concrete issues in health care and
bioethics which require urgent ethical decision-making and "pragmatic"
solutions from a cross-cultural perspective.

Under the influence of the Enlightenment, modernity is characterized
by the search for a common universal law, both in nature and in the
common world, in ethics as much as in bioethics. But increasingly, there
are doubts that the universal ethical code will ever be found. Instead of a
shared common morality in the field of bioethics, it is claimed that there
are numerous moral visions, each with its own value content. The
increasing reception of various systems of Eastern ethics in recent
decades have cast further doubts on the search for a comprehensive
unitary ethical system which can be shared by every human being and
community. They reflect intense differences in theoretical perspectives
and moral commitments which often involve un-bridgeable
disagreements, although this need not imply that any morality is but a
local (and temporary) custom and that sources of morality are purely
accidental and contingent.

At the same time, there are those who argue for a thin notion of global
ethics instead of a thick notion characterized by the search for a single
moral vocabulary and a single set of moral beliefs which claim universal
objectivity and validity. They defend a weak notion of global ethics
defined as the search for moral principles that can be shared cross-
culturally. On this understanding, global bioethics would mean nothing
more than acknowledging that "modern medicine all over the world has
generated the same kinds of moral issues that need to be addressed
through reason and argument" (Becker, p. 108, in this volume). The belief

*Julia Tao Lai Po-wah (ed.), Cross-Cultural Perspectives on the (Im)Possibility of Global
Bioethics, 1–18.*

is that much common ground can be secured in the midst of moral diversity through an open and rational discourse aimed at clarifying "the factual issues by removing misunderstandings and precisely defining the moral issue(s) at stake" (Becker, p. 108, in this volume).

Contributions within this volume are organized into five parts to explore five major themes central to the important debate on the possibility and impossibility of a global bioethics. The first part explores the moral foundations of bioethics, and contemporary reflections on the nature and role of moral theory. In particular, it examines the rival claims between the universalist and particularist views of morality. The second part analyses the character of bioethics in post-traditional, post-modern societies. It explores the impact of moral loss and moral diversity, and the increasing difficulty of appealing to care and community to guide important life passages and health care decisions. The third part evaluates current debates on the possibility and impossibility of a global ethics, and its implications for consensual solutions of specific bioethical problems across cultures. The fourth part addresses moral conflicts and ethical dilemmas in everyday health care practice, and analyses the way they reflect competing moral visions regarding the permissible treatment of humans by humans, the limits of biotechnological interventions, and the nature and goals of medicine. The volume concludes by examining several alternative philosophical approaches and moral perspectives in the hope of opening up new modes of self-understanding, and expanding space for new strategies for bioethical exploration, unconstrained by demands of universal moral consensus, and unbiased by claims of moral relativism.

I. THE FOUNDATIONS OF BIOETHICS IN A POST-TRADITIONAL, POST-MODERN WORLD

The volume begins with Engelhardt's penetrating analysis of the different plausible understandings of the foundations of bioethics and their implications for aspirations to a global bioethics that transcends particular societies and communities. He examines three competing visions of bioethics and analyses how each of these three perspectives makes plausible a different understanding of bioethics based upon their different conceptions of the nature and source of moral autonomy. His analysis shows that within the cosmopolitan liberal framework, all morality must be chosen authentically and autonomously by moral agents who contract with each other. Autonomous individual liberty therefore constitutes the ground of moral authority. For libertarian liberals, the only source of common authority among moral strangers is the consent of individuals. In traditional understandings of human flourishing, authority is grounded in religiously rooted moral visions or in cultural commitments to enduring orderings of values. The conflicts between these competing moral visions de facto count against the possibility of claiming the existence of a moral consensus to support a global bioethics. Universalization of morality can in effect mean silencing moral impulse and constricting moral life. How ought one to respond? Engelhardt offers two suggestions. On the one hand, it is important to resist claims on the behalf of a global bioethics such as that represented by the cosmopolitan liberal moral vision. On the other hand, it is important to take moral diversity seriously by nurturing the conditions under which it can flourish. This implies affirming the importance of exploring regional and cultural bioethics, as well as the richness and goodness of those traditional moral visions that allow the ample flourishing of the human good (Engelhardt, 2000).

II. CARE AND COMMUNITY: LOST PERSPECTIVES?

The two chapters by Tao and Joseph in Part I extend Engelhardt's analysis of the impact of moral loss and moral diversity on making care and community less accessible in post-modern, post-traditional societies. In her chapter, Tao welcomes the liberating opportunity offered by post-modernity to search for a different mode of self-understanding and an alternative bioethical perspective neither dominated by the mainstream Western individualistic paradigm of human relations, nor committed to an understanding of morality as impartiality as defined in conventional impartialist ethical theories. Such an emphasis on individualism and impartiality has made it increasingly difficult to see why we should care about others. It has also encouraged a mode of moral reasoning in bioethics which emphasizes general norms and which tends to overlook the contextual nature of decisions in patient care and the importance of a healthcare professional's character. Her analysis shows how in the face of the dominance of impersonal ethics and the alienating and depersonalizing forces of modern society, a reductionist response has set justice and care as irreconcilable perspectives. She argues for a new bioethics that is more sensitive and more responsive to the ideal of "just caring." For this, we need a picture of a just community that is far more than a picture, where rules and principles are followed with impartiality. She examines three alternative ethical frameworks: feminist care ethics, agent-based virtue ethics, and Chinese Confucian ethics, for a different understanding of the relationship between justice and care. Her analysis supports a new bioethical approach to allow context, particularity and relationship to have a central place in our understanding of complex bioethical issues, and a re-energized moral discourse which seeks to balance justice and caring for the sake of human flourishing in the post-modern world.

Joseph's chapter offers further analysis of the post-traditional character of the lives and deaths of many people across the world. His essay is focused on the theme of moral loss and the challenge it poses to bioethics and medicine in the new century where one can no longer experience a sense of care and community to guide important life passages. Drawing on his own experience as a priest visiting patients in the hospital, he shows how individuals living in large-scale societies in a fully secular, pluralist moral context, no longer possess a point of moral orientation that can provide definite moral guidance for decision making on life and death issues in health care. The moral confusion and disorientation which result from the absence of support and context once provided by family and community, often lead to the abandonment of the moral discourse, and a shift from the substantive to the procedural, when addressing substantive decisions at the end of life. The absence of a definitive point of moral orientation can imply a contraction of the sphere of human moral experience by making moral choices a matter of individual creations. The consequence is that in a post-traditional world, we can admit at most a vestigial "minimalistic" morality, limited solely by the demand for tolerance. To avoid further moral and spiritual fragmentation, and if we are not to get stuck with a procedural morality which deeply impoverishes moral life, Joseph urges us to examine ways in which to invigorate the roles of family and particular moral communities so that they can continue to serve as important moral networks within which individuals find their moral bearings.

III. THE POSSIBILITY AND IMPOSSIBILITY
OF A GLOBAL BIOETHICS

Part II examines a number of theoretical issues raised by concrete examples of biotechnological applications which bear importantly on contemporary debate between the possibility and impossibility of a global ethics. Through exploring these moral dilemmas and cultural conflicts, the essays re-evaluate the controversy between the view that there is a common morality accepted by all in the field of bioethics (see, e.g., Beauchamp and Childress, 1979, 1994), and the opposing view that there are different moral visions and moral rationalities which means that a comprehensive unitary ethics is an impossibility.

It begins with Qiu's chapter, which explores the challenge posed for genetic intervention and reproductive human cloning by Chinese Confucian moral tradition. He draws attention to Confucian values of filial piety and ancestor worship and how their moral requirements can create ethical dilemmas and cultural conflicts in the context of modern biomedical technology. He questions the claim that there are some foundational moral values that transcend particular cultures, and that the foundations of global ethics are independent of the norms of any particular culture. He reminds us that a global bioethics can be a dream of reconciled humanity or a nightmare of cultural domination of world bioethics by one set of moral assumptions and style of reasoning. He argues in support of a weak notion of global ethics, which presupposes humility about our ethical wisdom, based upon an awareness of the limits of rational argument and of the prejudice which distorts our claims to ethical impartiality.

In a similar vein, Hu's chapter begins with a critique of individualism, the main doctrine which underlies Western ethical, political and social thought, and which is also the dominant value system central to the enterprise of bioethics as it has developed in the west. He is concerned that in the light of this doctrine, the individual is emphasized as the focus, with both ontological and methodological priority. This has in turn led to an analogy between physical atoms and moral individuals under a kind of reductionism whereby individuals are perceived to be playing the same role in society as atoms in the physical world. In contrast to this unhealthy atomism and reductionism, Xu argues that a person is a product of a set of relations, and cannot be reduced to his genes. He proposes a relational paradigm as the basis of a new bioethical approach. Under the

relational paradigm, a person is understood to be at one and the same time
a solitary being and a social being. As a solitary being, she attempts to
protect both her own existence and the existence of those closest to her, to
satisfy her personal desires, and to develop her innate abilities. As a
social being, she seeks to gain the recognition and affection of her fellow
human beings, to share in their pleasures, to comfort them in their
sorrows, and to improve their conditions of life. In this sense, it is
impossible to think of her, or to understand her life, outside the
framework of society. It is also in this sense that human beings have both
a biological life and a biographical life. In conclusion, he argues for a
relational paradigm, drawing in part on the resources of Confucianism, to
replace the individualistic framework in bioethics. His analysis shows
that there are indeed foundational disagreements between different moral
traditions which render suspect the vision of universal moral consensus.

In his reappraisal of the foundations of bioethics, Lee also argues for
the need for development of new perspectives which are more sensitive to
the reality of difference across persons and cultures to guide bio-medical
decision making. He examines the main objections raised by
contemporary bioethnographers to mainstream principalism and to the
contemporary bioethics of autonomy. He also notes the increasing
divisiveness that appears among bioethical practitioners and theoreticians,
particularly with regard to issues about justice and self-identity on the one
hand, and about incommensurability and relativism on the other. In
conclusion, he also turns to Confucian moral philosophy for contribution
to a new ethical framework which can take difference and particularity
seriously. He argues that the Confucian theory of universal moral human
nature, with its emphasis on equal moral worth and on human self-
cultivation through living in culturally meaningful practices is able to
support a new bioethics which emphasizes care without sacrificing
individuality, which accepts uniqueness without undermining community,
and which respects difference without being committed to moral
relativism.

Becker's chapter takes a different approach and emphasizes the
importance for morality to aim at objectivity and universal validity for its
assessment. While acknowledging that the goals of global ethics may
seem rather elusive, and that ethics is always contextual, he argues
against the temptation to give up claims to universality in ethics and settle
for "pragmatic" solutions. Using prenatal diagnosis as an illustration, he
argues for the importance of establishing some central values to guide

moral evaluation. He suggests that this might prevent state policies and legal supports of prenatal testing from using the technology as a way to relieve society of the burdens of caring for those with Down's syndrome. He proposes two fundamental moral values as the common ground for framing the bioethical discourse on prenatal diagnosis: (1) human life, and (2) autonomy. However, Becker also agrees that global ethics should not be exclusively understood as the existence of globally shared sets of moral values and principles, but rather as "the open and methodological process of inquiry into specific moral issues from which nobody is excluded." For Becker, it is the requirement of the moral point of view to step into the shoes of others and look from their perspectives that raises the hope of consensual solutions to moral problems.

Holliday's chapter also explores the possibility of a universal moral framework to guide bioethical discourse, arising from developments in modern biomedical technology. His optimism is based upon the implications of continuing advances in genetic engineering for expanding and re-structuring theories of justice in Western societies. He argues that in the age of biological control, it is evident that the range of things about which we can seek to be just has substantially expanded. Because we now have the ability to decide many aspects of individuals who are to populate the just community, the question "Who are the members of the community?" takes priority over the question "What interests do people have?" in formulations of theories of justice. At the same time, the practice of human genetic engineering also raises urgent questions of justice regarding: Who should have access? Who should have a say? What should be done about those for whom responsibility is borne by others? How accountable are those who make decisions on behalf of others? While it is expected that liberals and communitarians will have different answers to these questions, Holliday is hopeful that advances in human genetic engineering will favour a universal rights-based theory of justice to protect the integrity of the person. The consequences are an expansion of the scope of justice, and the introduction of an element of universalism into theories of justice in the West. He argues for a cosmopolitan approach to determining the contents of a rights-based account of global bioethics, embracing communitarian inputs and some local variability. Such a claim, he admits, is not uncontentious, as evidenced by the ongoing dispute between liberalism and communitarianism. However, alternative systems of ethics developed outside the experience of the Western world (e.g., the Confucian system), which offer quite different understandings of the central values that frame bioethical discourse, will be the real test for this vision of a universal rights-based notion of justice.

Delkeskamp-Hayes's chapter draws a different conclusion for the prospects for a universal rights-based ethics, which is also mirrored in a global bioethics. She analyses the aspiration of contemporary bioethics to establish in international law, conventions, and policy a single, globally guiding understanding of moral principles that can shape health care policy across the world. She shows how the origin of this aspiration to universal validity can be traced to the Enlightenment project that found its influential expression in the thought of Immanuel Kant (1724-1804). The same commitments have also inspired the human rights movements. They have led to attempts to combine deontological respect for autonomy rights with protection of a teleological notion of human dignity as a universalizing model for international bioethical cooperation. But the result is self-defeating. Using the Convention on Human Rights and Biomedicine as a case example, she examines the conceptual inconsistencies which result from the aspiration to combine respect for autonomy rights with protection of a human dignity that is conceived in a content-rich, Christianity-inspired sense. It muddles the distinction between leaving people to choose as they will and restricting them to choose as they should. It compromises the fundamental forbearance rights of individual choice and diminishes the political space for moral difference when such a model is politically enforced on national, international and global levels. Ironically, it is the Kant-inspired understandings of autonomy, with their thick commitment not only to human choice but also to a particular good held to be intrinsic for the moral life, which diminish the political space for moral difference and which have turned out to be hostile to the peaceful co-existence of diverse moral communities. At the same time, the regulative principles which could be derived from this reasoning turn out to be decidedly less helpful than one might have expected, once applied to the courses of action opened up by technological advances in modern medicine (such as the selling of organs). This leads Delkeskamp-Hayes to argue that the most one could achieve for the securing of peaceful social interactions, both within and across political communities and societies, is securing respect for a purely formal human dignity exhausted by respect for autonomy. She concludes that such a meta-moral view of human autonomy offers a more viable way of securing greater tolerance of cultural difference for national and international cooperation in the field of bioethics.

IV. MORAL DILEMMAS AND COMPETING VISIONS

The essays in Part III examine a number of concrete moral dilemmas in health care and bioethical practice arising from new biomedical technology and competing moral visions in post-traditional societies. They evaluate to what extent universal standards such as autonomy, rights, justice and objectivity are adequate to guide moral dilemmas involving the selling of organs, the unequal distribution of health, the permissibility of sex reassignment surgery, the management of medical information, the use of reproductive technology by homosexuals, the making of treatment decisions for neonates, and the practice of evidence-based medicine. Their analyses show that not only are the bioethical questions we face becoming increasingly difficult, but they arise in circumstances devoid of a moral orientation in terms of which to know concretely what should count as a good answer. At the same time, there are also increasing questionings about the role and place of traditional cultures and beliefs, and the extent to which we might be able to draw on our own cultural heritage to solve problems and resolve conflicts in bioethical issues which transcend any single culture.

Yu's chapter examines the principle of self-ownership and the rights-based bioethics which it offers as a candidate for providing a universal standard to guide the resolution of moral dilemmas. In many ways, the principle seems to provide clear and unambiguous guidance for settling moral dilemmas in health care situations involving decisions over the ending of life, the control of one's body, and the disposal of one's body parts. But Yu's analysis shows that the principle is in fact morally inadequate for assessing bioethical issues because it can only ground a thin notion of autonomy as self-ownership, and it can be indiscriminately applied to endorse all forms of assisted suicide and all kinds of organ transactions between consenting adults. A global bioethics grounded in such an individualistic principle of self-ownership, together with the narrow understanding of autonomy it implies, can easily ignore the social and material pre-conditions necessary for the exercise of self-control and self-mastery embedded in a richer notion of autonomy. It cannot sufficiently take into account in its moral deliberations the requirement of a just environment for making a free and autonomous choice. In this sense, it is inadequate for settling contemporary debates over fundamental bioethical issues which are highly concrete and contextual. Yu concludes by arguing for a more encompassing notion of autonomy to replace the principle of self-ownership as the foundation for a rights-based bioethics. But this would involve a thick notion of autonomy, beyond a minimal notion of forbearance. It would also involve appeal to a standard beyond autonomy itself to determine the contents of the social and material conditions of a just environment within which autonomy will be realized.

The chapter by Marchand and Wikler further examines issues of distributive justice in relation to the distribution of health in a population, and the inadequacy of appealing to abstract principles of justice to resolve the problems. Their central concern is the socioeconomic inequalities in health within a society. Their research shows that in nearly all countries, those at the bottom of the socioeconomic scale suffer anywhere from 2-4 times the mortality and morbidity rates of those at the top. This holds true regardless of differences in the material position of the worst-off, and even in the wealthiest countries where the wealthiest groups enjoy better health and longer life than those in the next position down the socioeconomic scale. What is startling is that this gradient persists even in countries with universal access to health care, where health care resources approach a fair distribution. High-cost medicine does not appear to promise as much by way of improving health as we might expect from the adoption of certain socioeconomic arrangements, such as narrowing income inequalities in a society. This suggests that justice may require more than expanding access to health care – it may require a just distribution of health as well. Also, "class prospects," understood as "life prospects" attached to various socioeconomic classes constitute an even more important unit of moral concern. They further examine three egalitarian perspectives for assessing the justice of a society's distribution of health: "the strict equality view," "the maximin view," and "the urgent needs view." But these three perspectives are equally constrained by the problem of the lack of some justifiable notion of a "standard life-expectancy and standard level of health," which can command universal acceptance, to guide assessment of health inequalities across cultures and societies. This leads them to conclude that medical reality is socially constructed; local understandings have an important role in determining the contents of a standard life expectancy or a standard level of health. Their analysis underscores the need to make explicit and to debate openly the normative assumptions which shape notions of evaluating the justice of a society's distribution of health, instead of merely appealing to abstract principles of justice which have no specific contents.

The chapter by Chan and Fung analyses the moral dilemmas surrounding the management of medical information in a pluralistic world. They argue that moral dilemmas can only be *managed*, not resolved, and that the appropriate way to handle a moral dilemma is highly dependent upon the context in which the dilemma occurs. Drawing on examples arising from the management of medical information in post-modern societies, they challenge the view that there are universal procedures or principles that will enable us to resolve moral dilemmas in all possible contexts (universalism). They argue that the way in which moral issues present themselves regarding information management is largely a function of how societies, East and West, respond to the loss of relationships and communities, as we move from thickly bound communities to individualistic societies. It is impossible to affirm a specific set of global rules about the relationship among the interests of individuals, communities and societies regarding the management of medical information. They argue for open dialogues and public deliberation by representatives in review boards and ethics committees as a more viable way of arriving at culturally sensitive, compassionate, and informed decisions regarding reasonable courses of action in the face of difficult circumstances. These boards and committees should be small and cohesive; their members should be chosen to offer informed opinion, but not to represent every shade of opinion; and their positions should be public, consistent, and defensible. In this way, they are also more likely to help us re-establish the willingness to communicate and to re-establish a meaningful dialogue that respects the moral pluralism of contemporary society.

Schmidt's chapter provides another illustration of conflicting moral visions in bioethical discourses. He reviews current debates about sex-reassignment surgery and the profound differences of opinion which exist among individuals, communities and societies about understandings of male/female differentiation. It shows that in a post-modern world, even in the most fundamental matter of human coexistence, there is no definite universal answer to the question of what constitutes gender. Such differences bear importantly on conflicting views about the permissibility and non-permissibility of particular medical treatments such as sex reassignment surgery. They indicate that there is no longer any fixed point of orientation from which to guide thinking and judgement in health care and bioethics. Different ethical positions reflect conflicting views between seeing sex-reassignment as a necessary form of therapy to relieve suffering and seeing it as an immoral interference with nature. They also reflect conflicts of views over priority between the right and the good in moral decision making. The need to re-think the nature of human identity becomes an urgent task for individuals, societies and communities, and so does the need to re-examine received opinions on questions about who should initiate medical treatment, define treatment, or determine the goals of treatment in this pluralistic age. Instead of looking for perfect consensus, which he believes can never be achieved in our post-traditional world, Schmidt concludes by underscoring the importance of recognizing and establishing structures that offer scope for dissent within society.

Sze's chapter examines a different but equally controversial debate in health care practice. This relates to the debate about access of homosexuals to reproductive technology to procreate children and organize a family. Arguing from a rights-based perspective, Sze defends the position that homosexuals should enjoy the same rights as heterosexuals to use reproductive technology, which like any other form of technology exists to compensate for the inadequacies of humankind. Homosexuals, like heterosexuals in our society, should enjoy a basic universal right to use reproductive technology. Following the liberal rights tradition, he argues that the only reason their rights could be constrained would be if they infringed upon the rights of others. He agrees that children brought into the world through this kind of technology might have to cope with a lot of unfair pressure and discrimination. But this is more the fault of society than the fault of such children's homosexual parents, and it would be unfair to restrict on this basis the right of homosexuals to use reproductive technology. However, as he goes on to consider the case of single parents or single persons, Sze supports, for prudential reasons, stringent screening to restrict their right to use reproductive technology. He refuses to grant them equal rights on the ground that statistically, single-parent families tend to have more problems than families with both parents. This, however, creates conflicts within the universal rights perspective itself, since it allows a basic universal right to be compromised for prudential reasons. It also implies a kind of discrimination against single parents in refusing to recognize that the problems of single parent families are also largely a result of society's discrimination and lack of adequate support. More fundamentally, such a position raises questions about who should set the standards for screening and how those standards should be set. What values should the standards uphold? These questions challenge the adequacy of a standard of universal rights for settling complex disputes in health care and bioethical issues. They also indicate that the answers may have to be local, and not global.

Khushf's chapter analyses the complex realities of treatment decisions for severely compromised neonates in the context of modern high-tech medicine. He examines how treatment decisions for severely compromised neonates involve multiple levels of assessments, involving family, physicians, culture, religion and society, which often have competing values. They involve variable value assessments and cost/ benefit calculations that are not independent of deeper commitments regarding the nature and purpose of human life. According to Khushf, such assessments take place at the intersection of the practical domain of medicine, on the one hand, and the rich interpretive frameworks of purpose and meaning of living and dying on the other. They cannot be resolved by a simple ethical principle, and the current debate between the principles of 'sanctity of life' and 'quality of life' fails to capture the problem of decision-making in its full scope and complexity. Neither can such decisions be made on the basis of medical science alone because "technology cannot do moral work." The ethical issues associated with treatment of premature infants cannot be resolved simply by appealing to medicine since no scientific basis would enable a physician to say that a person in a vegetative state should not live, and it is therefore inhumane to treat such a person. In conclusion, Khushf argues that these issues are better addressed by a process of negotiating different values to achieve a balance between social policy, medical norms, and parental discretion, taking into account the role and responsibility of the individual child, family, medicine, religious and cultural community, society and the state. We should avoid adopting a reductionist response or utilitarian system that opts out of genuinely ethical deliberations.

Au's chapter examines the implications of a new paradigm of evidence-based medicine for global bioethics in the new century. Hailed as a "new engine of health care," evidence-based medicine holds out the promise of stamping out variations in practice and liberating medicine from subjective bias and personal opinion. It offers a vigorously standardized and formatted methodology for application to clinical practice, management and resource allocation. Au explores a number of deep issues and questions raised by this new paradigm in medicine. First is the danger of a tendency to deify science; second is the concern about justice for minority patient groups, and third is the question whether all medical interventions can be evaluated by evidence-based medicine. These concerns suggest that it may not be ethically and clinically sound to eliminate all variations in clinical practice. At the same time, there is the emergence of a countervailing force from within medicine itself under the new initiative of "narrative-based medicine." Under this paradigm, heterogeneity of illness stories is respected and assimilated into clinical decisions. It emphasizes that context and meaning are important in health care and medicine. Appropriate solution or intervention depends upon understanding the meaning of the situation and knowledge of the personal, cultural and social context of the patient. Given these fundamental differences, Au argues that an important breakthrough in the development of health care and bioethics will depend upon finding a way to bridge the gap between evidence-based medicine and narrative-based medicine, rather than suggesting that either one of them ought to be completely replaced by the other.

V. GLOBAL BIOETHICS: NEW POSSIBILITIES AND NEW STRATEGIES

Should global bioethics be thick or thin? Should it emphasize convergence or divergence? In light of these substantive differences and deep disagreement between moral visions, should we continue to pursue global bioethics or should we abandon the project? The essays in Part IV offer different suggestions for opening up new strategies and perspectives for bioethical explorations in the new century.

Cutter's chapter echoes the claim that because of moral pluralism, a common procedural morality seems to be the only option we are left with as we enter the new century. She believes that we can share a context-rich communitarian bioethics only with those who are committed to the same fundamental moral premises and value rankings. Concepts that frame medicine are local because they involve descriptions, values and social commitments, all of which involve difficulties in commanding global agreement. Her analysis of developments in molecular genetics also shows that we will increasingly understand disease as a local as opposed to global concept. There will be an increasing realization that one organism's disease is another's adaptation. Genetics also highlights the complex relation between normal and abnormal, or health and disease. This will further challenge the legitimacy of any single set of standards or classificatory scheme of disease in medicine. It is also expected that health care professionals will increasingly use genetic tests to match drugs to an individual patient's body chemistry. The results will likely lead to a revolution in understanding and treating disease in the new century. The important insight for bioethics is to acknowledge that differences within and across societal and communal boundaries must be taken into consideration in framing disease categories. Furthermore, individual differences are important to understanding the disease expressed by a particular organism. In this way, Cutter draws the conclusion that genetics offers additional support for a local, as opposed to global account of bioethics.

Cheng's chapter argues for a holistic understanding of humanity as the basis for a new bioethics. He begins by affirming the values of recent developments in bio-medical technology and bio-genetic engineering. But, he also points out that there is a need to recognize the limits and boundaries of the values upon which norms and rules of action are to be formulated, adopted and justified. He argues that as we re-examine bioethical principles and their foundations, we also have to examine our understanding of humanity, in order to assess the values of technology and new norms of action. He points out that not only are modern societies fractured, but also our understanding of humanity. There is a need to develop a holistic and deeper understanding of humanity which makes appropriate distinction between a human individual and a human person. Drawing on the moral resources of Chinese Confucian tradition, which emphasizes the moral cultivation of the individual and the thickly bound community, he proposes a process-agency approach to re-conceptualize our understanding of humanity and its relationship with community. He further proposes two basic standards of valuation of biomedical technology to guide global bioethical discourse. The first is a holistic principle of the human person. It requires biotechnology to conform to the potential and inherent ends of a whole person and not just the ends of a biological individual. Second, he suggests a universalistic principle of the human community that requires biotechnology to conform to the potential and inherent ends of the whole community of human persons, and not just the ends of a single or selected community of biological individuals or human persons. For Cheng, the pursuit of a global bioethics is dependent upon the development of such a deeper and holistic understanding of humanity and the human person.

The chapter by Sakamoto takes the discussion in a different direction. He explores the possibility of building a new bioethics based upon a new foundational philosophy, methodology, and policy approach. He begins by examining the genesis of bioethics in the west and its early roots in "modern humanism" which he argues is a kind of "human-centricism." But since the 1980s and 1990s, there have been serious challenges to this mainstream perspective with its emphasis on "informed consent," "self-determination" and "patients' rights." The challenges are posed by rapid developments in genetics, and in Asian and environmental approaches to bioethics. Sakamoto argues that Asian worldviews are characterized by the lack of natural-artificial dualism that underpins much of the Western understanding of the relationship between nature and human beings. The Asian ethos supports a holistic view over an individualistic view of bioethics that emphasizes social harmony and values nature, society, community, neighborhood and mutual aid. In the views of Sakamoto, the new bioethics, instead of being principle-oriented, should be characterized by the search for some sort of holistic harmony among the antagonists of competing moral traditions, guided by social techniques of bargaining and compromising. Instead of referring to universal principles, it should seek to maintain "bargain dialogues" for reaching acceptable "bargain consensus" in the face of moral diversity. He argues that the only possible way to realize the Asian ideal of harmony is to accept that every ethical and moral code is essentially relative to its period and region. The role of bioethics is to maintain a dialogue in order to reach some kind of acceptable consensus among bargaining parties. It achieves its function not by applying any set of invariant principles, but by drawing on and enhancing universal feelings of compassion and common humanity which already exist in human nature.

The final chapter by Fan offers a new strategy for bioethical exploration in order to understand the diversity and plurality of bioethical accounts more deeply. His central concern is that the role of moral theories in bioethical accounts has been exaggerated. A moral theory can contribute a thin structure to a bioethical account in which concrete moral substance is arranged, but it is insufficient for drawing any concrete conclusions. Moreover, because principles or doctrines identified by moral theories are abstract formal statements, they can easily obtain the support of the intuitions of individuals, creating a false consciousness of moral consensus and overlooking deep disagreements and incommensurable differences. He introduces the notion of a "moral perspective" to capture the moral commitments and convictions which constitute the content-full moral substance of a bioethical account. The source of these commitments lies in the ways of life that are shared by members of a moral community. A moral perspective in this sense is concrete, canonical and content-full, unlike a moral theory which is general and comprises formal statements that are inevitably vague, ambiguous and underdetermined without further interpretation. As an illustration, Fan further uses the case of Peter Singer vs. David Friedman to analyse the tripartite interplay among moral theory, moral account and moral perspective, to show how they develop incommensurable bioethical accounts and draw different bioethical conclusions even though they both use a utilitarian theory to construct their account. Fan concludes that bioethical investigations in the new century should focus on disclosing moral perspectives, rather than debating formal principles and abstract statements of moral theories. Such an approach will enable us to achieve a deeper understanding of different bioethical commitments. It will help to unveil different moral perspectives, dispel illusions of a global bioethics and support respect for difference and particularity in bioethical discourse.

VI. GLOBAL BIOETHICS OR GLOBAL DIALOGUE ON BIOETHICS?

The essays in this volume offer different approaches and perspectives on the pursuit of global ethics in the new century. But their conclusions all point in one direction. We should avoid practicing moral parochialism under the mask of promoting universal ethics. As we enter the new century, we are increasingly deepening our understanding of how moral discourse, moral narratives and moral commitments take different shape within particular cultures and traditions. As we become aware of our pluralistic world, we are increasingly committed to learning ways to respect and appreciate the richness and diversity of human values and the ever-evolving bioethical discourse that they support. Such richness and diversity in human moral aspirations and practices deserve space and support for their pursuit of sustaining a continuing global dialogue on bioethics in order to meet the challenge of complex bioethical problems in the 21st century. Instead of seeking to establish a comprehensive

JULIA TAO LAI PO-WAH

unitary global bioethics, cultures and societies can benefit from mutual learning through the more modest project of creating a continuing global dialogue on complex and urgent bioethical issues which know no border, and which generate the same kinds of moral conflicts and dilemmas, requiring solutions through reason, argument and compassion. Such a global dialogue should be based on a sense of genuine humility about our human rationality. It should be conducted on the basis of a full recognition of the need to respect difference as well as the need to negotiate and to justify difference, carried out through open and rational discourse, and from which no one ought to be arbitrarily excluded.

Department of Public and Social Administration
City University of Hong Kong
and
Centre for Comparative Public Management and Social Policy
City University of Hong Kong
Hong Kong

REFERENCES

Beauchamp, T.L. & Childress, J.F. (1979) (1994). *Principles of Biomedical Ethics*, 1st edition & 4th edition. New York: Oxford University Press.
Engelhardt, Jr., H.T. (2000) *The Foundations of Christian Bioethics*. Lisse: Swets & Zeitlinger Publishers.

H. TRISTRAM ENGELHARDT, JR.

MORALITY, UNIVERSALITY, AND PARTICULARITY: RETHINKING THE ROLE OF COMMUNITY IN THE FOUNDATIONS OF BIOETHICS

I. LIBERALS, LIBERTARIANS, AND CULTURAL/RELIGIOUS CONSERVATIVES

An assessment of bioethics at the threshold of the new millennium must begin by acknowledging the numerous competing visions of the field, as well as their divergent foundations. There is a received authorized account with a kind of governmental establishment in much of North America and the West. It regards issues both moral and bioethical within the framework of liberal cosmopolitan assumptions: morality is seen to be grounded in autonomous moral agents who contract with each other in the realization of social structures and who should favor autonomous free choice and fair equality of opportunity over particular, especially traditional familial and communal moral commitments. There is also the libertarian liberal insight that, when moral strangers meet, they will have no source of common authority other than their own consent. By default, the authority of their common undertakings as moral strangers must be understood as drawn from the permission of moral agents, not from God or from a content-full authoritative understanding of moral rationality. Unlike the cosmopolitan liberal, the libertarian liberal recognizes that the circumstance of moral pluralism within which permission is the source of authority does not support the conclusion that autonomous individual choice should have a value over other goods, including those celebrated within the context of traditional moral communities.

Over against these liberal visions, there are traditional understandings of human flourishing, which ground their authority neither in a supposed prior value of autonomous individual liberty, nor in the permission of moral agents, but rather in a connection with God, nature, or an experience of reality that discloses a possibility for human realization and flourishing not justifiable in discursive, rational terms. Such communities range from those within which traditional Christians, Moslems, Jews, Parsees, Daoists, and Buddhists live their lives to the religious/cultural communities one finds embodied in Confucianism and Shintoism. Within

Julia Tao Lai Po-wah (ed.), Cross-Cultural Perspectives on the (Im)Possibility of Global Bioethics, 19–38.
© 2002 *Kluwer Academic Publishers. Printed in Great Britain.*

these communities there may be understandings of the appropriateness of abortion, third-party-assisted reproduction, physician-assisted suicide, and euthanasia, which will be on the one hand at odds with cosmopolitan liberals and on the other hand much more content-full than what is available to libertarian liberals who must regard all that persons fashion as moral strangers as creations of human choice, not as disclosures of fundamental reality or enduring orderings of values.

Each of these perspectives makes plausible a different understanding of bioethics. Within the context of liberal cosmopolitan commitments, the libertarian liberal is a challenge to aspirations to all-encompassing health care policy, while the cultural/religious conservative is a challenge to the particularity of the cosmopolitan liberal. The cultural/religious conservative gives evidence for holding that the cosmopolitan liberal also belongs to a particular community of faith, however secular. The cultural/religious conservative will also have grounds for suspicions concerning the cosmopolitan liberal in that cosmopolitan liberal convictions have undermined robust, traditional communities in the West and throughout the world. Where all morality must be chosen authentically and autonomously, one is invited to step out of history and tradition, thus losing the orientation provided by the community of one's ancestors. The libertarian liberal, in contrast, is pleased peaceably to make space for as many communities as there are persons to fashion them. Libertarian liberals take the project of robust communities seriously without commitment to any in particular. Finally, cultural/religious conservatives understand why their particular community carries a truth unknown to others and therefore why it has special access to the human good and human flourishing.

This essay will address the dialectic among these three moral perspectives. It will first lay out how the cosmopolitan liberal perspective arrogated to itself the status of the moral vision acclaimed by consensus. The essay will then turn to the grounds for the absence of consensus in bioethics and for the presence of the moral pluralism that de facto defines bioethics across the world. The conflicts between the moral visions compassed by this pluralism will be described under the category of culture wars to indicate the conflicts between the cosmopolitan liberal moral vision and those who understand its particularity and arbitrariness. Finally, the character of communities will be explored in greater depth, leading to the conclusion that the bioethics of the next millennium will find itself plural in character and in its foundations.

II. THE MIRAGE OF CONSENSUS

Much is made of consensus (Bayertz, 1996; ten Have, 1998). From the perspective of *Realpolitik*, this is understandable. The more one can convince people that they share a common moral vision, the more governable a society is likely to be. It is in the interest of rulers to create in the governed a belief that all share foundational moral commitments. It is even more useful to embed the lives of all citizens in a common moral narrative so that all experience themselves bound in a thick community of moral sentiments. Even when such a communality does not actually exist, the false perception or false consciousness of such common commitments can be useful. Toward the end of governance, one would expect that rulers would employ intellectuals, including philosophers, to develop arguments in support of the favored consensus, as well as arguments aimed at showing that apparent disagreements are without substance.

Given the ever greater importance of health care for modern societies, one would anticipate that there would be a significant social and political interest in manufacturing a consensus in bioethics for health care policy. One would even expect to find intellectuals and especially philosophers hired to aid in this project.[1] Following Karl Marx, it is reasonable to anticipate that philosophers would be engaged as "conceptive ideologists, who make the perfecting of the illusion of the [ruling] class about itself their chief source of livelihood" (Marx and Engels, 1967, p. 40). It is not hard to find bioethicists who are examples of such conceptive ideologists. They are engaged in affirming, manufacturing, and creating the appearance of moral consensus in order to move towards uniform health care policy. They garner prestige and power by affirming a moral vision, which allows them to participate in governmental commissions aimed at sustaining the illusion of consensus.

Even at the international level, there is a search for global consensus expressed in the attempt through philosophical reasoning to establish a global ethics (Hoshino, 1996). Many hope to establish internationally one content-full moral perspective, thus closing out opportunities for the pursuit of divergent understandings of bioethical probity in different countries. Usually, examples of wrong conduct are chosen whose wrongness does not depend on a particular moral account, indeed, on any content-full moral understanding, but rather on the use of persons without their permission.[2] Moreover, when defending the establishment of a global ethics, little attention is paid to the slaughter of millions in this century in the pursuit of universal justice and fairness.[3] If anything, the history of this century provides a chronicle of atrocities associated with the attempt internationally to impose particular content-rich understandings of moral probity, justice, and fairness on the basis of scientific reasoning and supposed sound rational argument.

The plausibility of consensus can be buttressed by noting the ways in which commissions and ethics committees can work together despite seeming disagreements. Given the social dynamics of the creation of commissions and committees, this is to be anticipated. First, those who create commissions and committees do so not to support philosophical analysis of moral and philosophical differences or disputations regarding which views should be affirmed. Instead, the goal is collaboration, in particular, the production of common moral understandings, often with very specific policy objectives in view. It would be a misunderstanding of the political function of ethics commissions and committees were one to appoint persons with truly divergent moral and philosophical understandings. Such appointments would not create an ethics commission or committee, but the basis for an ongoing and likely engaging philosophical seminar. Second, those appointed to such commissions and committees, since they are not likely to be drawn from communities with substantively different moral and ideological commitments, will in the process of their deliberations tend to experience the discovery of common moral ground, and may experience the disclosure of a consensus.

Consider, for example, what will likely occur if one appoints to a committee a utilitarian and a deontologist who share common ideological understandings. If they are able adequately to reconstruct their original pre-theoretical commitments within their different theoretical frameworks, they will discover that their middle-level principles in fact have materially the same force. Although they may give different justifications for principles such as autonomy, beneficence, non-maleficence, and justice, they will discover that they can work together, that they can share common middle-level principles and can come to common agreements about particular cases. They may even have an important eureka experience: consensus is real. They may also conclude that moral theory is not particularly important (Beauchamp and Childress, 1979), thereby driving an artificial wedge between moral theory and normative judgments, which in turn is used to justify the discounting and dismissing of deep theoretical disagreements. Of course, the outcome will be different if persons begin from different pre-theoretic moral and/or political commitments. For example, persons with social democratic and libertarian moral and ideological commitments will discover how deep their disagreements are when they both appeal to a middle-level principle of justice to analyze particular case examples. Where one group will see needs generating rights to welfare, the other group will see needs generating temptations to violate the property rights of others. Also, if persons are divided by theoretic disagreements carrying moral substance, they will encounter similar points of dispute. A Kantian and an act utilitarian, when confronting a case where lying will save a person's life, will have quite different views of proper conduct as well as of beneficent action.

Even if there were a broad consensus, what moral conclusions could be drawn? What grounds would one have to conclude that a moral consensus corresponded with moral truth rather than mere moral fashion? Why should one assume that the *vox populi* is the *vox dei?* That is, even if a consensus could be established, one would need to show its moral authority. Moreover, how much of a consensus proves which normative claim regarding the content of morality? At this point, one is driven to foundational issues. How does one know when one knows truly what one ought morally to do? The contingent sentiments and moral fashions of the age may bring people to hold that they know truly how they ought to act. Still, the foundational question arises: how can one in such important matters as morals know one has indeed come to the correct conclusion? After all, unlike conclusions in matters of science, conclusions in matters of morals lead to authorizing coercive public action.[4]

III. THE CULTURE WARS: BATTLEGROUNDS IN BIOETHICS

The field of bioethics is marked by substantive disagreements about most issues that matter. There are substantive disagreements regarding the moral probity of genetic engineering, third-party-assisted reproduction, abortion, physician-assisted suicide, euthanasia, and the requirements of justice in health care. In part, these disagreements are embedded in different moral visions, many of which are religious. Many moral disputes regarding the meaning of sexuality, reproduction, birth, suffering, dying, and death have roots in traditional moral understandings that appreciate a substantive significance in these passages of human life.[5]

The points of controversy among religiously rooted moral understandings and those of the various secular bioethics will not go away. They are anchored in different understandings of how one can know moral truth and what the nature of that moral truth is.

Beyond the culture wars that engage religious visions against those secular, there are contrasting secular understandings of justice, fairness, and appropriate polity. For example, is a society best organized when it gives greater value to civil liberties than to equality, and only then considers matters of security and prosperity? Or is the human condition better enhanced by giving first ranking to security, then prosperity, and only then to liberty and last of all to equality? Different thin theories of the good will ground the contrasting moral visions underlying a social democracy or a softly authoritarian capitalist state, as illustrated, for example, in comparisons between the governing ideology of Cambridge, Massachusetts, and that of Singapore. To gauge the moral differences involved, including the moral plausibility of the cultural differences between Singapore and Cambridge, Massachusetts, one might imagine the public policy makers of a large developing country reflecting on the Soviet Union's attempted transition to a social democracy with the conclusion that a softly authoritarian capitalist state would have been preferable to the pursuit of a social democratic polity.[6] The policy makers might choose a softly authoritarian capitalist state because they hold that a direct transition to a social democracy is fraught with great risks not only to security and prosperity, but to liberty interests as well. Such policymakers might in addition hold that a softly authoritarian democratic society is preferable because of the positive support it could give to traditional family structures, commitments to responsibility, and the development of moral character, all of which they consider endangered in a social democracy. This conviction might be further enforced by commitments to particular values such as those structured within Confucian cultural commitments. Given different thin theories of the good, even ones that differ only in the respective ranking of liberty, equality, security, and prosperity, not to mention different understandings of human flourishing, possible risks and benefits will be differently assessed making the pursuit of a social democracy either morally obligatory or morally to be avoided.

Or to take another example, consider the contrast between those who favor autonomous individualism and those who would give moral priority to family life. Those who regard autonomous individualism as the presumptively appropriate relation among persons would require any deviations to be established by explicit statement and agreement. For example, patients would be presumptively treated as autonomous individuals willing and committed to choosing on their own, unless they explicitly demanded to be regarded and treated within a traditional family structure. The decision to live within a traditional family structure that does not value autonomous authenticity would need to be tolerated by a cosmopolitan liberal as morally deficient but still grounded in self-determination. The cosmopolitan liberals, those committed to authentic autonomous choices and fair equality of opportunity, as well as to a global community of liberated individuals, will hope that through state controlled, ideologically directed education the choice to remain in a traditional community can be shown to be morally inappropriate and thus progressively discouraged.

On the other hand, if one considers life within a traditional family structure as the presumptively appropriate relation among persons, the burden of proof shifts. Persons are approached as nested within the thick expectations of traditional family structures, unless they explicitly state that they wish to be regarded and treated as isolated individuals. Those who understand human flourishing as realized within the thick embrace of a robust community will recognize the cosmopolitan liberal alternative as morally impoverished. From an external perspective, both approaches can be justified as social structures deriving their authority from the permission of those involved. In this way, each would satisfy libertarian liberal requirements by grounding their practice in permission. But from the internal perspective, the grounding of family-oriented approaches need not be in permission. Instead, it might be sought from a commitment to traditional, for example Confucian, understandings of family structure. That is, the justification from the perspective of those living within a tradition will be in terms of that comprehensive moral vision.

At stake in the contrast between these approaches are different moral assumptions regarding social structures, as well as different views as to who should bear the burden of defeating which standing presumption. The presumption in favor of treating patients as autonomous individuals grows naturally out of the cosmopolitan liberal endorsement not only of permission as a source of authority but of autonomy as a value. There is a category shift from permission as the source of moral authority among moral strangers to autonomous individual choice as the cardinal value. The cosmopolitan liberal does not merely recognize free choice as the general justification to moral strangers of life (Engelhardt, 1996), but also affirms liberty as a value that must be given priority in the good life. As such, liberty becomes the cardinal value for an emerging, cosmopolitan, liberal community committed to authentic, autonomous choice and opposed to non-cosmopolitan communities unstructured by liberal moral commitments, that is, moral views that do not give priority to liberty and equality. Once committed to achieving personal liberty, they may be committed as well to securing the material resources necessary for its realization and therefore will come to affirm equality in all areas bearing on liberty and opportunity.[7] In a high-technology society, exemplar lifestyles are defined in terms of the possibility of entering into self-rewarding careers. Anything that impedes such careers is suspect, including traditional gender roles in the family. Personal success and fulfillment is defined in terms of economic success as well as societal power and status. Most importantly, because liberty as a value is given priority, the choice to set one's liberty aside in the pursuit of a traditional family structure or set of cultural commitments is always brought into question as illiberal, as establishing a moral vision where liberty does not have primacy.

With a libertarian liberal understanding, both the choice to live as an autonomous individual and the choice to live within a traditional family structure have equal standing. In each case, there is authorization through permission. The choice to pursue human flourishing within a traditional family structure that does not give prior lexical ordering to liberty and equality is for the libertarian liberal on a par with the choice of the cosmopolitan liberal to live a life committed to autonomous authenticity, as long as both are grounded in permission. Finally, the commitment to a traditional moral understanding can seek to ground the primacy given to family structures not in permission, but in the human flourishing seen to be possible only in terms of a traditional way of life. Such a project should be regarded as illiberal by libertarian liberals only when the permission of the participants in this way of life cannot be presupposed. The issue is not whether the people within a way of life seek to justify their moral commitments in terms of permission, but whether they can be regarded as having acted with the consent of those involved.

There are, in short, substantive moral disagreements that de facto count against the possibility of claiming the existence of a moral consensus to support a global bioethics. Some of these disagreements are rooted in metaphysically anchored religious understandings of the deep meaning of human life and the purpose of the universe. Since these understandings are nested in conflicting transcendent claims, it will not be possible to resolve such controversies by sound rational argument. Even commitments to different thin theories of the good or to different assumptions regarding the moral appropriateness of social structures will lead to alternative moral views separated by different initial premises. Since the initial moral premises will differ, such controversies cannot be resolved by sound rational argument. Although in some of the cases of the contrast between individualistic and family-oriented approaches to patients both sides may agree that individuals are the source of moral authority.[8] there will still be disagreements as to how individuals should realize their own best interests, as well as under what circumstances one has obligations to speak one's own mind. There will also be different views as to which costs are worse, having practices that may lead to ignoring the desire of some individuals to act within their own families, versus practices that may lead to ignoring the desire of some individuals to act independently of their own families. Which circumstance is understood as worse will depend on antecedent rankings of values and conceptions of the good life. At stake are conflicting moral rationalities embedded in different foundational understandings of human flourishing.

The same difficulty lies at the root of disputes regarding different understandings of justice. A socialist and a libertarian may agree with the adage that justice is "the constant and perpetual wish to render everyone his due."[9] At issue is what is due to whom and under what circumstances. There are deep disagreements regarding the circumstances under which needs generate rights and the extent to which welfare claims can defeat claims to the ownership of resources. Such disputes concern the nature of property rights, as well as the circumstances under which one must surrender one's property to the needs of others. One might think here of the contrast between an account of such matters as defended by John Rawls in his *Theory of Justice* (Rawls, 1971), and what is defended by Nozick in his *Anarchy, State, and Utopia* (Nozick, 1974). In this contrast, Rawls' account bears the heavier burden, for his account requires justifying a particular vision of justice or fairness. In contrast, Nozick provides what is tantamount to a default position, in that his account can be seen to rely on permission alone.

The difficulty of resolving moral controversies by sound rational argument lies in the circumstance that, to make a moral judgment, one must already have a moral or value perspective. If one attends to the consequences of different moral visions, one cannot produce a comparison without begging the question. One must already agree as to whose moral standard to use. If, for example, one narrowed moral concerns to liberty, equality, prosperity, and security, one would not be able to determine which moral vision maximizes the realization of the good unless one already knows how one ought to compare liberty, equality, prosperity, and security. The matter is somewhat like trying to determine the value of a wallet full of American dollars, Australian dollars, Canadian dollars, New Zealand dollars, Singapore dollars, and Taiwan dollars. One must already know the exchange rate in order to make the calculation. *Mutatis mutandi*, one must know the proper measure of comparison in order to determine how to compare different goods.

This problem is not solved by invoking a particular substantive view of moral rationality. To endorse in a principled fashion one thin theory of the good, to select one cluster of principles, judgments, and intuitions in reflective moral equilibrium, to endorse one understanding of moral rationality, one must already have prior moral guidance. At the outset, one must have in hand the moral foundations for what one is seeking: a guiding value perspective. In short, the moral pluralism we face is not merely a de facto diversity of values. The plurality has its origin in our character as moral knowers. In our human condition as we find it, we do not have a noetic capacity for eidetic moral knowledge (Kontzevitch, 1988). As a result, we cannot advance conclusive discursive arguments regarding the correct, canonical moral vision without begging the question or engaging in an infinite regress because we do not share common moral views, common weightings of intuitions, or a common understanding of the good. Because we do not agree regarding the specific character of the good, we encounter substantive moral controversies with significant public policy implications. Without a common background measure by which to resolve such controversies, they are interminable. Moral pluralism is thus real and moral consensus at best a pious, secular hope. At best, we can recognize that in the face of diverse views of God and of moral rationality we can still draw commonly recognizable moral authority from the permission of persons. The appeal to persons as a source of moral authority rather than to God or to an account of moral rationality becomes the default position constituting libertarian liberalism. This position is at peace with a diversity of moral visions and communities.

IV. COMPREHENSIVE DOCTRINES, MORAL NARRATIVES, AND ROBUST COMMUNITIES

It is not just that the moral world is fractured by numerous, competing moral visions. In addition, many experience profound moral loss. They find themselves deprived of a substantive moral community. This experience of loss often occurs against the backdrop of a religious or religious/cultural commitment once embraced, but now abandoned. It is especially in terms of religious commitments and the communities they sustain that many have an image of a thick fabric of common experience. Such a particular community provides a reference point for substantial community life in (1) affording metaphysical orientation through disclosing the deep meaning of reality, (2) providing narrative structure by interpreting all life occurrences, transforming seemingly surd events into happenings with significance, and (3) sustaining a thick commitment of caring through well-structured webs of obligations that are independent of prior particular agreements. Such robust communities are generally (1) inclusive in the sense of encompassing all elements of life in their interpretive framework and (2) exclusive in requiring unqualified commitment be given only to that community. Those communities that allow membership in more than one religious community either are not robust[10] or make such an allowance from the perspective of a cultural community that has a polytheistic world vision permitting the embrace of a plurality of religious understandings integrated or at least lodged within an overarching moral and metaphysical framework.

Robust communities provide (1) metaphysical grounding, (2) axiological direction, (3) social context, (4) exemplars of successful moral conduct, and (5) a thick narrative in which both to embed the individual and to sustain the individual's relationship with others, as well as with the cosmos. Within a robust community, one has an account of its structure and its moral commitments as rooted in the very character of reality. One is provided with an ordering of values and right-making conditions, as well as with a social structure within which one can achieve the human moral good. In such a community, there are usually not only persons who are moral experts about the good life, but also persons who are in authority to resolve moral controversies. That is, within well-functioning moral communities there are not only persons who have special knowledge but also persons who have special moral authority. To take an analogy from the law, there are both the equivalent of those who are experts regarding the moral life, as there are experts regarding the law (e.g., lawyers), as well as those who are in moral authority as sheriffs are in authority within a legal system. There will be theologians or inspired moral experts, on the one hand, and persons who may be in authority, like abbots of a monastery, on the other.

In contrast with life within the thick web of such communally recognized moral reality, the life of the cosmopolitan may appear remarkably free. Cosmopolitan liberals are at liberty individually to make presumptively appropriate choices regarding how to approach the major passages of their lives, such as reproduction, birth, suffering, and death. In such "private" areas of life,[11] cosmopolitans are doubly autonomous in the sense of being responsible both for framing rules for their own conduct as well as for applying them as they judge best. In addition, from a cosmopolitan liberal perspective, free choices are fully affirmed morally only when they endorse liberty and equality. On the one hand, the individual acts responsibly through investigating and articulating the moral law; on the other hand, the individual acts with authentic responsibility in applying the moral law. Although there will be persons who are experts regarding the complex geography of moral concerns and agreements, there will not be persons who are in moral authority, simply from their status in a society. Through choice and agreement, particular fabrics of obligations can be fashioned, some of which may bring people into moral authority. From a liberal point of view, such authority can only be derived from the consent of those who derive their authority not from special claims rooted in the nature of being or God's will, but from permission. As a consequence, the cosmopolitan liberal moral world will be appreciated as a social construction rather than a structure given to and constraining humans as such.

In contrast, those who live within a robust community will recognize the foundational elements of their social structure as givens. The family and traditional family structures, as well as taken-for-granted orderings of values and right-making principles, will bind not simply because of the permission of those who wish to live in their terms, but most significantly because of their correspondence with the good and the true. It will be clear, for example, to an Orthodox Christian that, despite the consent of all competent parties, abortion is tantamount to murder, artificial insemination by donor a form of adultery, and assisted suicide self-murder. In addition, many such communities will claim special access to moral knowledge available only through participation in the thick web of moral assumptions and commitments which frames that community's way of life. Knowledge will be given through living in a particular fashion. Because there are different understandings of the good and the true, and different views of how to resolve epistemic claims in these matters, there will be a plurality of communal perspectives that will endure.

Between the level of individual choice and the overarching framework of society and the state, there will be varying genres of communities. There will be limited communities directed to different dimensions of life, including corporations, universities, political parties, as well as partially secularized religious communities. Such communities will generally not require presumptive, primary loyalty. They will not supply a thick narrative both orienting and anchoring the individual, as well as providing intimate communal support. At best, they will provide care from strangers, welfare support, counseling, and therapy from dispassionate, value-neutral professionals. Those in need will not generally be aided out of love by selflessly committed persons. The care provided will generally have neither the metaphysical significance of charity[12] nor the affection of an individual responding out of personal concern.

Such limited communities will be neither (1) all-inclusive, nor (2) exclusive. Neither will they supply (1) metaphysical orientation, (2) a fully directing moral narrative, or (3) a thick fabric of caring. Indeed, one of the characteristics of the emergence of limited moral communities is that individuals find themselves embedded in numerous moral narratives, none of which is recognized as having deep, pre-emptive, moral and metaphysical purchase. That is, such communities do not provide (1) metaphysical grounding, but instead (2) only limited axiological direction, often reducing the ethical to the aesthetic. They will not offer (3) a robust sense of social context, but instead (4) social contexts in the plural. As a result, (5) moral exemplars will always be limited in the force of their authority, as well as (6) because of the limited character of the narratives they provide.

The more communities lose their robust character, the less individuals will have an unambiguous sense of their own moral and metaphysical orientation. In addition, the less it will be possible for individuals to understand their lives as embedded within a thick moral narrative or context of significance. Cosmopolitan liberals in particular will not be able to make unambiguous sense of, or at least provide a deep meaning for sexuality, reproduction, birth, suffering, dying, and death. The passages of human life will be encountered as material to be woven into aesthetic narratives or compositions. Cosmopolitans will not experience themselves as told into a narrative grounded in being itself. It will be difficult for cosmopolitans to take metaphysical account of the tragedies of life in general and of their own suffering and death in particular. Tragedy will rather be a function of the seemingly surd character of reality, which appears deaf to human concerns. Tragedy simply happens. Suffering has no momentous, final redemptive significance. From the perspective of those who live their lives within robust moral communities, the reproduction, birth, suffering, dying, and death of cosmopolitan liberals will have a shallow character.

For cosmopolitan liberals, there will be no final perspective from which an account can be given of the universe or of those who live in it. As a consequence, the life narratives of liberal cosmopolitans will have a post-modern character not only in being plural, but also in lacking a narrative integration around a deep sense of self or the meaning of reality. Such narratives will at best be tentative and at worst only an affirmative story of self liberation, which in the end can have no enduring content. The moral task will be to endeavor to establish the legal, cultural, and material conditions for the possibility of persons autonomously assembling for themselves and with consenting others the mosaic of achievements and experiences they will find personally fulfilling. The task of choosing freely without external constraints will bring into question all content-full moral accounts other than that of the affirmation of liberty, thus progressively evacuating moral content or transforming it into the aesthetic.

The cosmopolitan liberal should find none of this to be unexpected or wrongheaded. Human reason, morality, and community must be grounded in human concerns, not divine or metaphysical truths, if they are to be open to all in the public forum and in terms amenable to public reason. Moral community in this sense is a creation of persons and their reason, not a revelation of God. Indeed, insofar as the moral community is to bind all persons in terms open to the public forum, it must be articulated in claims that can be assessed by all. As a consequence, the moral discourse of the general community of humans must avoid overly particular cultural commitments; most importantly, it must avoid claims of special moral insight. Moral commitments must be justified and framed in terms defensible to all. Comprehensive doctrines, which provide the foundations of robust moral communities, are acceptable only if those who embrace them are required to prescind from their claims before entering into the public forum. To engage in public debates one must affirm a kind of public reason denuded of deep metaphysical claims and particular moral differences.

Robust traditional communities are, as John Rawls recognizes, the enemy of social democracy, which seeks to achieve moral community at the societal level. The difficulty is that, when community is sought at this general level, it is neither able to provide a thick moral narrative and a commitment to caring, nor will it tolerate robust moral communities with their all-inclusive explanations and their commitments to exclusive membership. Such robust moral commitments threaten the dignity of others. Rather than simply judging that alternative life-styles represent different approaches to living life, some of which will be right for some persons and wrong for others, robust religious communities support strong judgments that life-styles that appear as alternatives to living rightly are simply wrong. Robust, traditional, religious communities when considering matters such as abortion, alternative sexual life-styles, and euthanasia may consider other lifestyles as not simply different, but as morally misdirected. Moreover, they may identify such moral choices and lifestyles as morally misguided by appeal to a moral viewpoint grounded in experience beyond the liberal cosmopolitan ethos of secular morality. The dignity offense is then double. Not only are the free choices of others condemned, but also the basis for the condemnation lies beyond the bounds of general cosmopolitan culture.

Rawls acknowledges this point when he recognizes that the world of the social democrat, that of the cosmopolitan liberal, is anchored in an idea of public reason that "neither criticizes nor attacks any comprehensive doctrine, religious or nonreligious, except insofar as that doctrine is incompatible with the essentials of public reason and a democratic polity" (Rawls, 1997, p. 766). Rawls unites the life of the autonomous individual with the social fabric of the welfare state, which conveys both financial welfare support and protects dignity rights. In such a context, it becomes politically incorrect for a member of a religious community who understands the deep evil of abortion to denounce abortionists as murderers or those who seek physician-assisted suicide as those who would be self-murderers. Yet, for those who live within a robust moral community, they will have clear understandings that many of the autonomous individual choices of cosmopolitan liberals are deeply misguided. Rawls, however, will not willingly support space for them to pursue freely their robust and illiberal view of human flourishing.

For Rawls, there are no robust rights to privacy, no exclaves within which consensual communities can act out views of human flourishing that may collide with the values of the cosmopolitan liberalism realized within a social democracy.

> A domain so-called, or a sphere of life, is not, then, something already given apart from political conceptions of justice. A domain is not a kind of space, or place, but rather is simply the result, or upshot, of how the principles of political justice are applied, directly to the basic structure and indirectly to the associations within it. The principles defining the equal basic liberties and opportunities of citizens always hold in and through all so-called domains. The equal rights of women and the basic rights of their children as future citizens are inalienable and protect them wherever they are. Gender distinctions limiting those rights and liberties are excluded. So the spheres of the political and the public, of the nonpublic and the private, fall out from the content and application of the conception of justice and its principles. If the so-called private sphere is alleged to be a space exempt from justice, then there is no such thing (Rawls, 1997, p. 791).

The implication is that even when a community organized around illiberal commitments is consensual, it must be regarded as unjust and to be abolished, if only through tax incentives, burdensome constraints, and education. A robust community may not have moral commitments that are at odds with the society of cosmopolitan liberals.

V. THE SEARCH FOR COMMUNITY: LOOKING FOR MEANING IN THE WRONG PLACE

As we look into the next millennium and the future of bioethics, we find the field in a paradoxical condition. On the one hand, it often advances itself as united in a consensus that neither exists in fact nor can be established in principle. Moral plurality is not only real. It cannot be set aside by sound rational argument without begging crucial premises or arguing in a circle. Yet, a consensus transcending this plurality is invoked to legitimate health care policy. Although the thick moral commitments of cosmopolitan liberalism cannot be established by sound rational argument, they are nevertheless endorsed as providing the moral structure and content for a community that demands to be global. This demand stems from a conflation of the necessary role of permission as the source of authority when moral strangers meet and liberty as a value to be realized in one's life through authentic autonomous choices. This project of authentic autonomous individuality constitutes a view that persons should on the basis of an examined life affirm only that moral law they can discursively show to be well established and which they can without undue cultural influence apply to their lives. In affirming this moral view, one moves from recognizing permission as the source of secular moral authority to considering free choice as the source of secular moral meaning. This focus on meaning derived from free choice leads to affirming the conditions for such liberty and therefore to endorsing free equality of opportunity as a condition for the relationship among persons. Also, because all that has meaning must be in concert with the cardinal commitment to authentic autonomous choice, the constraining hand of tradition and of the past is brought into question, evacuating moral content and transforming the moral into the aesthetic.

At the same time that this cosmopolitan liberal ideology is affirmed and embraced as the foundation for bioethics and health care policy, there is the nagging sense that there is a lack of a community of meaning and caring once experienced by many. This hunger for meaning then often leads to a more intensified embrace of cosmopolitan liberalism without recognizing that, by its very character, cosmopolitan liberalism brings into question and removes the basis for the strong community that many seek. Nor is there a recognition that the only justifiable goal for the liberal is a neutral social structure within which diverse consensual communities can flourish on their own terms. The result can only be a public policy that the libertarian liberal will understand to be misguided and the true communitarian will understand to be an enemy of the integrity of robust communities.

How ought one to respond? First and foremost, one must recognize that cosmopolitan liberalism is but one among other particular views of human flourishing, and as such is on similar footing as other substantive views such as particular religions and cultural accounts. It would be a serious error to regard cosmopolitan liberals as a morally necessary consequence of recognizing that persons are the source of moral authority when moral strangers meet. Not only should one recognize that the content of cosmopolitan liberalism does not carry a moral necessity, but also that one is free to affirm the richness and goodness of those traditional moral visions that one understands as allowing the ample flourishing of the human good. On the one hand, this will require resisting claims on the behalf of a global bioethics. On the other hand, this will require affirming the importance of exploring regional and cultural bioethics, as well as those embedded in particular religious understandings. This will mean taking seriously as well the competitive plausibility of other political structures within which health care policy can be nested. Since the cosmopolitan liberal worldview has no more right to be imposed on unconsenting persons than thick traditional moral views, it is thus worth indicating in closing the equal plausibility of softly authoritarian, capitalist political views. After all, security and market liberties have as much a claim to enforcement as those thick civil liberty rights of the cosmopolitan liberal that in fact constrain free choice among consenting persons[13] through the imposition of a particular view of authentic autonomous flourishing. As we go to the future, we must learn to take moral diversity seriously and to nurture the conditions under which it can flourish.

The lesson of libertarian liberalism is that by default we have no authority to deny space for robust communities, at least when their structure can from the outside be seen to rise out of the permission of their participants. This is the case even when individuals join not out of a commitment to authentic autonomous choice, but to submission, obedience, and humility. As we go to the future, we must take moral diversity seriously and provide substantial place not only for individual, but for communal privacy within which robust communities can have their own moral space. Those who realize that a robust traditional community discloses moral truths which discursive arguments cannot secure will need to nurture the conditions under which such a robust community can flourish.[14]

Department of Philosophy, Rice University
Houston, Texas, USA

NOTES

[1] For a candid assessment of the role of bioethicists in giving authority to a particular health care proposal, that of President Clinton, see O'Connell, 1994 and Secundy, 1994.

[2] One of the major concerns has been with respect to the use of humans for research without adequate consent. Insofar as adequate consent is absent, this would count as a moral violation not only for the cosmopolitan liberal, but for a libertarian liberal understanding of morality. That is, its status as a violation would not depend on a particular view of human flourishing, but would only involve a recognition of persons as the source of moral authority when moral strangers meet [Engelhardt, 1996].

[3] The Soviet Union, Maoist China, and Cambodia under Pol Pot, in the pursuit of the rights of workers and peasants, as well as the realization of universal justice, slaughtered tens of millions of individuals. In their commitment to a content-rich understanding of liberation and progress, they came to endorse terror, if not as an intrinsically valuable instrument, then at least as a useful means for achieving what they took to be a more humane future. Consider, for example, the remarks of Merleau-Ponty:

> It is certain that neither Bukharin nor Trotsky nor Stalin regarded Terror as intrinsically valuable. Each one imagined he was using it to realize a genuinely human history which had not yet started but which provides the justification for revolutionary violence. In other words, as Marxists, all three confess that there is a meaning to such violence—that it is possible to understand it, to read into it a rational development and to draw from it a humane future [Merleau-Ponty, 1969, p. 97].

[4] Conclusions of scientific investigation may lead to the coercive imposition of particular policies, but this is usually only after such findings have been placed within and interpreted by a set of moral and value interpretations. It is the non-epistemic values, in particular moral values, that lead to the imposition of a particular policy in the name of justice and right action. See Engelhardt and Caplan, 1987. For a study of the ways in which the fashioning of empirical data has played a role in the achievement of political purposes, see Graham, 1981.

[5] Although many of the arguments raising questions regarding the moral allowability of abortion are robustly embedded within religious moral and metaphysical assumptions, many are not. See, for example, Brody, 1975.

[6] A softly authoritarian state is a non-totalitarian state that suppresses political criticism of its governance, but otherwise provides rule of law, freedom of religion, and even freedom of speech in matters political not bearing directly on that state's governance.

[7] For an example of an argument connecting equality and liberty, see Nielsen, 1985.

[8] Family-oriented approaches may differ as to whether they regard the family as the source of both authority and value, or instead regard individuals as ultimately the source of moral authority, but families as the content for human moral flourishing.

[9] "Justitia est constans et perpetua voluntas jus suum cuique tribuens" (Justinianus, 1970, 1.1, p.5).

[10] Communities are understood to be robust only when they provide metaphysical orientation, strong narrative structure, and a thick community of caring.

[11] Many decisions regarding reproduction, birth, suffering, and death are considered private insofar as they involve the consensual collaboration of moral agents while eschewing state enforcement on persons not party to the collaboration.

[12] Charity may by many be recognized as a way of turning from self to God and others and thus to salvation, while at the same time allowing the recipient to learn humility and advance in virtue.

[13] Consider the fact that current fair-housing legislation in the United States forbids Orthodox Jews from forming a community with restrictive covenants so that only Orthodox Jews may

live in the proximity of a synagogue. Such laws make it difficult for communities of like-minded individuals to live together, especially when these communities are based on common religious commitment. For an elaboration of similar arguments but with regard to religious discrimination in employment, see Epstein, 1992.

[14] The reader should know that the author belongs to a robust moral community: he is an Orthodox Christian. See Engelhardt, 1995.

REFERENCES

Bayertz, K. (1996). *The Concept of Moral Consensus*. Dordrecht: Kluwer.

Beauchamp, T., & Childress, J. (1979). *Principles of Biomedical Ethics*. New York: Oxford University Press.

Brody, B. A. (1975). *Abortion and the Sanctity of Human Life*. Cambridge: M.I.T. Press.

Engelhardt, H. T., Jr. (1996). *The Foundations of Bioethics*. 2nd ed. New York: Oxford University Press.

Engelhardt, H. T., Jr. (1995). Moral content, tradition, and grace: Rethinking the possibility of a Christian bioethics. *Christian Bioethics, 1(1)*, 29-47.

Engelhardt, H. T., Jr., & Caplan, A. L. (Eds.) (1987). *Scientific Controversies*. New York: Cambridge University Press.

Epstein, R. (1992). *Forbidden Grounds: The Case Against Employment Discrimination Laws*. Cambridge: Harvard University Press.

Graham, L. R. (1981). The multiple connections between science and ethics. In: D. Callahan & H. T. Engelhardt, Jr. (Eds.), *The Roots of Ethics* (pp.425-438). New York: Plenum Press.

Hoshino, K. (Ed.) (1996). *Japanese and Western Bioethics*. Dordrecht: Kluwer.

Justinianus, Flavius Petrus Sabbatius (1970). *The Institutes of Justinian*. T. C. Sandars (Trans.). Westport: Greenwood Press.

Kontzevitch, I.M. (1988). *The Acquisition of the Holy Spirit in Ancient Russia*. Platina: St. Herman of Alaska Brotherhood.

Marx, K., & Engels, F. (1967). *The German Ideology* New York: International Publishers.

Merleau-Ponty, M. (1969). *Humanism and Terror*. J. O'Neill (Trans.). Boston: Beacon Press.

Nielsen, K. (1985), *Equality and Liberty: A Defense of Radical Egalitarianism*. Totowa: Rowan and Allenheld.

Nozick, R. (1974). *Anarchy, State, and Utopia*. New York: Basic Books.

O'Connell, L. (1994). Ethicists and health care reform: An indecent proposal? *Journal of Medicine and Philosophy 19*, 419-424.

Rawls, J. (1997). The idea of public reason revisited. *University of Chicago Law Review 64(Summer)*, 765-807.

Rawls, J. (1971). *A Theory of Justice*. Cambridge: Harvard University Press.

Secundy, M. (1994). Strategic compromise: Real world ethics. *Journal of Medicine and Philosophy 19*, 407-417.

ten Have, H., & Sass, H. (1998). *Consensus Formation in Healthcare Ethics*. Dordrecht: Kluwer.

PART I

COMMUNITY AND CARE:
LOST PERSPECTIVES?

JULIA TAO LAI PO-WAH

IS JUST CARING POSSIBLE?
CHALLENGE TO BIOETHICS IN THE NEW CENTURY

I. INTRODUCTION

This paper begins by reviewing some main concerns about the
inadequacies of dominant theoretical approaches to bioethics in the last
century. These concerns have led to increasing pressure to search for
alternative frameworks and new paradigms to guide heath care and
biomedical decision making which do not cast caring and justice as two
oppositional and irreconcilable moral requirements. The paper examines
three alternative frameworks: feminist care ethics, agent-based virtue
ethics, and Chinese Confucian ethics and analyses how they can provide a
different orientation for framing and developing alternative systems of
bioethics. It concludes that these three ethical perspectives can provide
foundations for a new bioethics which is more sensitive and more
responsive to the moral ideal of 'just caring' than current mainstream
impartialist or principle-based approaches. The paper argues for the
importance for bioethical discourse to engage seriously with these three
ethical perspectives in order to move away from the dualism and
reductionism of the current American dominant paradigm of bioethics,
which cannot be the global bioethics for the new century.

II. MORAL LOSS, MORAL DIVERSITY AND CONTEMPORARY
BIOETHICS

Living good and dying well is thought to be increasingly difficult for
want of moral consensus and moral guidance in post-traditional and post-
modern societies which characterize the new century. If however moral
loss and moral diversity are indeed the main features of moral life in our
post-traditional, post-modern world, we have reasons both to
commiserate and to celebrate. We commiserate the moral disorientation
which is unavoidable, but we celebrate the moral liberation which is
inevitable. The former is the source of moral perplexity and confusion;
the latter is the source of opportunities for innovations and paradigm

*Julia Tao Lai Po-wah (ed.), Cross-Cultural Perspectives on the (Im)Possibility of Global
Bioethics*, 41–58.

shifts. Such emancipation is particularly welcome in the field of bioethics which has been under great pressure to search for new approaches to understand and to respond to breathtaking and controversial developments in biotechnological applications in divergent societies and cultures in the new century.

Much of today's medical ethics and received principles of bioethics reflect the influence of ethical theories which employ, in the main, a modern western philosophical framework which is then applied to issues such as consent, euthanasia, surrogacy, organ transplantation, etc. Such an approach requires critical examination, particularly in light of the requirement of pluralism and in the context of non-western cultural traditions. Furthermore, the ethos of modern bioethics is much dominated by values and methodologies that are closely associated with American liberal political ideology. Many of today's received principles of bioethics, with their emphasis on individual choice and patient autonomy, can be traced to the patients' rights movement of the 1960's in the U.S. The central goal is to affirm patients' rights, in particular, the right to decide about their own health care, and to uphold equal treatment and due process. Congruence with American liberal political ideology explains the success of health care ethics in the second half of the previous century. As pointed out by Callahan,

> The final factor of great importance [to the acceptance of bioethics] ... was the emergence ideologically of a form of bioethics that dovetailed very nicely with the reigning political liberalism of the educated classes in America. Politically America has always been a liberal society, as manifested by the market system economically and by a great emphasis on individual freedom in our cultural and political institutions. Bioethics came along with the kind of intellectual agenda that was wholly compatible with that of liberalism. (Callahan 1993, S8)

It is not surprising that contemporary bioethics discourse is dominated by an American-dominant bioethics with its central values of individual choice, personal autonomy and informed consent. However, recent expansions in biotechnological applications and the increasing reception of eastern systems of ethics have thrown up challenges which cast serious doubts on the reliability of conventional theories and dominant approaches to provide satisfactory answers or adequate solutions. They raise questions about the limits of autonomy in patient care, the role of family in medical decisions, the place of care in moral reasoning, and the importance of the moral character of medical professionals.

III. UNIVERSALIST THEORIES AND PRINCIPLE-BASED APPROACHES

These doubts and questioning have further led to much criticism and re-evaluation of universalist moral theories and mainstream principle-based bioethics which mirror these theories and which constitute the paradigm of modern bioethics. In particular, they reflect three major concerns. First is their over-emphasis on morality as impartial rule-following. Second is their exclusive focus on general abstract norms in moral reasoning. Third is their excessive individualism in defining the goals of medical ethics.

Moral theories in the west are generally characterized by their concern with universalism, expressed in the articulation of general rules and principles, or "criteria" of right and wrong to guide moral judgment. For example, consequentialist philosophers such as Hobbes (1651) conceive of morality as a set of rules based on collective interest, or a system of contractual reciprocity, which if everyone follows, everyone benefits. But what makes moral rules normatively binding is the agent's own interest. Mutual advantage and agreement bring the individual agent's good and the collective good together. What morality requires of me is what it would require of any person in a situation like mine. Moral agents are conceived of as isolated, abstract individuals in the sense that they are both independent of others and free to choose what relationships to have with others. The model of interaction is contractual – the moral agent chooses to whom she will be related and the conditions of the relationship. Their moral obligations are spelled out in abstract rules, rules that are general enough to bind all others similarly situated.

In a similar vein, liberal rights philosophers such as Rawls (1971), who defend an ethic of justice, conceive of morality as constituting a fair or just system of rules for resolving disputes among individuals who are self-interested and mutually disinterested. In Rawls's vision, the ethic of justice is also premised on a presupposition which defines self and others in universal or general terms. It understands morality to be impartial in treating all as separate but free and equal individuals, and to be concerned with upholding the primacy of universal individual rights to protect free choice and autonomy.

Under the influence of these universalist theories, contemporary bioethics emphasizes a view of justice which takes as its central point a notion of impartial treatment. It also supports a view of moral reasoning as the development of the understanding of rules and rights. It holds that some general norms or action guides are central in moral reasoning (see for example Beauchamp and Childress, 1989; Childress, 1998). There is little place for moral emotions and care in moral reasoning which emphasizes the understanding of relationships and responsibility within this paradigm of bioethics.

Increasingly, there is much objection to these universalist theories and the bioethics paradigm which they support. The major concerns are their inadequate treatment of personal (and family) relationships, and their tendency to lead to an objectionably minimalist conception of a good life. The consequence is that they tend to leave us with an impoverished moral vocabulary and a highly constricted moral vision because they tend to evaluate all actions in terms of 'right', 'wrong', 'obligations,' or 'permissible' (Oakley, 1998, p. 86). Because of their conception of justice as the following of rules with impartiality, they cannot see any significant place for care in the moral life, or in bioethical decision-making. In fact, care is often viewed with suspicion because it is seen to undermine the requirements of justice and impartiality. Their reductionist approach and dualistic perspective tend to set justice and care as two irreconcilable perspectives which are invariably opposed to each other. Neither can they recognize that impartiality and relationships are both central to moral life and moral deliberation.

There are also strong objections to conventional principle-based approaches to bioethics because of their excessive abstractions, and their tendency to overlook the multiplicity of features which can have moral relevance in different contexts and which vary across different cases. They do not adequately capture the contextual nature of decisions in patient care nor do they give due recognition to the moral importance of a healthcare professional's character in their conception of bioethics. As a consequence, they have difficulty in responding sensitively to real life bioethical issues. In particular, they are found to be either inadequate or irrelevant to address issues which entail highly complex caring responsibilities and commitments in a wide range of contexts, involving family, friends, and distant strangers, and in complicated practices involving patients who are in long term care, who are incompetent, or who are demented.

These concerns and critical reflections have led to the search for new modes of self-understanding and new perspectives for framing and interpreting issues in health care and bioethics which can restore caring in human life and contemporary bioethical accounts. This essay will examine three alternative ethical systems from both east and west, and explore their potential contribution to the development of new approaches to guide bioethics in the post-traditional and post-modern world.

IV. ETHICS OF CARE

The first proponents of a care ethics, Gilligan (1982) and Noddings (1984), and other feminist writers such as Held (1993) and Hirschmann (1992), understand moral agents to be embedded in particular social contexts, relationships and personal narratives. They argue a different thesis that persons are not fundamentally separate and isolated, but rather are literally constituted by the relationships of which they are a part. The starting condition of all human beings is an enveloping tie. Each of us enters into a complex web of relationships, including family, friends, neighbours, colleagues, fellow community members, fellow citizens, and so on in our life which provides the basis for developing individuality and particularity. Relationships of these different kinds involve different forms of care and concern. Our place in the network of relations is thus also a locus in a network of different forms of care. Contractual relations are not to be ruled out, but they would not be regarded as paradigmatic of human relations.

A care ethics as developed by these writers sees ethics as fundamentally concerned with how properly to care for the particular others to whom we are related within the various different relations of care and concern we share with them. Morality, on this view, is about responsibility and responsiveness within relationships. What morality calls for is fundamentally a matter of what to do within these specific relations. The form in which ethical issues present themselves is thus not a matter of what someone should do in a situation like mine where a person to whom I am related in such-and-such a way will be affected in such-and-such a manner. Care, as emphasized by care ethicists, is particularistic. Ethical issues appear in a particularistic form. This calls for increasing recognition of the significance of contexts, relationships and obligations for understanding human needs and health care choices.

1. Process and Particularity

In particular, two essential features are distinctive of an ethics of care. They have important implications for a different understanding bioethical issues and moral decision-making. First is the emphasis on a process approach to ethical deliberation. In contrast to conventional principle-based approaches, a care ethics pays much attention to procedural questions of ethical deliberation and emphasizes a process of ethical thought and discussion involving the participants. It is committed to involving all relevant others in thinking through their respective problems and about how their input is itself relevant to what one should do in a moral situation. A good example would be when someone has to decide whether to continue life-support for patients. This is not a technical decision that only doctors are capable of making. Nor is it a personal decision which affects the individual patient alone. It is an ethical decision on which patients and others may have views no less defensible than those of doctors. For care ethics, the preferred way to solve a moral dilemma is not to set up a hierarchy of principles, and have one type of consideration trump another. It recognizes that one's response in a moral dilemma involves the more delicate and demanding task of weighing the various moral considerations bearing on the situation. The way to react in a moral dilemma is to think contextually and make discrete decisions, instead of appealing to coarse-grained principles. There is no hard and fast rule generally applicable for all situations, for all people. Such an approach provides justification for greater consideration of contexts, relationships, and biographical histories in assessment processes for health care decisions. It also legitimizes greater involvement of family members, personal friends, and professional care-givers in the decision-making and care giving process.

The second essential feature of the ethics of care is the emphasis on particularity. It is the essential element which defines the distinctiveness of interactions in close personal relationships. Particularity implies irreplaceability and non-substitutability (Chan, 1993). As Chan explains, "when we conceive of a person or an object as a distinct particular, we mean that the person or the object cannot be replaced or substituted without changing the nature of the relationship we have with the person or the object."

Chan further distinguishes between a weaker and a stronger notion of irreplaceability (1993, p. 114). For the weaker claim, a person is irreplaceable, because each individual has her own individual traits, qualities and idiosyncratic elements which creates a unique complex wholeness that make special contribution to a relationship. Her role therefore cannot be replaced by another person without changing the nature of the relationship. For the stronger claim, it is not in virtue of her distinctive qualities that a person becomes irreplaceable. It is the person herself qua herself that is irreplaceable. The individual is in principle non-replicable, not just because she possesses qualities that are de-facto non-replicable. It is the person herself, in which particularity truly inheres.

The particularity perspective of the care ethics enables us to see the person who is related to us as the irreplaceable particular. It also enables us to see the relationship we share as a unique particular good. We do not see persons as mere bundles of qualities. One's particularity is not so much due to the particular qualities one possesses as a person, whether these be rationality, or capacity for autonomy. Rather it is to be accounted for by one's participation in the common good of relationships.

It is in this sense that Charles Taylor (1989) describes a relationship as a common enterprise, a common good. A common good is not a mere combination or coordination of separate actions made by each individual person. This gives a common good a special sort of value. It is the commonness that constitutes its value, the value of the good. A relationship is a common good because there are common actions and meanings and there is sharing. The agents are the co-authority in the interpretation of the relationship. Partaking in a common good confers a special particularity to the participants in the common good. Relationships are important in the view of care ethics because they confer particularity on the participants. Relationships which capture our core commitments are central to making one the kind of person one is, constituting one's basic orientation to the world, and shaping the meaningfulness of one's life. They are core commitments because they are identity-conferring commitments (Blustein, 1991, p. 53).

2. Personhood and Relationship

A care ethics conception of person emphasizes that every person is a unique particular. It emphasizes that the intrinsic conception of personhood includes particularity, and not only rationality or autonomy. In this sense, the other is not just a bearer of claims or rights or rational capacity; the other is also a bearer of particularity in a world of sociality with whom the self is connected. It is the value of particularity in the sense of non-substitutability and irreplaceability which constitutes the basis of personhood.

Such a conception of personhood understood in terms of particularity is in sharp contrast to the notion of personhood in either consequentialist or deontological moral theories which underpin the paradigm of modern bioethics. In utilitarianism, the worth of persons is derivative from their status as loci of desires, pleasures and pains. On Kant's view (1987), human beings are ends-in-themselves and they have dignity in virtue of their rational will or lawmaking capacity. But care theorists argue the view that persons have value just because of the irreplaceability of persons.

In the face of the dominance of impersonal ethics and the alienating and depersonalizing forces of modern society, care ethics' emphasis on particularity in human relationships has important implications for assessing bioethical issues. Such an insight allows us to move away from the deeply entrenched view in western ethics that "rationality" or "rational capacity" is the central important quality in defining personhood. It also reminds us that patient autonomy does not have to be the overriding principle to guide ethical and bioethical decisions. In real-life consideration of bioethical issues, such a point of view can enable us, for example, to move away from the traditional over-emphasis on "higher-brain functions" as the single most important criterion to define human personhood when assessing the mental capacity of brain-injured or demented patients. The danger of equating loss of rationality with loss of personhood is that it can lead to the conclusion that loss of rationality implies failure to count as a person and consequently less entitlement to care, resources and treatment (Au, 2000, pp. 212-215). The claim to less entitlement is often justified on the ground that since personhood is defined in terms of rationality, if rationality is lost, the worth of an individual's life is lost. In this way, the labeling of "loss of personhood" can have the grave consequence of supporting a nihilistic attitude that in turn reinforces the loss of personal identity of the patient, making "non-person" a self-fulfilling prophecy (ibid).

Similarly, the danger of viewing patient autonomy as the primary value to guide health care decisions can lead to the conclusion that certain patient groups, such as those in long term care, and those who are incompetent or suffering from dementia, will count only as "minimal human persons" who require only "minimal moral consideration"

(Pullman, 1999, p. 33). A care ethics is much more relevant to guide bioethics in the context of long term care than the dominant ethics of patient autonomy. It is more able to justify equal moral consideration for these patients since it does not make moral consideration proportionate to the degree of rational capacity or autonomy present in the recipients of care. The care ethics perspective enables us to appreciate that each person is an irreplaceable particular, who, even in her utmost neediness of care, or severe loss of rational capacity, as in the case of Alzheimer or demented patients, may still be contributing to the on-going nature of human relationships which are a value in themselves. In this kind of caring, one is meeting the other person in herself, not as bearer of universal properties.

3. Contextual Thinking, Moral Reasoning

For care ethics, the moral point of view is the caring point of view. It emphasizes contextual thinking over impartial thinking in our approach to bioethical issues. Contextual thinking is thinking which hinges on experiencing actual and imagined situations. Placing oneself in concrete situations, actual or imagined, facilitates one in picking up salient features of a situation. Judgments are made on the basis of the salient features. It is a way to treat every situation as a discrete situation warranting discrete solutions. The emphasis on process also is an emphasis on inclusion and on hearing people's voice their own concerns. In this way, care ethics make respect for difference and particularity the genuine requirements of a defensible moral perspective. A care ethics enables us to recognize that "long after the capacity for autonomy has diminished or vanished, relationship ties continue to exert normative force" (Pullman, 1999, p. 26).

In summary, a care ethics is distinguished from traditional universalist theories by its emphasis on the context and process of ethical deliberation, on the role of emotions in moral reasoning, and on understanding personhood as being dependent upon connectedness in particularistic relationships, rather than being dependent merely upon the capacity for rationality. In health care and bioethical decisions, it gives much greater recognition to the place of family, friends, and professional care-givers than principle-based approaches. A care-based moral perspective poses five important questions to the prevalent mode of moral

thinking which has come to dominate bioethical decision-making in this century:
   (1) Is every case analyzable in terms of right?
   (2) Is autonomy the highest value in every case?
   (3) Is rationality sufficient for defining personhood?
   (4) Is sufficient attention given to the acknowledgement of the fundamental human condition of interdependence?
   (5) Is the patient the only responsibility of the doctor?

V. CRITIQUE OF THE ETHICS OF CARE

Critics (see for example, Marilyn Friedman, 1993; Daryl Koehn, 1999) have also raised three major sets of questions about the care ethics :
   (1) What limits should be placed on our care obligations? How to avoid abuse of care? How should care be constrained by other values? How to prevent care from becoming oppressing, exploiting and enslaving?
   (2) How is care to have an impact on public life? Does care have anything to say about inequality and unjust institutions that could provide the basis for collective action? How to move from the personal to the social and the political?
   (3) Does care have anything to say about strangers and distant others? How to induce the agent to widen the scope of her concern and attention beyond the sphere of intimates? Does admitting care in our moral and political framework mean the displacement of justice and moral judgment made on the basis of justice?

The central theme running through these three sets of questions seems to be the larger question of "Is Just Caring Possible?" The question of "just caring" also brings into focus another important concern which is that a care ethics would still need to be grounded in some conception of the good life in order to justify what kind of relations qualifies as an ethically good action.

VI. AGENT-BASED ETHICS

In response to these concerns, Michael Slote (1998) argues for an ethic of caring to be conceptualized as a form of virtue-ethics. He regards the morality of caring as a form of agent-based virtue ethics according to which it is the agent's motivation or motive which becomes the ultimate basis for evaluating actions, institutions, laws and whole societies as well. The moral goodness, rightness, and wrongness of actions is a function of whether they exhibit or reflect a (sufficiently) caring attitude (or motive sufficiently close to caring) on the part of the moral agent. But Slote also emphasizes that caring only counts as caring in the fullest sense if it aims at and takes steps to promote the well-being (or virtue) of those within its orbit of concern.

According to Slote, "caring is admirable precisely because it is directed toward the welfare of another and thus because of the kind of *intentional state* it intrinsically is" (174). Caring conceived as a form of agent-based virtue ethics emphasizes the value of caring as a function of the *inner life* of the agent. It focuses on agents and on admirability and moral goodness of actions. However, notwithstanding the view that the virtue/motive of benevolence is fundamentally and inherently admirable, it does not entail that the virtuous, benevolent individual can simply choose any actions she pleases without the admirability or goodness of her actions being in any way compromised or diminished.

Slote argues, on this view, that the laws, customs, and institutions of a given society express the motives (and knowledge) of the social group in *something like* the way actions express an agent's motives. National public institutions and law are just when they reflect (enough) such concern on the part of (enough of) those who create (or implement or maintain) them or at least do not reflect a (great) lack of deficiency of this motive. On this view, the social virtue of justice is a function of individual virtue(s), i.e., of the virtuousness of the individuals who constitute a given society. In this revised ethics of caring, it is argued that self-concern should be in some measure or degree balanced against concern for others *considered as a class,* and this would much reduce the danger of anyone's becoming a slave of or to others. Obligations to strangers and distant others are recognized, because the morally most attractive or virtuous individual is one who has *deep* concern for particular others, but who is also *broadly* concerned with the well-being (and moral development) of other human beings. Slote believes that this conception of justice within a care ethics is a practically attainable ideal if sympathy with and concern for others come naturally to people and can be cultivated by proper education and forms of social life.

Slote's reconceptualization of the care ethics would allow the ethics of caring to widen its concern to social morality and to address our obligations to people we do not know from within a virtue-ethical framework. Such an approach to social justice is in marked contrast to Rawls's approach according to which a theory of the justice of the basic structure of society has to precede any account of individualistic moral norms or virtues. Furthermore, Slote's account of agent-based care virtue ethics supports the need for bioethics in the new century to re-focus attention on the moral importance of a health professional's character, since reference to character is *essential* in the justification of right action.

This is particularly relevant to the realities of long term care which entails paternalistic intervention from the outset, and requires ever-increasing degrees of such intervention as physical and mental functions deteriorate. While this fact indicates a need for vigilance in regard to potential abuse, it is argued that in the end it is the moral character of the caregivers rather than some tenuous conception of patient autonomy that is at issue (Pullman, 1999). This is not to say that a virtue-ethics approach is incompatible with the idea of patient autonomy as an important value. But we must have confidence that health care professionals will strive to serve the on-going interests of their patients to the best of their abilities. It is important that the potential for abuse regarding paternalistic intervention should never be under-estimated or ignored. However, the issue for concern is often not paternalism per se, but rather the character of the agent who exercises it.

A virtue-ethical framework can also provide an important countervailing force to the rise of the consumerist movement in health care. Under this movement, health professionals have little moral independence from the patients who they serve. This could lead to a kind of 'deprofessionalization' of health care workers. A virtue ethics emphasis on the importance of the moral character of professionals can provide some remedy to this situation, and restore confidence that they will strive to serve the on-going interests of their patients to the best of their abilities.

VII. CHINESE CONFUCIAN ETHICS

Chinese Confucian ethics has played a significant role in shaping interpretations of, and responses to, bioethical issues in many eastern societies. It is a virtue-based ethics which emphasizes a relational understanding of personhood, a commitment to caring obligations, and a conception of *jen* or humaneness as the highest moral virtue. *Jen* or humaneness as the highest moral ideal pervades all human relationships, including the patient-doctor relationship. The Confucian moral tradition is often characterized as the philosophy of Humaneness *(Jen)*. It is no coincidence that medicine in the Chinese context is often referred to as the "Art of Humaneness" (Qiu, 2000, p. 292)

1. Humaneness and the Foundation of Medicine

When Confucians argue the thesis of humaneness as the art of medicine, what they put stress on is that the principle of medicine is the principle of humaneness. They are noted to have said, "Whoever has no chance to work as a good prime minister, may work as a good physician" (Unschuld, 1979, p. 98). The reason is that both a prime minister and a physician are expected to practise humaneness or *jen*.

Jen as humaneness is the key moral concept which guides human relation in Confucian society. The concept has two senses in the Confucian account. Taken narrowly as a particular virtue, *jen* refers to the virtue of benevolence and altruistic concern for others. As a general virtue, it stands for the highest moral ideal and encompasses all the other particular virtues such as humaneness, benevolence, reciprocity, propriety, filial piety, loyalty, etc.

The Confucian account locates the source of *jen*, humaneness, in the natural 'heart sensitive to the sufferings of others' which all humans possess. It is a virtue that could emanate from inside oneself to all aspects of life. Mencius explains a human being's inherent potential to have compassion and sympathy for others in this way:

No man is devoid of a heart insensitive to the suffering of others. When I say that all men have a heart which cannot bear to see the sufferings of others, my meaning may be illustrated thus: suppose a man were all of a sudden to see a child about to fall into a well, he will without exception experience a feeling of alarm and distress. He will feel so, not because he wants to gain the favour of the child's parents, nor to seek the praise of their neighbours and friends, nor from a dislike of the cry of the child. The heart of compassion is the germ of *jen (Mencius* 2A:6).

The claim that is being made here is that the heart of compassion is the seed of humaneness, *jen*. It is the genesis of the cardinal virtue of *jen* which we all possess in virtue of being human. Confucius himself firmly believes that "*Jen* can only come from the self – how could it come from others?" (*Analects* 12:1).

Compassion is important because it signifies attentiveness to and engagement with others. It enables us to know about others and to be others-directed. Thus the development of one's heart of compassion is the route to become a person of humaneness, or *jen*. The way to achieve this ideal is through a process of self-cultivation and moral transformation. It is therefore not surprising that Confucian ethics puts great expectations on the cultivation of the moral character of medical professionals who are supposed to be guided by the virtue of humaneness, *jen*, in their relationship with patients. As the Chinese saying goes, "Physicians have the heart of parents," suggesting that physicians should be as caring and committed to their patients as parents. In return, they receive the same trust and respect as parents deserve. Patient self-determination is a weak concept in Chinese medical ethics. Instead there is even today strong endorsement of the duty of physicians to engage in paternalistic interventions which are considered to be generally justified if motivated out of concern for protection or enhancement of the best interest of the patient.

2. The Place of Family in Bioethics

Confucian ethics emphasizes a moral duty to nurture and develop the heart of compassion to achieve the ideal of humaneness, *jen*. Such moral development is dependent upon the practice of care where family is the natural starting point of one's caring obligations. Confucian care ethics rejects universal love. Instead, Confucius himself also urges that a person practising *jen* should start from one's parents and siblings and then extend to other people. The process is always from those who are immediate to those who are far off, or from the role-related others to non-related people in general, from those we love to those we do not love. *Jen* begins in the family and in family relationships, but its final destination is the general other, a process which Mencius describes in this way:

Treat the aged of your own family befitting their venerable age and *extend* this treatment to the aged of other families; treat your own young in a manner befitting their tender age and *extend* this to the young of other families... In other words, all you have to do is *take this very mind here and apply it to what is over there.* Hence, one who *extends his bounty (tuen en)* can bring peace to the four seas, one who does not cannot bring peace even to his own family.... (*Mencius* 1A:7).

Similar to western feminist care ethics, Confucian ethics recognizes relationship and connectedness as the basis of personhood, as well as our moral starting point. Weaving through the Confucian care ethics are the twin doctrines of "love by gradation" and "care by extension" (Tao, 2000, p. 225). There is no requirement to treat everyone equally with the same impartial treatment. There is however the imperative to move from relationship-love to general love through self-cultivation, extension and transformation which is what distinguishes Confucian ethics from feminist care ethics. But they share the same insight that human relationship, rather than consent or right, is the basis of morality. Our obligations are defined by our relationships. In important health care matters and treatment decisions, family and family relationships are given special importance because of their centrality in Confucian ethics. Family decision-making and family consensus are valued no less than individual decision-making and independent choice. The self-determining individual is not the ideal in the Confucian framework. It challenges western biomedicine for its assumption of individuality and rationality as the basis of personhood. It emphasizes the relational basis of personhood and mutual dependence in the family which provides an important corrective to the excessive individualism in western biomedicine.

3. Care and Justice in Moral Judgment

Confucian ethics recognizes that moral dilemmas are inevitable when confronted by conflicting caring obligations, or tensions between impartial requirements and particular responsibilities. The Confucian ethical account offers no hard and fast rule which can be generally applicable for all situations, for all people. It does not emphasize setting up a hierarchy of principles, and have one principle trump another. Responding to moral dilemmas involves the delicate and demanding task of weighing the various moral considerations bearing on the situation and forming judgment. Confucius says:

> A man of humaneness (*jen*) in dealing with the world is not for anything or against anything. He follows righteousness (*yi*) as the standard (*Analects* 4:10).

The Chinese character "*yi*" for the English word "righteousness" means "appropriateness." In moral judgment, there is no requirement of giving equal weight to all, or treating like cases alike in Confucian ethics. But there is a requirement to be sensitive and responsive to all moral claims and to emphasize considerations of context, particularity and following the principle of appropriateness or righteousness to come to some kind of considered judgment in resolving moral dilemmas. A virtuous person is expected to possess the disposition to judge correctly. In this sense, Confucian ethics shares with feminist care philosophers the view regarding the limitations of universal principles and the ideal of impartiality by emphasizing the process of ethical reasoning. It also emphasizes moral judgment by moral characters. We learn from the following story from Mencius that even when care and justice are in direct conflict, Confucian ethics does not require the sacrifice of one for the other:

> Tao Ying asked, "When Shun was emperor and Kao Yao was the judge, if the Blind man (Shun's father) killed a man, what was to be done?" (Mencius said) "The only thing to do was to apprehend him." "In that case, would Shun not try to stop it?" "How could Shun stop it? Kao Yao had his authority from which he received the law." "Then what would Shun have done?" Shun looked upon casting aside the Empire as no more than discarding a worn shoe. He would have secretly carried the old man on his back and fled to the edge of the sea and lived there happily, never giving a thought to the Empire (*Mencius* 7A:35).

Shun, when forced to make a decision, chose to take the side of his father and abandon the throne. Instead of using his authority to bend the law in favour of his father, or sacrifice the integrity of the law for the sake of his father's life, he gave up the position which solidarity with his father would incur a violation of the law. In fulfilling his particularistic caring responsibilities, he observed his moral responsibility not to willfully subvert justice or to corrupt the law. The case is a good illustration of

Confucian moral reasoning which involves weighing the various moral considerations bearing on the situation and forming judgment to strike an appropriate balance between the different moral claims.

VIII. CONCLUSION

There is of course more to moral life than justice or caring. What Confucian moral reasoning emphasizes are the virtues of judgment, wisdom and the capacity to strike a balance, to find the mean between impartial requirements and particularistic claims. Bioethics in the new century should try to avoid reducing moral reasoning to a contest between caring and justice. Neither should it allow moral parochialism to be elevated as a kind of ethical universalism. Moral progress does not have to imply moral convergence. But it does imply developing a perspective that allows us to see the complexity of moral situations and to make appropriate judgment. It also implies developing moral character to strike a reasoned balance between the different demands that characterize moral life, and to respond sensitively to ethical dilemmas in bioethics and health care matters. As Steven Toulmin (1981) reminds us, practical reasoning is always a matter of judgment, never a matter of formal theoretical deduction. It is true that "in the absence of virtuous moral agents, we can never be insulated from unethical decisions and actions, no matter how many careful qualifications are made about if and when a principle can be justifiably invoked" (Pullman, 1999, p. 37).

Bioethics in the new century will need to be grounded in new thinking about foundational issues of morality which will reformulate our conception of morality and redefine our perspective on issues of approach, method, motivation and standard of morality. Chinese Confucian ethics, feminist care ethics, and virtue ethics are useful frameworks within which to look for the necessary resources and values for forging alternative approaches to enrich debates and to address issues of diversity, pluralism, and difference within the overall ideal of "just caring" in the 21st century.

Department of Public and Social Administration
City University of Hong Kong
and Centre for Comparative Public Management and Social Policy
City University of Hong Kong, Hong Kong

REFERENCES

Au, D.K.S. (2000). Brain Injury, Brain Degeneration, and Loss of Personhood. In: Gerhold Becker (ed.), *The Moral Status of Persons: Perspectives on Bioethics* (pp. 209-218). GA: Amsterdam-Atlanta.

Beauchamp, T. & Childress, J. (1990). *Principles of Biomedical Ethics*, 3rd Ed. New York: Oxford University Press.

Blustein, J. (1991). *Care and Commitment: Taking the Personal Point of View*, New York: Oxford University Press.

Callahan, D. (1993). Why America Accepted Bioethics. *Hastings Centre Report*, 23 (Suppl.), S8-S9.

Chan, S. Y. (1993). *An Ethic of Loving: Ethical Particularism and the Engaged Perspective in Confucian Role-Ethics*, Ph.D. dissertation, University of Michigan.

Childress, J. (1998). A Principle-based Approach. In: Helga Kushe & Peter Singer (Eds.), *A Companion to Bioethics (pp. 61-71)*. Oxford: Blackwell.

Friedman, M. (1987). Beyond Caring: the Demoralization of Gender. In: M. Hanen & K. Nielsen (Eds.), *Science, Morality, and Feminist Theory* (pp. 87-110). Calgary: University of Calgary Press.

Gilligan, C. (1982). *In a Different Voice*. Cambridge: Harvard University Press.

Held, V. (1993). *Feminist Morality: Transforming Culture, Society and Politics*. Chicago: University of Chicago Press.

Hirschmann, N. (1992). *Rethinking Obligation*. Ithaca: Cornell University Press.

Hobbes, T. (1651)(1991). *Leviathan*, R. Tuck (ed.). Cambridge: Cambridge University Press.

Kant, I. (1785) (1987). *Fundamental Principles of the Metaphysics of Morals*, T. L. Abbott (trans.). Buffalo: Prometheus Books.

Koehn, D. (1998). *Rethinking Feminist Ethics: Care Trust and Empathy*. London: Routledge.

Lau, D.C. (trans.). (1979). *Confucius: The Analects*. London: Penguin.

Lau, D.C. (trans.). (1988). *Mencius*. New York: Penguin.

Noddings, N. (1984). *Caring: A Feminist Approach to Ethics and Moral Education*. Berkeley: University of California Press.

Oakley, J. (1998). A Virtue-ethics Approach. In: Helga Kushe & Peter Singer (Eds.), *A Companion to Bioethics (pp. 86-97)*. Oxford: Blackwell.

Pullman, D. (1999). The Ethics of Autonomy and Dignity in Long-Term Care. *Canadian Journal on Aging*, 18 (1), 26-46.

Qiu, R. (2000). Medicine – the Art of Humaneness: On Ethics of Traditional Chinese Medicine. In: Robert M. Veatch (Ed.), *Cross-Cultural Perspectives in Medical Ethics* (pp. 292-307), (2nd ed.). Sudbury: Jones and Bartlett Publishers.

Rawls, J. (1971). *A Theory of Justice*. Cambridge: Harvard University Press.

Slote, M. (1998). The Justice of Caring. In: E.F. Paul, F.D. Miller & Jeffrey Paul (Eds.), *Virtue and Vice* (pp. 171-195). New York: Cambridge University Press.

Tao, J. (2000). Two Perspectives of Care: Confucian *Ren* and Feminist Care. *Journal of Chinese Philosophy*, 27 (2), 215-240.

Toulmin, S. (1981). The Tyranny of Principles. *Hastings Centre Report*, 10 (2), 31-39.

Unschuld, P. U. (1979). *Medical Ethics in Imperial China – A Study in Historical Anthropology*. Berkeley: University of California Press.

FR. THOMAS JOSEPH

LIVING AND DYING IN A POST-TRADITIONAL WORLD

I. INTRODUCTION

Individuals increasingly find themselves alone in large-scale societies that do not have the moral and metaphysical resources to aid them in finding meaning for their lives and deaths. Between these individuals and their surrounding societies there are increasingly fewer well-founded moral communities that can indicate the possibility for meaning and moral orientation. People who approach end-of-life decisions increasingly lack sufficient value commitments in order unambiguously to guide their choices. They find themselves in a state of anomie without direction or moral community. The result is that difficult choices, if they are to have a principled normative character, become impossible: decision-makers at the end of life often no longer possess a point of moral orientation that can provide definitive moral guidance. They know they have important choices to make, but they do not know how to characterize what is morally at stake, and therefore no idea of how to make a good choice. As a consequence, life-and-death choices in health care can lead to endless puzzles about what treatments to accept, what procedures to decline, and what forms of treatment are morally appropriate. When people no longer possess a clear view of the morally good life, they will have no clear view of when it no longer makes sense heroically to struggle against death. That is, they will lack an understanding of when attempts to postpone death are morally dangerous as well as what should morally be secured while dying, so as to have a good death.

In this sense, the challenges of bioethics and health care policy are made more difficult by the post-traditional character of the lives and deaths of many people around the world. Decisions are made outside of a tradition in the sense of outside of a taken for granted context of values, social roles, and exemplars of successful action. Increasingly, people lack the family structure, cultural values, cultural authorities, and moral narratives that have traditionally directed choices. Having abandoned a traditional moral context, they approach life and death decisions in the absence of a new source of structure, values, or authority. It may be unclear who is in authority to make decisions. It will surely be unclear

Julia Tao Lai Po-wah (ed.), Cross-Cultural Perspectives on the (Im)Possibility of Global Bioethics, 59–67.
© 2002 *Kluwer Academic Publishers. Printed in Great Britain.*

what moral content should guide decisions. Often the families of patients themselves are fractured. In addition, patients are frequently alienated from their religious commitments, and they have no one to give them direction. At best they may remember that they once possessed a sense of moral orientation which they no longer claim. Between this memory and their current state there may be a feeling of loss, even discomfort. As a result, although the patient and family have the sense that the end of a person's life is a moment of some significance, they have no moral, metaphysical, or cultural resources by which to characterize this significance.

This state of affairs is the very opposite of that for which an Orthodox Christian would hope. Where an Orthodox Christian would expect to find himself facing decisions within a family bound in love and commitment to God, the post-traditional person is no longer clear as to the nature of the family, or how to understand the meaning of life, death, and the universe. He lacks moral authorities as well as sources of moral content. This engenders a loss of fundamental orientation in many patients encountered by physicians and health care professionals, as well as by priests. Many people no longer possess what would be required to understand or to be able to make sense of their lives, their suffering, and their deaths: they lack an intact moral narrative or a coherent account of the good life and death.

In this essay, I draw on my experience as a pastor caring for my parishioners in the church and in their homes. Most especially, I address the changes I have seen take place in health care from the perspective of a priest visiting patients in the hospital. Here I speak not so much of my parishioners, but of others I encounter in the hospital, the person in the next bed who lacks a traditional moral context from which to acquire points of orientation. Even in my brief experience, I have witnessed the weakening of moral and spiritual orientation, and the results this has had on patients and their families. In short, I offer a selective phenomenology of the moral life world of many contemporary patients in order to indicate some implications of the loss of moral community: there is now often a lack of moral orientation when facing major passages of life.

II. AFTER THE FAMILY AND DEAF TO GOD

Patients come to health care without families. The problem can be merely geographical. For example, people retire in Florida, placing patients hundreds if not thousands of miles away from their families. Even where extended families could still in principle function well, geography disperses them so they cannot function effectively. They cannot give each other the support, orientation, and guidance they might want and seek. Often, the difficulty is not simply geographical, but structural. Members of families are alienated from one another. Over their life histories, they have married, reproduced, divorced, remarried, and reproduced again, leaving complexly fractured family circumstances, marked by distance and at times estrangement if not bitterness. Often matters are so complicated that one would need a map to determine who is related to whom, sharing what values, beliefs, and commitments. Appeals to the family in such cases will disclose a geography of problems rather than provide solutions. Rather than finding a structure that can help develop appropriate medical decisions in the face of suffering and death, one encounters in the family a source of conflicts and confusion.

The contemporary decrepitude of the family does not show that we should not take the family seriously. Even in its broken and strained character, the ideal of the family remains and should remain. It offers a natural social structure which, when intact and functional, can help in achieving moral orientation by providing a community bound in love and informed by a common moral understanding or religious vision. Either common love and commitment, or common moral or religious orientation by itself will help focus moral choices in the face of suffering and death. When these exist together in a family one has a foundational natural institution for facing suffering, disability and death. Yet, one may be tempted to retreat to a default position of autonomous individualism, rather than attempt to recapture the strengths of families within traditional cultural and religious contexts. One of the difficulties in attempting to reinvigorate the family is one of its very strengths: families bring with them concrete substantive moral commitments. Substantive commitments divide a society. They ingrain a moral pluralism. In addition, some critics are especially concerned about the family's connection with religious faith because of their commitments to rendering society ever more secular (Rawls, 1997).

Monotheistic religious concerns will in this regard be especially powerful and divisive by recognizing that if one does not orient all that one has, including one's family, to the God Who is the source of all, everything to some extent will go shy of the mark. Such a religious perspective also offers a diagnosis of contemporary difficulties in medical decision-making, especially end-of-life decisions. Traditionally, a unique ontological and axiological perspective is recognized. As one becomes deaf to God's word, it becomes ever less clear what the purposes of families are, what their role in caring for others during sickness and illness should be, and what values should guide medical decision-making. One is at the very least deprived of an account of human sexuality, reproduction, disability, suffering and death with metaphysical depth for both its narrative and its account of the good. A monotheistic religious account, by rooting its vision of appropriate moral decision-making in a creator God, offers an account which provides a harmony of the good and the right, as well as a location for the meaning of human suffering in terms of the ultimate meaning of things. As one steps away from such an account, one is confronted with an array of intriguing choices, but without a substantive basis from which to sort out the right answers.

III. ANSWERS REQUIRE CONTEXTS

Bioethics is full of puzzles. The newspapers and national commissions (e.g., the National Bioethics Advisory Commission) address an array of questions that appear for many unanswerable. Is cloning wrong? Is surrogate motherhood immoral? When is physician-assisted suicide an improper way to treat terminal illness? These are surely very important questions, but the specification of an answer requires a context. Because one once had definitive answers to such questions within a traditional moral context, one may falsely assume that one can have analogously specific answers within a fully secular pluralist moral context. To determine what a correct answer would be, one needs to know something of real moral content such as that provided through traditional religious accounts of the meaning of reproduction, sexuality, marriage, life, suffering, and death. This, of course, is exactly what one lacks when one asks such questions inside a fully secular, pluralist post-traditional moral context. One does not have a source from which to draw specific guidance for an answer. Instead, one has numerous answers and no principled way to sort out the right answer.

Here in illustration I offer a contrast between two stereotypical axes which bring together my experiences from encounters with a number of patients. One axis involves the availability of family and community support, the other involves the availability of a point of moral and metaphysical orientation. The two cases offered bring together and render anonymous details from a number of cases. My encounters with patients are often in semi-private rooms where one of the patients is an Orthodox Christian and the other not. In this context I witness a conflict between cultures. Take, for instance, the example of two patients, both of whom have serious illnesses likely to lead to death over the next six months to a year, where the question at issue focuses on whether aggressive treatment should be used or foregone. Aggressive treatment would postpone death and allow the patient then to be cared for at home. Withholding treatment would allow the patient to die in the near term, thus avoiding the need for nursing and other care at home as well as for likely numerous repeat hospitalizations. How should the family choose on behalf of the patient, presuming the patient, at least temporarily, is incompetent? Granting that in many jurisdictions the family surrogate decision-makers are supposedly reporting the wishes of the patient, realistically they will be giving an account shaped by their, not necessarily by the patient's, view of a good death.

In such circumstances, when I come to visit Orthodox Christian patients, I often encounter in the same semi-private room two patients superficially not that different from others in our culture. The Orthodox like others have been seduced by their culture often to the detriment of their family and their personal lives. Yet, in facing death they and their family in different ways come to recognize that they must now take dying and living seriously. Religious Orthodox Christians hope for an opportunity for reconciliation with family members and true repentance before God. They come to see the broken character of much they have done. To the amazement of the other patient's family in the semi-private room, the rituals of Orthodoxy envelope the ill parishioner in prayers that seem never to end. If anyone has devised a way of fashioning long prayers and rituals, the Orthodox have discovered a way to make them longer. Out of all of this there is usually a personal and familial, psychological and spiritual turn from the trivialities of life to the seriousness of dying well.

For the Orthodox family there is a last opportunity to take care of the dying person. Postponing death may allow not only the dying person through repentance to be reconciled to God and family, it will also allow the family to love unselfishly and turn to the patient with love. There is an understanding of a common task binding them in love to God and to each other. On the other hand, when the dying person has made peace with God and those around him, it will often be important not to attempt to postpone death. This will especially be the case when postponing death will cause one's life and dying to be enveloped in medicine and not in love to God and others. For the Orthodox Christian, death has its meaning in a larger sense of life. The process of dying is thus recognized as having a crucial significance for the patient as well as the family.

While all of this is taking place with the Orthodox patient and family, I frequently overhear the puzzled remarks of the other patient and family in the room. They have less of a sense of the need for repentance, reconciliation, and final acts of love. They are more concerned with a death with dignity, with comfort, and without mess and bother. They are often deeply conflicted and torn between different loyalties. After a short period of time, death with dignity for the family means a relief from the burdens of caring for a family member who will die in any event. On the one hand, they may feel a sense of loyalty to their dying family member. On the other hand, they are tired and exhausted with a labor that seems to them to have no ultimate purpose. Decision making for the patient then often becomes decision making in terms of the comfort of the family. Under such circumstances it becomes easier to sedate the patient and to hope that the patient will die with dignity where this often comes to mean a death without too much bother for the family and without the need of nursing care at home. In moments when the patient is incompetent, the two families report the past wishes of patients framed within quite different background moral metaphysical assumptions. When the patients are competent, they have radically different moral resources to draw on for end-of-life decision-making.

These examples of two quite different families represent different poles of two spectrums, one of family and community functionality and the other of moral and metaphysical orientation. Though there are even clearer contrasts, there are many cases in between. The poles of the spectrum indicate a contrast between the living and dying that embeds the family in a narrative with ultimate purpose versus a living and dying where there are only immanent, primarily aesthetic and comfort concerns.

The patient in the latter case cannot articulate enduring bonds of commitment nor a clear sense of the importance of dying well. Nor does that person have a community that can give psychological and social support, not to mention moral and spiritual guidance. Under these circumstances, one is confronted with two general alternatives. First, one might simply concede the day to the forces fragmenting families, traditions, and moral communities. Under this circumstance, one will need to look for surrogate structures to bring and sustain values. Confronting the next millennium would then involve trying to create new mechanisms to support individuals through attempting to give their lives meaning, if one wishes for a context of coherent meaning.

However, there is another alternative. One can instead return to family, tradition, and God and attempt to understand anew these ancient sources of meaning. Here, the challenge is to explore how particular communities can sustain nurturing and caring families, maintain commitments to values, and orient their energies to the God Who is the Source of all, while still living in a larger, indeed increasingly global society which is secular, and at best indifferent to religion. In the light of this option, which I take to be the only true option, the challenges of the next millennium take on quite a different character. One is challenged to recapture for traditional moral and religious communities certain elements of the life of the ghetto, in the sense of a traditional community sustained within as well as set apart from and against an increasingly global society marked by secular moral understandings. One will need to find means to sustain family and community in the larger context of an indifferent, if not hostile society.

IV. BIOETHICS AND THE FRAGMENTATION OF CULTURE

This analysis of end-of-life decision-making is meant to lead to three observations offered as heuristic points of departure for bioethical reflections concerning the vanishing roles of family and community in providing moral orientation. The first observation is a very practical one drawn from my observations as an Orthodox Christian priest giving pastoral care in hospitals: once family structure is broken, values lost, and orientation to God abandoned, it is very difficult to put things back together as they were. Moral and spiritual fragmentation has real consequences. My first observation is one of caution: one must examine the ways in which one can invigorate the roles of family and particular moral communities. It makes more sense to turn to the next millennium as an opportunity to maintain important social structures, moral commitments, and orientation to God rather than to ask whether one can fabricate useable surrogates. The hope to create in one generation a surrogate for what has functioned for centuries can at best be a dangerous temptation.

The second observation is that feasible surrogates will not easily be forthcoming. My position here is obviously grounded in my Orthodox Christian religious commitments and my understanding that God is the Center of all. As one steps away from God, from Christ, things do not work as they should. Much of the broken character of the lives of patients attests to this, as does the confusion about how to make choices for one's own life in the absence of any sense of the meaning of one's life or one's death. Only deep meaning will prevent suffering and death from being deprived of moral significance, and of having instead only aesthetic significance. If one does not recognize an ultimate moral significance to life, suffering and death, living well becomes an aesthetic not a moral project. In this circumstance, one will abandon moral discourse when one addresses substantive decisions at the end-of-life. The focus will shift from the substantive to the procedural.

This leads to my final observation: in a morally fragmented, pluralist society bioethical discussions are doubly burdened. Not only are the questions themselves difficult, but they are asked in circumstances lacking the moral orientation or spiritual guidance to indicate what should count as an appropriate answer. On the one hand, one may have a cultural memory of when there was a framing moral significance which could give a definitive place to moral and aesthetic concerns. On the other hand, in the absence of a definitive point of moral orientation, moral choices become individual creations with a fully immanent meaning. Those who know there is a transcendent truth will recognize such surrogates as radically inadequate to human existence. Those who do not recognize such a truth can but experience a contraction of the sphere of human moral experience, a retreat from a Truth once seen at its horizon.

Bioethical decision-making requires points of moral orientation and direction. This is especially the case when the decisions being made involve issues once held to be of great moral moment, such as those involving the process of dying. When there is no longer a sufficiently thick moral account of what is at stake in living and dying, there remains only a sense that there is something of importance to do, but without a clear sense of what exactly that might be. The result is moral confusion and disorientation. This confusion and disorientation may border on anomie when compounded by the absence of the support and content once provided by family and community. After having lost metaphysical orientation to God, in the absence of intervening moral structures between the individual and the larger society, the death of patients may be marked by moral vacuity, if not chaos.

Director of Pastoral Research
International Studies in Philosophy and Medicine
North St. Petersberg, Florida, USA

REFERENCE

Rawls, J. (1997). The idea of public reason revisited. *University of Chicago Law Review 64 (Summer)*, 765-807.

PART II

THE (IM)POSSIBILITY OF GLOBAL BIOETHICS

REN-ZONG QIU

THE TENSION BETWEEN BIOMEDICAL TECHNOLOGY AND CONFUCIAN VALUES

Since the Asian economic crisis and the new Balkan War, the world has become much more uncertain. It is hard for me to predict how the third millennium will be, and also how the bioethics of the third millennium will be, so I limit my paper to two issues: the value of traditional cultures in the 21st century and the third millennium, and the possibility of the pursuit of global bioethics or universal ethics.

I. THE DESTINATION OF TRADITIONAL VALUES: THE MUSEUM OR THE RUBBISH HEAP?

Knowledge is power. Modern technology is a great and fearful power. Wherever it invades, tradition seems to give way to it. It creates new culture everywhere. Almost every country with a traditional culture, after being defeated by imperialists, tries to modernize. They try to introduce, apply and develop modern technology in their country. However, this in turn causes the conflicts between tradition and modernity, or between traditional values and modern values.

The Chinese may never forget that the history since the Opium War in 1840 is a history of humiliation, exploitation and aggression by foreign powers. These powers used advanced weapons, attractive commodities and ingenious technologies to knock China's door open, and impose their values on the Chinese people. The first response from the Chinese was to try to repudiate all of these powers outside the door. The dismantling of a railway near Beijing by the government of the Qing Dynasty and the 1900 Yihetuan Movement,[1] in which traditionalists tried to use *qi gong* to fight against aggressors, are events symbolic of the response. But the counter-action to the latter was the further aggressive invasion by the Eight Powers.[2] The attempt to keep tradition unchanged and to reject modern values failed. Since then, many attempts have been made to reform the country following the example of Japan, but these attempts were guided by a principle called "Chinese learning as a fundamental structure, Western learning for practical use" (*zhong xue xi yong*). But

Julia Tao Lai Po-wah (ed.), Cross-Cultural Perspectives on the (Im)Possibility of Global Bioethics, 71–88.
© 2002 *Kluwer Academic Publishers. Printed in Great Britain.*

these attempts at reform failed too. Eighty years ago there was a May 4 Movement, the objective of which was to invite Mister Sai (science, modern science including modern technology) and Mister De (democracy) to China, along with the slogan "Down with Confucianism" *(Da dao Kong Jia Dian)*. This objective has become a perpetual dream for Chinese intellectuals and laymen. Regarding the slogan "Down with Confucianism," should the traditional Chinese culture, with its core of Confucianism, be relegated to the historical museum or should it be discarded upon the rubbish heap? Or can it still can play some important and even indispensable role in the era of modern science and technology in the 21st century and the third millennium?

The same questions can be asked of modern biomedical technology, which can efficiently treat or prevent human diseases in an unprecedented way, including altering the human genome, human body, human mind, human relations and human species. In this case also, cultural conflicts and moral dilemmas emerged. The responses to modern biomedical technology in China vary. In a Shanghai case, grandparents and their family rejected a boy born by AID (artificial insemination by donor). In a Beijing case a patient's family sued the doctor for taking their kin's cornea postmortem for transplantation. In both cases Confucian values, which include extending ancestors' lives and maintaining bodily integrity before burial or cremation, were neglected in the practice of modern biomedical technology. Some scholars, including a Taoist philosopher and Heidegger expert at our Institute, argue for the abandoning of all research into the nucleus of the cell and the exploration of the secret of life. Others try to reinterpret Confucianism or Taoism in order to adapt it to modernity. Given this latter strategy, gene therapy, human cloning and other controversial projects are acceptable to Confucianism and Taoism. In this part of my paper it will be argued that Confucianism and Taoism have to be reinterpreted in a way in which they can adapt to modernity, but that there is a hard core that cannot be interpreted in different ways or reinterpreted away. A proper interpretation or reinterpretation of Confucianism will not impede the application or development of biomedical technology, and meanwhile it can play a very important normative role in it.

Human genetics and cloning technology are being developed in great China. In a society where Confucianism has deep-rooted influence on people's mind and social conventions, many challenging issues have emerged and demand a response. In recent years there have been hundreds of articles in professional journals, popular magazines, newspapers and books, and programs shown on TV on the mainland of China, Hong Kong and Taiwan on human cloning and the human genome project. In 1998 and 1999 symposia on ethics and bioethics were held on the mainland of China, Hong Kong and Taiwan at which many participants made presentations on human cloning and the human genome project. A number of articles addressed ethical issues in genetic intervention and cloning from Confucian and Taoist perspectives.

1. The debate on genetic intervention and reproductive human cloning

The approaches and conclusions of these articles are diverse. Interestingly enough, some proponents of Neo-Confucianism and Neo-Taoism favor these technologies, including reproductive human cloning. S-C Lee[3] argues that for Confucians there may be something in the course or change of nature that is not in accordance with human moral judgment, and that humans have natural moral requirement to change it. It is a kind of making up for the inadequacy of nature that violates the *dao* of nature. The argument is that: "Humans can carry forward the *dao*" (*ren neng hong dao*). If humans do so, it actually realizes the Confucian ideal that humans participate in the creation and transformation of nature. So in this sense Lee sees no problem with "playing God" for Confucianism. In other words, we are required to play God together with God. If there is something wrong in human reproduction, we can make up for it with negative eugenics, gene therapy for treatment, and even human cloning. But positive eugenics and gene therapy for enhancement is not expanding the *dao*, but violating it (Lee, 1998). But I think Lee's Confucianism is too new to adhere to the hard core of Confucianism. Confucian classics reiterate of the *dao* that one is *yin* and the other is *yang*, and that only *yin* or only *yang* alone cannot produce or reproduce. Confucius warned us to stand in fear of it (*wei tian*) because *tian* is omnipotent and the warning implies that humans should not play *tian* in a similar sense with playing God.

Another argument is raised by D-H Chen.[4] He argues that Neo-Daoists will not reject gene repair, gene therapy, organ cloning and human cloning because they are not against the law of nature. "It is the *dao* of nature (*tian*) that the excessive be reduced and the deficient supplemented" (*Dao De Jing*, Chapter 77). Neo-Daoists must only reject medical domination, commercial oligarchy and the abuse of eugenics

(Chen, D.H. 1998). However, that they are 'not against the law of nature' is not a good argument for the use of these technologies. When nuclear energy was used in war, it was also not against the law of nature. On the contrary, its success showed that the bomb makers knew the law of nature. The hard core of Daoism is 'inaction' (*wu wei*), or 'do nothing unnatural'. For Daoism all things, including humans, in the universe are and should be the natural outcome of interaction between *yin* and *yang*. It is hardly possible for Daoists to justify the practice of reproductive human cloning or even therapeutic organ cloning.

Most Chinese ethicists argue that Confucian values are in conflict with the use of these technologies. Their arguments mainly focus on familial integrity or the orderly familial relationship, although they don't exclude other arguments. P-S Pang[5] argues that there is an essential difference between IVF (*In vitro* fertilization) and reproductive human cloning. In the latter there is no conjugation between elements from the child's parents. It follows that the child cannot grow up in the mutual love and union of its parents. Thus, reproductive human cloning seriously erodes the most fundamental ethical basis of human relationship in a society. He also warns that it cannot be ignored that human cloning will bring about negative consequences for the cloned child, human relationship, familial structure and social stability (Pang, 1998). His argument seems to be that reproduction without conjugation entails that the cloned child cannot grow up with love. It is implied that parents take a different attitude toward a child being made, from a child being begotten. Or, as P-K Ip[6] argues, human cloning will lead to the objectification of the cloned child because it is made or even manufactured (Ip, 1998). In the same vein, Vincent Shen[7] argues that human cloning jeopardizes the harmony of the human relationship (*ren lun*) because it is made without sex and love between the parents (Shen, 1998).

The first point in the argument that human cloning will jeopardize the orderly familial relationship is that human cloning will lead to the objectification of the cloned child and a shortage of love between parents and child, due to the absence of conjugation. Here I want to point out that if we observe the relationship between an adopted child and its parents or between a child of IVF and its parents, we shall observe that the child can grow up with love even if it was not born as a result of the interaction of its parents' *yin* and *yang*. The same would seem to apply if it were made in a test tube.

The second and most important point in this argument involves the indeterminate status and role of the cloned child (Ip, 1998). T-J Kuang[8] also argues that reproductive human cloning will make the human relationship (*ren lun*) confused and puzzled (Kuang, 1998). I will discuss this contention later.

Some colleagues even argue against reproductive human cloning from the viewpoint of human dignity. This argument is also related to the familial integrity argument. R-P Fan[9] argues that reproductive human cloning will do harm to the human parent-child relationship, and hence do harm to human dignity. Reproduction without conjugation will destroy normative or orderly relationships between parents and children and thereby undermine human dignity. For Confucianism, human nature is nurtured and developed in an orderly human relationship, the core of which is the familial relationship – especially the father and son, husband and wife relationships, which are the basis of the society and country. If these relationships are abandoned or destroyed, the good human nature will be harmed (Fan, 1998b). He also argues that Confucians should not accept genetic interventions that would jeopardize orderly human relations, especially familial relations (Fan, 1998a). S-B Liao[10] and his colleagues also argue against reproductive human cloning due to the possibility that it will bring the familial relationship into jeopardy (Liao *et al.*, 1997).

Some other arguments provided by colleagues in great China are interesting, but more controversial. Jonathan Chan[11] (Chan, 1998a, 1998b) argues that unless there is evidence to prove that reproductive human cloning will do physical or mental harm to the cloned or others, human cloning should not be prohibited. The problem with this argument is that we can get this kind of valid empirical evidence only after cloning a child. Professor Shen also argues that reproductive human cloning is not good for the development and diversity of human beings, and that the cloned lack free will (Shen, 1998). However, Parfit's argument (Parfit, 1984) will neutralize his first point: a cloned child may think that her/his coming into the world with some mutated genes is better than her/his non-existence. And all children lack the free will to decide whether to be born.

2. The challenge from Confucianism to genetic intervention and reproductive human cloning

The challenge from Confucianism to genetic intervention and reproductive human cloning may focus on three issues: Do they violate personal bodily integrity? Do they jeopardize familial relations or integrity? And do they jeopardize the harmony between nature and humans or the integrity of nature?

(1) Personal bodily integrity

From a Confucian viewpoint, the human is a conjugation of the two finest energies, called *yin* and *yang*, which come from father (heaven) and mother (earth) respectively. Keeping personal bodily integrity is a value, for each body is endowed from her/his parents and cannot be harmed in any way, otherwise it would violate a primary ethical principle called *xiao*, or filial piety. In some Confucian classics like *Xiao Jing* (*Book on Filial Piety*) it is even said: "Skin and hair that are endowed by parents cannot be damaged." Actually, the body or any part of the body can be damaged only in the following two circumstances:

1. To meet her/his parents' needs. There are twenty-four models of the son with filial piety in traditional China. In one such model the son cuts one piece of flesh from his leg to feed his parents because they wanted to eat meat but were too poor to buy it.
2. To sacrifice himself for practicing *ren*. Confucius once said: "You should not preserve your life damaging *ren*, instead you should sacrifice yourself for practicing *ren*" (*The Analects of Confucius*).

Anatomy and surgery have never fully developed in traditional China, because dissection of a dead body and operation on a body will violate the principle of filial piety. It is an ethical requirement of the principle of filial piety that a child treat her/his parents' illnesses and care for them, and for the same reason each has to make an effort to treat his own illnesses in order to recover the balance between *yin* and *yang*, i.e., health. Although operation on a body and dissection after death will violate personal bodily integrity, this constraint is too strict to be accepted by most modern Chinese. They have to give up the fundamentalist interpretation of filial piety and reinterpret it so as to adapt it to modernity. The hard core of the principle *xiao* is to love, respect and care for one's parents, and *xiao* is the beginning of *ren* – Confucianism's central principle. For example, the donation of the body or an organ postmortem is practicing *ren*, doing good to others, in cases where the benefit to others (*ren* is loving, caring and doing good to others) will outweigh failing to keep bodily integrity. The same goes for operation on a body: otherwise the harm of being dead will outweigh a body not perfectly integral. In the same vein, manipulation of a gene or cell will not conflict with a reinterpreted Confucianism if it is for the treatment or prevention of disease. But if the intervention is for enhancement of a certain favored trait, it may cause an imbalance between *yin* and *yang* in the body and jeopardize personal bodily integrity. This is unacceptable even to our reinterpreted Confucianism unless there is evidence to prove otherwise.

(2) Orderly familial relations and familial integrity

Keeping familial integrity and orderly familial relations may be even more important than keeping bodily integrity for Confucianism. Each person in the family has her/his own certain status and proper role that are not permitted to be omitted or violated. To extend the life of one's ancestors is a requirement of the principle of filial piety, but this extension should be by the conjugation between *yin* (from wife) and *yang* (from husband), and not by only *yin* or only *yang*. In Confucian classics like *Yi Jing* (*The Book of Changes*) and *Li Ji* (*The Book of Rites*), it is reiterated that:

> Only *yin* or only *yang* alone cannot generate;

> One *yin* and one *yang* are called the *dao*;

> Gentlemen (*jun zi*) pay great attention to the marriage that unites two families into one in order to serve ancestors in the temple of family and extend the family to future generations. Man and woman are different than each other, so there is affection between husband and wife, then there is kinship between father and son, then there is the political relation between a king and his subjects.

So for Confucianism conjugation between husband and wife is the basis of the family and the country. Any non-conjugal reproduction will weaken the blood tie between the members of a family and lead to its instability. A case in Shanghai in 1989 showed that a traditional family could not accept a boy infant born by AID, because the child did not come from the family's bloodline. However, Confucianism can be given a weaker interpretation wherein the focus is on an affectionate tie rather than a blood tie between parents and child. Together with the concession that the balancing of values differs in different contexts, most Chinese people can accept AID, in vitro fertilization, surrogate motherhood, and genetic intervention for treatment and prevention of a disease, but not reproductive human cloning and genetic intervention for enhancement.

A cloned child will have an uncertain status in the family even though the law may help to stipulate it. Family members will have to face the child everyday, and won't know whether he is the father's brother or son, the mother's son or brother-in-law, or the siblings' uncle or brother. And the cloned child could grow up confused about her/his social identity and role. The orderly familial relationship would be shaken. Even if we give a weaker interpretation to Confucianism, a modern Confucian has to admit that reproductive human cloning would weaken the affectionate tie between the members of a family, because of the confusing status of the cloned in the family, and lead to the instability of the familial relationship. For Confucianism, the family is the basis of a society, so these results would also have a destructive impact on the society at large.

(3) The harmony between nature and humans

One of Confucianism's normative concepts is the harmony between nature and humans. Nature and humans are ontologically common: both of them are a product of the interaction between yin and yang, and they are common in nature, insofar as both possess a continuing creativity. 'The dao of heaven (nature)' and 'the dao of humans' are same, but 'the dao of humans' should follow the 'the dao of heaven'. Nature is an exemplar for humans. Human action should follow the dao of nature. Humans can carry forward the dao, but cannot transcend the dao. No action can be justified if it is done unnaturally or against the dao of nature.

One of consequences an action done against nature is that it disturbs the harmony between nature and humans in that yin and yang must be balanced. In this case the identity, integrity or diversity of nature is disrupted. When we talk about the human genome, we say that it is the common heritage of mankind.[12] The human genome will change for various reasons, but its 'identity' or 'integrity', which makes Homo sapiens as it is, has no change. The same can be said of nature. Nature has changed and will change forever, but its 'identity' or 'integrity' remains intact. The identity or integrity of nature means that nature is an entity that can sustain the survival and reproduction of most present and future species, and preserve the diversity of species of plants, animals and human beings (Qiu, 1998). Confucianism would not generally object to the use of technology. But it would object to its use when it would cause such a serious adverse impact on nature that it would become a totally different nature that no longer could sustain the survival and reproduction of most present and future species. In a word, Confucianism objects when the diversity of species of plants, animals and human beings would be destroyed. Reproductive human cloning would probably reduce the diversity of human beings, and genetic intervention for enhancement would probably disturb the harmony within the human body and between humans and their environment. Genetically-enhanced or cloned humans may constitute a new species of human beings, which would affect the harmony between nature and humans. Human action that would harm the harmony between nature and humans, disrupting the balance of *yin* and *yang*, should be discouraged, unless there is evidence to the contrary.

Confucius taught us to respect nature or stand in fear of nature (*wei tian*), and, accordingly, we have to think that there are many things we cannot predict. As S-C Wu[13] has argued, the human genome could be a holistic, non-reductionistic and non-linear complexity system. There is a possibility that it will have a butterfly effect somewhere. The fact that some positive consequence can be predicted and that the intention is good is not sufficient to guarantee the absence of a catastrophe. So for gene therapy for enhancement and reproductive human cloning, we should *wei tian*, in a sense similar with "heuristics of fear" (Wu, 1998).

The conclusion of this section is that whether Confucianism favors genetic intervention and human cloning or not depends on how we interpret Confucianism and adapt it to modernity. Confucian ethics is good-oriented and contextualistic. There are some values in it that can be interpreted in one way as favoring genetic intervention or human cloning but interpreted in other way as not favoring them. But there are also some hard core values that it is difficult to interpret in different ways or just interpret away. A proper interpretation or re-interpretation of Confucianism will not impede the development of new technology, but can play some normative role in its application. The same applies to other traditional cultures in other parts of the world.

II. HOW IS GLOBAL BIOETHICS POSSIBLE?

The character of modernity is to pursue universal law(s) or principle(s) both in nature and in the human world without exception in ontology/epistemology and ethics/bioethics. This modernization is the strategic aim of the government of many developing countries including China. But it has turned out to be a Westernization of the whole country to some degree, while traditional values persist. When these countries turned to the market economy, they became involved in globalization that in turn enhanced Westernization (Tangwa, 1998a). Modernization and globalization seem to be an expansion of Western values to the whole world. While the proponents of traditional values are crying out in alarm that China will be lose its identity because of Westernization, some Westerners and international bodies that are controlled by Westerners feel that it is necessary to hasten the process.

Some international bodies are very enthusiastic to look for a universal ethics or global bioethics. Ironically enough, in a Symposium on Universal Ethics that was held in Beijing in 1998 and sponsored by UNESCO, many Chinese Confucian philosophers argued for the universality of Confucian values. It seems to follow that the universal ethics, at least in part, should be Confucian ethics. At the fourth Congress of IAB (International Association of Bioethics), which was held in Japan in 1998, the theme was global bioethics. However, one who was not present argues that a canonical, content-rich, secular bioethics cannot be justified, and its international imposition would be immoral. He is H. Tristram Engelhardt, Jr. (Engelhardt, 1998a). A. Campbell,[14] who was present, expressed his doubts, while still admitting the existence of global ethics. Campbell wonders whether global bioethics is a dream or a nightmare (Campbell, 1998). I will argue below that in Japan global bioethics is a dream, but may become a nightmare.[15] In what follows I would like to examine the arguments for and against global bioethics in different ways.

1. Arguments for global bioethics

One argument for global bioethics is that bioethics is the love of life, and love is a normative principle as well as a universal value across communities and cultures (Macer, 1998). It was argued that when we translate four basic principles into the language of love, they will become universal too: non-maleficence is self-love, beneficence is love of others, respect is loving life, and justice is loving good. There are two rather convincing counter-arguments I would like to cite here:

1. Love is too complex, generic and diffuse a concept, and one with too many problematic associations and connotations, to conveniently and economically carry our characterization of Bioethics. If we describe bioethics as love of life, how does it accommodate at one and the same time one's love of crickets and termites, indifference to glow-worms and butterflies, fear of snakes and scorpions and aversion towards vultures and chameleons (Tangwa, 1998b).

2. 'Love', as a normative concept is not only vague, but it can be misleading, can be abused, and is regularly misunderstood, which, in effect, makes it an unreliable and even dangerous principle. ... The paradoxical conclusion is that, exactly from an attitude of 'Love', we should reject 'Love', not as the highest good, but as a normative principle in the language of bioethics (Doering, 1998).

The second argument for global bioethics is on the basis of the concept of human right. It was argued that if human rights is a legitimate concept, at least some ethical values must be universal. There seem to be two alternatives before us: Either we must altogether deny the legitimacy of the concept of human rights, or accept the possibility of global bioethics. A major proponent of this argument has admitted that different countries or regions of the world may assign different priorities to different values, but still insists that all must accept certain fundamental values within the framework of human rights (Maklin, 1998). I would like to point out that in some cultures there is not even a word for 'right' in the relevant language. In a Confucian framework, there is no place for universal human rights; instead rights are related to role. The so-called framework of human rights is one that came mainly from the Western cultural tradition. I have argued that right-talk is not a panacea that can solve all the issues facing bioethics. At a meeting devoted to discussing the reaffirmation of the slogan "Health for All," I said that if we don't do anything to develop areas that are underdeveloped, right-talk is useless. Health as a right is a positive right. You have to provide products and services to guarantee the exercise of the right to health. If these do not exist in an area, i.e., there is nothing to eat or to wear, the slogan "Health for All" is empty. Now we see some regional organization taking action on the basic universal value of human rights, but the consequences show that there is no such universal value in the region as well as in the world. For some Western experts and politicians the rights of chain murderers seem more important than the rights of their many possible victims, and civil and political rights seem more urgent than social, economic and cultural rights. These so-called experts and politicians don't care how those victims were or will be killed or raped, or how to get rid of poverty for hundreds of millions of people in developing countries. It seems that they do not think protecting victims or getting rid of poverty is a human rights issue.

The third argument is that there are some foundational moral values that can transcend particular cultures, so the foundations of global ethics are independent of the norms of any particular culture. A proponent of this argument further claims that the universality of these norms permits criticism of local actions and practices that violate the norms even if these actions and practices are dominant in the culture (Beauchamp, 1998). The problem is how can any moral values be independent of or transcend any particular culture. Instead, the more foundational a moral value is, the more dependent on a particular culture it is. Any moral value or any ordering of values presupposes a moral vision that is embedded in a culture.

A similar argument claims that any bioethics must become explicit about its foundations, and articulate its rock-bottom grounding. It has been suggested that some systems that ground their norms in nothing more than the norms of a special group, country, ethnic group, or religion are not really ethics, ultimate underpinnings for standards of moral conduct, but merely social mores. The conclusion is that any true ethics, and therefore any true bioethics, must have universal foundations, that is, must be grounded in a source that applies to all humans regardless of culture (Veatch, 1998). The problem is where these foundations come from. Can it be believed that these foundations can come from nowhere, i.e., from an authority that does not come from any culture? It is required that bioethics must also make clear its epistemological assumptions – who can claim authority for knowing the moral content found at those foundations. It means that the 'who' or knower can transcend any community or culture. So the knower must be an infallible God rather than a fallible human being who is unavoidably embedded within a community or culture.

Bioethicists from developing countries have expressed their skeptical attitudes toward global bioethics, and their skepticism is plausible. One skeptic has asked how a 'global bioethics' can be possible in a world inhabited by different cultural groups whose material situations, power, ideas, experiences and attitudes differ rather markedly and who are not, in any case, equally represented in globalization efforts (Tangwa, 1998b). Another skeptic has worried about ethical imperialism. On the one hand there is concern that universal ethical standards should be adhered to for all research on humans worldwide. On the other there are concerns about the imperialistic imposition of requirements that are culturally specific, and about the potential for exploitation by researchers from industrialized countries. It is clear that imperialism was not only a political a force which existed during the age of Empire, but that it is an ongoing feature of hegemonic world views, civilizations and religions (Benatar, 1998). The skepticism about global bioethics also comes from bioethicists in developed countries. One such skeptic has asked whether global bioethics is a dream of reconciled humanity, like the famous dream of Martin Luther King, or a nightmare of a homogenized lowest common denominator ethics. Or, worse, does it represent the cultural domination of the world of bioethics by one set of moral assumptions and one style of reasoning? Even the term *global bioethics* can be accepted only if it is be based on a respect for the richness of human culture and the fragility of the physical and social environments that sustain human life. The fundamental value of global bioethics ought to be humility about our ethical wisdom, based on an awareness of the limits of rational argument and of the prejudice which distorts our claims to ethical impartiality (Campbell, 1998). This position seems to accept a weak version of global bioethics, but reject its strong version.

At the Symposium on Asian Bioethics after the Congress I examined the arguments for and against global bioethics. My conclusion is as follows. The weak version of global bioethics, i.e., moral values shared by different communities or cultures, can be justified. Even so it reserves space for different communities or cultures to interpret and apply these shared values differently, and to develop those values which are not shared by others and which are community-relative or culture-relative. But its strong version, i.e., a single overarching bioethical theory that is invented by somebody somewhere and claims that its theory, principles, rules or values hold true to all moral communities or cultures, may lead to undesirable consequences and even ethical imperialism (Qiu, 1999b).

2. Arguments against global bioethics

Dr. Engelhardt and his colleagues have already elaborated arguments against global bioethics that I think are very convincing (Engelhardt, 1996, 1998a, 1998b; Fan 1998c). In what follows I would like to add some further arguments.

One argument against global bioethics is that everybody, including ethicists and bioethicists, is embedded in a certain community or culture. He/she cannot transcend the community or culture and her/his moral vision cannot come from nowhere. When it is argued that the foundational moral value should transcend any particular culture, or that the value of human rights is universal, these values are definitely embedded in moral communities and one kind of Western culture. It follows that the 'global' bioethics supported by these values is a bioethics that comes from Western culture but pretends to be universal to all moral communities and all cultures. Only God's moral vision comes from nowhere. But God himself is embedded in Judeo-Christian culture, and has no moral authority in Confucian culture.

The second argument against global bioethics is that the morality or ethics between different moral communities or cultures is incommensurable. Thomas Kuhn argued that the terms 'mass', 'space' and 'time' seem to be the same, but their meaning is different in Newtonian physics and Einsteinian physics. The phenomenon is similar in ethics. Values seem to be shared by different moral communities or cultures, but their meanings differ among them. Autonomy in Western culture means decisions made by an independent person, but in Confucian culture it means decisions made by a family incorporating the person. Bioethical paradigms in different moral communities or cultures are incommensurable. This means that there is no meta-criterion or meta-criteria with which to judge or evaluate the relative merits of different bioethical paradigms. It also means that there is no common moral authority, no common moral norm, and no common vision with which to evaluate an action taken by a person or an institution from different moral communities or cultures.

The third argument against global bioethics is that when bioethics pretends to hold true for all moral communities or cultures, it may lead to ethical imperialism. A so-called global bioethics may borrow some canons of morality from existing bioethics or invent something new. The fact is that Western bioethics is dominant in the world. So, the newly articulated global bioethics may be tainted with a strong color of Western culture, or may be just an another version of Western bioethics within the clothes of global bioethics. When it is imposed on the non-Western communities or heretic communities within Western culture, this constitutes ethical imperialism. Now the warning has become reality. When some country or a number of countries makes a military action somewhere using the excuse of protecting the universal value of human rights, it is claimed that it is an ethical war, but it is a war of ethical imperialism. We know from history that any aggressive war is in the disguise of protecting somebody's rights. Ethical imperialism has two characteristics. The first is a moral double standard – one moral standard is used to judge action taken by one group, another used to judge action taken by another group. The second is hypocrisy – for the ethical imperialist, not only is might truth, but might is also morality. While torturing individuals is a human rights violation, torturing a nation is not!?

All proponents of global bioethics neglect the advance of recent philosophical inquiry. At most they are at the level of logical positivism. Philosophy is now at a stage where universalism, foundationalism and essentialism are open to question. There is no rock-bottom foundation in epistemology or ethics. To quote Popper, the foundation is like stakes bumped into the sand. It is fallible. Experience is theory-laden. Deductive inference will have a high-level premise that comes either from intuition or from a higher level premise that will lead to infinite regress. The foundation cannot be secured in the essence of person either. Serious challenges have also been raised by feminism, postmodernism and multiculturalism. Confucius was right when he said: "Human nature is similar, but practice makes them far apart" (*Confucius' Analects*). Although the human genome may be the same, personhood cannot be reduced to the human genome. No person can get rid of her/his social and cultural context, just as nobody can pull her/his hair to leave the ground. There is no identical essence for persons in different socio-cultural contexts. There is not even an identical essence for men and women, or Chinese and American people. Everything that was claimed as universal has turned out to be local. The diversity of cultures, ethical values and moral communities can never be eliminated, so there is a limit to the pursuit of a universal or global ethics.

How properly to resolve the conflicts between different cultures, moral values and moral communities, and how to find a way to avoid ethical relativism and nihilism while accepting moral and cultural diversity will be a formidable task for bioethics, as well as ethics in general in the 21st century and the third millennium. The only way to resolve ethical conflicts is by dialogue, negotiation and the reaching of consent. In solving or resolving ethical conflicts, Confucianism emphasizes 'harmony'. The concept is heuristic because it means we have to respect other moral communities and cultures, be tolerant of what is we think wrong but members of other moral community think right, be patient in dialogue, negotiation and the reaching of consent, and never impose our values on members from other communities. I believe Confucianism and every other traditional culture will have its own position and play its own role in this task.

Program in Bioethics
Centre for Applied Ethics
Chinese Academy of Social Sciences
Beijing, China

NOTES

1 The Yihetuan Movement is known as the Boxer Rebellion in the West.
2 The Eight-Power Allied Forces were Britain, the United States, Germany, France, tsarist Russia, Japan, Italy and Austria. Many of them are now members of NATO.
3 Mr. Lee is a professor at the Institute of Philosophy, Central University, Taiwan.
4 Mr. Chen is an associate professor at Buddhist University, Taiwan.
5 Mr. Pang is an associate professor at Hong Kong Baptist University.
6 Mr. Ip is a professor at Hong Kong Open University.
7 Mr. Shen is a professor at Chengchi University, Taiwan and President of the Chinese Society for Philosophy, Taiwan.
8 Mr. Kuang is from East Sea University, Taiwan.
9 Mr. Fan got his Ph.D. from Rice University and is now assistant editor of the *Journal of Medicine and Philosophy* and assistant professor at City University of Hong Kong.
10 Mr. Liao is an associate professor and Chairman of the Department of Ethics, Institute of Philosophy, Chinese Academy of Social Sciences.
11 Dr. Chan is an associate professor at Hong Kong Baptist University.
12 Please see UNESCO's Declaration on the Human Genome and Human Rights.
13 Mrs. Wu is a professor at Buddhist University, Taiwan.
14 A. Campbell is the President of International Association of Bioethics.
15 What happened in the Balkans shows that the pursuit for a universal ethics or global bioethics may become a nightmare which leads to a disaster of mankind, as well as a danger in which it can be exploited as a tool to justify aggression by a new generation of imperialists.

REFERENCES

Benatar, S. (1998). Research ethics in 'developing' countries: Universalism, imperialism and context. *Global Bioethics: 4th World Congress of Bioethics*, 4-7 November, 1998 at Nihon University Hall, Nihon University, Japan, p. 97.

Beauchamp, T. (1998). Foundations of global bioethics. *Global Bioethics: 4th World Congress of Bioethics*, 4-7 November, 1998 at Nihon University Hall, Nihon University, Japan, p. 19.

Campbell, A. (1998). Global bioethics – Dream or nightmare? *Global Bioethics: 4th World Congress of Bioethics*, 4-7 November, 1998 at Nihon University Hall, Nihon University, Japan, p. 18.

Chan, J. (1998a). From Chinese bioethics to human cloning – A methodological reflection. *Chinese & International Philosophy of Medicine, 1(2)*, 49-72.

Chan, J. (1998b). Human cloning, harm and personal identity. *Proceedings of International Symposium on Bioethics*, vol. 1 (pp. 150-166). Tsung Li & Jia Yi, Taiwan, June 16-19.

Chen, D.H. (1998). Bioethical dimension of the secular Daoism: A reflection on cloning and genetic engineering. *Proceedings of International Symposium on Bioethics*, vol. 1 (pp. 73-86). Tsung Li & Jia Yi, Taiwan, June 16-19.

Doering, O. (1998). Can love be a language of bioethics? *Global Bioethics: 4th World Congress of Bioethics*, 4-7 November, 1998 at Nihon University Hall, Nihon University, Japan, p. 125.

Engelhardt, H. T., Jr. (1996). *The Foundations of Bioethics*, 2nd ed. New York: Oxford University Press.

Engelhardt, H. T., Jr. (1998a). Beyond a global ethics: Take moral diversity seriously. *Global Bioethics: 4th World Congress of Bioethics*, 4-7 November, 1998 at Nihon University Hall, Nihon University, Japan, p. 30.

Engelhardt, H. T., Jr. (1998b). Critical care: Why there is no global bioethics. *The Journal of Medicine and Philosophy, 23(6)*, 643-652.

Fan, R. P. (1998a). Human cloning and human dignity: Pluralist society and the Confucian moral community. *Chinese & International Philosophy of Medicine, 1(2)*, 73-94.

Fan, R. P. (1998b). Genetic intervention and the Confucian tradition. *Proceedings of International Symposium on Bioethics*, vol. 2 (pp. 1-14). Tsung Li & Jia Yi, Taiwan, June 16-19.

Fan, R. P. (1998c). Critical care ethics in Asia: Global or local? *The Journal of Medicine and Philosophy, 23(6)*, 549-562.

Ip, P. K. (1998). Human cloning, ethics and Asian values.' *Proceedings of International Symposium on Bioethics*, vol. 1 (pp. 18-30). Tsung Li & Jia Yi, Taiwan, June 16-19.

Kang, P. S. (1998). To clone or not to clone: The moral challenges of human cloning. *Chinese & International Philosophy of Medicine, 1(2)*, 95-124.

Kuang, T. J. (1998). Value dispute on human cloning and genetic engineering. *Proceedings of International Symposium on Bioethics*, vol. 1 (pp. 135-149). Tsung Li & Jia Yi, Taiwan, June 16-19.

Lee, S. C. (1998) On the social and ethical puzzles in human cloning: An analysis of applied ethics.' In: S. C. Lee (Ed.), *Ethics and Life-Death: Collected Papers of Asian Symposium on Applied Ethics* (pp. 105-122), The National Central University Press, Tsung Li, Taiwan.

Liao, S. B. *et al.* (1997). On human cloning. *Guangming Daily (March 14)*.

Lo, P. C. (1998). Introduction: How to think about human cloning? *Chinese & International Philosophy of Medicine, 1(2)*, 1-48.

Macer, D. (1998). Bioethics is love of life. *Global Bioethics: 4th World Congress of Bioethics*, 4-7 November, 1998 at Nihon University Hall, Nihon University, Japan, p. 20.

Macklin, R. (1998). Is global bioethics possible? *Global Bioethics: 4th World Congress of Bioethics*, 4-7 November, 1998 at Nihon University Hall, Nihon University, Japan, p. 105.

Manickavel, V. (1998). Are Southern countries missing bioethics or left out? *Global Bioethics from Asian Perspective, Asian Bioethics Symposium* (pp. 47-53). University Research Center, Nihon University, November 9-10, 1998.

Ni, H. F. & Liu, C. Q. (1999). Ethical issues in genetic and reproductive intervention. *Medicine and Philosophy, 1*, 50-53.

Parfit, D. (1984). *Reasons and Persons*. Oxford: Clarendon Press. pp. 351-441.

Qiu, R. Z. (1997). Cloning technology and its ethical implications. *Studies in Dialectics of Nature, 6*, 1-6.

Qiu, R. Z. (1998), Social and political impact of genetic engineering. *Genes the World Over* (pp. 81-91). Evangelische Akademie Loccum.

Qiu, R. Z. (1999a). Human genome project and ethics. *Journal of Dialectics of Nature, 1*, 70-79.

Qiu, R. Z. (1999b) How is global bioethics possible? A Chinese perspective. *Global Bioethics form Asian Perspective*, Asian Bioethics Symposium, University Research Center, Nihon University, November 9-10, 1998, pp. 5-10.

Sakamoto, H. (1998). Towards a new 'global bioethics'. *Global Bioethics: 4th World Congress of Bioethics*, 4-7 November, 1998 at Nihon University Hall, Nihon University, Japan, p. 18.

Shen, V. (1998). Is human cloning supported by any ethical argument? *Chinese & International Philosophy of Medicine, 1(2)*, 125-144.

Tangwa, G. (1998a). Globalization or westernization? Ethical concerns in the whole bio-business. *Global Bioethics: 4th World Congress of Bioethics*, 4-7 November, 1998 at Nihon University Hall, Nihon University, Japan, p. 37.

Tangwa, G. (1998b). Is bioethics love of life? An African view-point. *Global Bioethics: 4th World Congress of Bioethics*, 4-7 November, 1998 at Nihon University Hall, Nihon University, Japan, p. 124.

Veatch, R. (1988). Foundations of bioethics. *Global Bioethics: 4th World Congress of Bioethics*, 4-7 November, 1998 at Nihon University Hall, Nihon University, Japan, p. 29.

Wu, S. C. (1998a). Ethical issues in human genome project. In: S. C. Lee (Ed.), *Ethics and Life-Death: Collected Papers of Asian Symposium on Applied Ethics* (pp. 123-158). Tsung Li: The National Central University Press.

Wu, S. C. (1998). Genetic information and every-lifechoices: A view from the postmodern ethical perspective. In: *Proceedings of International Symposium on Bioethics*, vol. 2 (pp. 15-31). Tsung Li & Jia Yi, Taiwan, June 16-19.

Yu, Y.L. *et al.* (1998). Human genome project and humanistic responses. *Medicine and Philosophy, 7*, 346-350.

HU XINHE

ON RELATIONAL PARADIGM IN BIOETHICS

I. PUZZLES OF BIOETHICS AND THE RISE OF A RELATIONAL PARADIGM

1. Puzzles of bioethics

With the rapid development of modern industrial society, many ethical problems arise. A central one is the tendency of increasing social disintegration and loss of communal bonds. Common ideals of freedom, equality and fraternity which have had force since the Enlightenment are losing their social power. The centripetal force which once united us to make efforts for the same goals is losing the ability to bend us to the basic principles which served as moral foundations of our ethics.

These challenges are made more acute with the revolutionary advancements in genetic science and biomedical technology and the moral questions that they press; we confront the disturbing metaphor of genetic essentialism, and its attendant notion of pre-disposition, which is a paradigm that reduces all characteristics of a person, including smoking, alcoholism, even his criminal behaviors, to a genetic base. Such a paradigm of medicalization, even "geneticalization" of social problems has engendered problems in the area of law and criminal justice, employment and health care (insurance), as well as in the very concept of "patient" and "therapy" (Kegley, 1998, pp. 41-60). Another one is a case in which an HIV-positive male wants his doctor to keep his serological status confidential so that he can marry his girlfriend, or the notorious case in U.S., the Tarasoff case, in which a patient told his psychiatrist that he planned to kill his girlfriend (Qiu & Hu, 1998, p. 168). And many more of them, as indicated in the Third Theme of this Conference, "Bioethics and Concept of the Self", came from those vulnerable situations like the fetus in the abortion debate, the terminal patient in the euthanasia controversies, the Alzheimer patient with mental dysfunction, etc. The results of this circumstance are the centripetal forces which bring into question the possibility of better understanding of the relationship among individual, community, and society. A relational paradigm,

Julia Tao Lai Po-wah (ed.), Cross-Cultural Perspectives on the (Im)Possibility of Global Bioethics, 89–104.
© 2002 *Kluwer Academic Publishers. Printed in Great Britain.*

drawing in part on the resources of Confucianism, is offered as a possible new foundation for addressing this challenge.

2. Individualistic framework

Among those theories or principles at the root of these ethical problems and challenges is individualism. Individualism has become a main doctrine that underlies western ethical, political and social thought since John Locke. In the light of this doctrine, the individual should be emphasized as the focus, with both ontological and methodological priority. The consequence is that the principle of autonomy becomes foremost in an ethics of principlism. Here, the "ontologically" prior means that in the context of this view, the only real social reality is individual. Group and community are seen as nothing but a collection of individual persons. The individual, therefore, should have primacy over group, community and the state. Because it is the individual who has priority in time, it is the individual who makes a contract that serves as the basis of the state. It is the individual in whom reason was seen as an inherent endowment and in whose reason morality is based, thus giving the individual prior moral authority. As a consequence, the first function of the state is to protect the liberty of the individual in his pursuit of his interests, property and happiness.

As "methodologically" prior, the individual is a well-defined entity even without relating to his context, so that the individual could be used as the starting point of all theoretical analysis and construction (Watkins, pp. 270-271). Needless to say, such a well-defined entity would be an ideal model that implies some "self-evident" presuppositions, such as being rational, autonomous, self-interested or in pursuit of maximum benefit, etc. But each of these propositions predicts contraries, as in the debate about whether the essence of man is autonomous or plastic (Hollis, 1977). When proceeding on the basis of this model, even if we admitted it, we would run up against paradoxes at almost every step. One example is from the theory of choice. In a book with the same title, the authors give us a theoretical construction from individual choice to interactive choice, then to collective choice, on to a model of *Homo economicus* (Heap, 1992, p. xi). But as we know, there are paradoxes, such as Newcombe's paradox, the prisoner's paradox, Arrow's paradox, and the voter's paradox bedeviling each step of the construction. One explanation for these paradoxes is due to the model of *homo economicus*. And another one is due to their reductionistic tendency. The methodological individualism in social sciences follows a reductionistic approach that tends to reduce social events or facts to dispositions or psychology of persons and even reduces the characteristics of persons to genes. But as has been shown, reductionism cannot succeed even in much simpler cases such as in natural sciences.

As we know in the philosophy of science, the strong form of reductionism is physicalism, but it failed to accomplish its ambitious program. In physical science itself, a time-honored reductionistic doctrine is atomism. In fact, there is an analogy between physical atoms and moral individuals (Watkins, p. 270). The role played by individual selves in human society is similar to that played by atoms in the physical world. Genetic essentialism or genetic determinism shows a close analogy between atomism and individualism. In retrospect to this analogy, we can discern a similarity with Kant who treated physics and ethics in corresponding ways, regarding one in the pursuit of knowledge regarding laws of nature, and the other in the pursuit of knowledge concerning laws of freedom (Kant, 1949, p. 50). We can even trace it to Locke who admired Newton's work, including his modern-style atomism and the individual tendency in a moral doctrine. If so, what could we learn from the recent story of atomistic physics?

3. The rise of a relationalistic paradigm

The reductionistic strategy in physics seems to embody two directions since Locke. One of them is atomism which aims to reduce all physical objects to their ultimate constituents, such as atoms, fundamental particles, etc. The other is a doctrine called "the view of the primary quality" which directs reduction to the "primary qualities" of physical objects; these primary qualities are seen as the inherent, invariant essence of objects, which means they are independent of both their environment and our recognition. As a consequence, the reality of physical objects can be reduced to the reality of their primary qualities. However, we now know that atoms have an inner structure and "fundamental" particles are not so fundamental as we once thought them to be. The features of micro-particles are that they cannot be divided into even smaller particles. On the one hand, the parts may be bigger than the original particles as the shoe-strap theory suggests. The particles even cannot be divided into more basic components for they are inlaid in a very strong interactive, or relational structure, as the quark theory tells us. In the latter direction, since the revolution of physics in the beginning of this century, we knew that according to the theory of relativity, those basic properties such as length, duration and mass, etc., which were taken as invariant primary qualities, are frame-laden. Furthermore, according to quantum mechanics, even the most fundamental physical properties, wave property and particle property, are relative to the measuring equipment. Both quantitatively and qualitatively, the so-called primary qualities are not determinate except in relation to their surroundings. In both atomistic and primary quality-oriented directions, the reductionistic doctrine fails. From both the relativity and quantum theory, a relational paradigm of physical reality has emerged and has found "full corroboration in modern physics. At first, the theory of relativity revealed that the geometric-kinematic properties, such as position, time, and velocity (as well as length, size, duration, and mass), previously regarded as objective features, depend upon the frame of reference. Subsequently, quantum mechanics showed in addition that these properties are relative also to the means of observation" (Jammer, 1966, p. 404). A relational doctrine in philosophy of science which claims that those once taken as physical realities (such as particles, their "primary qualities", etc.) are in relation-restraints, has been justified upon these bases. The physical relations are realistic and the relations are a priori in many cases (Luo & Hu, pp. 359-379).

Moral individualism is a counterpart, even a derivative of physical atomism and shares the same reductionistic methodology. If this methodology could not work in the case of physics and has been replaced by a relationalistic paradigm, then the latter account may enlighten us as we approach moral puzzles.

II. ARGUMENTS FOR THE RELATIONAL PARADIGM

1. Transcendental argument

A relational paradigm emphasizes relations as the focus or as the a priori point of departure for ethical theory. In its light the individual self is relation-based. How then could we justify this relational paradigm and what role is played by relations for those individual selves?

The first approach is a transcendental argument. Here the word "*transcendental*" means a study which is not from experience but is universally suitable in all experiences, "which is occupied not so much with objects as with the mode of our knowledge of objects in so far as this mode of knowledge is to be possible *a priori*" (Kant, 1929, p. 59, A12=B25-26), which is to focus on the conditions of making these objects possible, conditions of making those objects what they are, conditions as H.E. Allison means in his notion of *epistemic condition* (Allison, 1983, p. 10). For Kant, his transcendental problem is "how is pure natural science, so exact and so universal as Newton Mechanics, possible?" And his task is to find the *a priori* conditions for such knowledge. We have a similar problem, "how is an individual self or a person, so familiar and obvious a fact, possible?" And our task should be to pursue the *a priori* conditions for a person.

How is an individual self possible? What makes a person what he is? What are the epistemic conditions for a person? Needless to say, it is a very complex problem. Renzong Qiu thought the necessary and sufficient conditions for an entity to be a person could be analyzed into three dimensions: biological, psychological and sociological. I think that view is rather reasonable (Qiu, 1998, pp. 2-10). I will also deal with the three dimensions, or necessary conditions, for being a person.

First of all, what makes a person what he is biologically? Intuitively speaking, it is his exterior features, his characteristics of behavior, etc. But in the sense of transcendental aesthetics, all these characteristics can appear only in a framework of the pure forms of sensuous intuitions, which are related to his genetic structure, his blood relations with his parents, etc. All of these serve as *a priori* conditions, predetermining his biological relations with the *Homo sapiens,* i.e., his ethnological status biologically. Therefore, an individual self is biologically possible only on the basis of a set of *a priori* biological relations.

Second, what makes a person what he is psychologically? Traditionally, the essentially psychological features of a person are his reason, consciousness, reflection and autonomy. Then, how is a rational or autonomous person possible? Where is the consciousness, especially the self-consciousness of a person from? The basic modes of consciousness of a person are sensation, understanding, and reason. In each of them, an interaction with surroundings is necessary. Any sensuous intuitions, judgments of the understanding and rational inferences are based on such interactions. As to self-consciousness especially, the interactions are those with other intelligent agents, through which the capacity of self-consciousness can be developed. So a child can posses the capacities of consciousness and self-consciousness only through continuous interactions with his mother's breeding, with his other family members and with much larger groups. The autonomy of a person is not innate. It is the consequence of a process of interactions. The consciousness and self-consciousness of individual is in fact an internalization of interrelations or social activities. As we all know, if a newborn were raised by wolves since his birth and never interacted with human beings, he would never develop those abilities that mark the human species, such as self-consciousness, self-reflection, etc. Therefore, an autonomous person is possible psychologically only through psychological interactions or social relations.

Last, but not least important, the social condition for the possibility of being a person involves the very essence of humans as a species. Following the psychological argument that a psychologically sound person is interaction-based, a person is possible sociologically only through a context of social communications. Similar to what N. Bohr once said: we are both audience and players on the arena of life. We might say that persons are both products and producers of social context, and a society is both a kind of relational institution which restrains a person's behavior and a collection of persons which is composed by all kinds of roles for persons.

Although a dominant doctrine is individualistic in social science since Locke and Kant, many philosophers emphasize the social dimension of a person as what he is. Feuerbach once said that an isolated individual does not possess the essence of man that lies in the species of man. Marx went a step further to say:

> Feuerbach resolves the essence of religion into the essence of man. But the essence of man is no abstraction inherent in each single individual. In its reality it is the ensemble of the social relations. Feuerbach, who does not enter upon a criticism of this real essence, is hence obliged: 1. To abstract from the historical process and to define the religious sentiment regarded by itself, and to presuppose an abstract – isolated – human individual. 2. The essence of man, therefore, can with him be regarded only as 'species', as an inner, mute, general character which unites the many individuals only in a natural way (Marx, 1976, p.7).

What he emphasized is the historical and social aspects of the essence of man. Durkheim emphasized the objectivity and reality of those social constitutions and ideologies, such as laws, language, custom, education and religious beliefs, etc., which are both external to individuals and endowed with coercive power over them (Durkheim, p. 245). All of this goes to show that an autonomous individual is not innate but a product of culture and society.

2. Empirical argument

Although the transcendental property of relation goes unnoticed in our daily life, especially in an individual-oriented society, a reflection of the behaviors of individuals, or a theoretical analysis of some of the difficulties manifested in an individualistic society, will always bring us back to the recognition of relations, a model of man existing in relations. "Real, practical and concrete men are all men in relations" (Xia, 1997, p. 7). Even "without deep reflection one knows from daily life that one exists for other people – first of all for those upon whose smiles and well-being our own happiness is wholly dependent, and then for the many, unknown to us, to whose destinies we are bound by the ties of sympathy" (Einstein, 1954, p. 8).

In fact, any ethical theory is at least partially a relational theory whose motives or objects are the relations between persons, or between a person and society, and whose aim is to establish some moral standards and principles as the guides for and restraints of human behaviors and activities. An individual-oriented theory can cause difficulties because its emphasis on the autonomy of persons inclines one to neglect one's dependence on other persons and society. I think that is why Kant ascribed the self-control, self-restraint and self-overcoming of reason to what he called "autonomy." Namely, he ascribed the external moral norms to the internal imperative of reason. But could we ascribe the moral norms solely to inner necessity as having nothing to do with external compulsion? Indeed, "whatever is morally important in our institutions, laws, and mores can be traced back to interpretation of the sense of responsibility of living individuals" (Einstein, 1954, p. 27). But could we say the ideas of these individuals are not the intellectual fruits of their time and society?

Here, I would like to cite a saying of Einstein himself as a proper answer to the question of relation between a man and society.

Man is, at one and the same time, a solitary being and a social being. As a solitary being, he attempts to protect his own existence and that of those who are closest to him, to satisfy his personal desires, and to develop his innate abilities. As a social being, he seeks to gain the recognition and affection of his fellow human beings, to share in their pleasures, to comfort them in their sorrows, and to improve their conditions of life.

Both efforts of these two kinds mould the characters of a man,

but the personality that finally emerges is largely formed by the environment in which a man happens to find himself during his development, by the structure of the society of which he grows up, by the tradition of that society, and by its appraisal of particular types of behavior. The abstract concept 'society' means to the individual human being the sum total of his direct and indirect relation to his contemporaries and to all the people of earlier generations. The individual is able to think, feel, strive, and work by himself; but he depends so much upon society – in his physical, intellectual, and emotional existence – that it is impossible to think of him, or to understand him, outside the framework of society. It is 'society' which provides man with food, clothing, a home, the tools of work, language, the forms of thought, and most of the content of thought; his life is made possible through the labor and the accomplishments of the many millions past and present who are all hidden behind the small word 'society'. It is evident, therefore, that the dependence of the individual upon society is a fact of nature which cannot be abolished (Einstein, 1954, pp. 153-154).

To show the relation between the moral crisis and the individualistic tendency, I want to cite another saying of Einstein:

Innumerable voices have been asserting for some time now that human society is passing through a crisis, that its stability has been gravely shattered. It is characteristic of such a situation that individuals feel indifferent or even hostile toward the group, small or large, to which they belong.

And talking about

the essence of our time, it concerns the relationship of the individual to society. The individual has become more conscious than ever of his dependence upon society. But he does not experience this dependence as a positive assert, as an organic tie, as a protective force, but rather as a threat to his natural rights, or even to his economic existence. Moreover, his position in society is such that the egotistical drives of his make-up are constantly being acccntuatcd, while his social drives, which are by nature weaker, progressively deteriorate. All human beings, whatever their position in society, are suffering from this process of deterioration. Unknowingly prisoners of their own egotism, they feel insecure, lonely, and deprived of the naive, simple, and unsophisticated enjoyment of life. Man can find meaning in life, short and perilous as it is, only through devoting himself to society (Einstein, 1998, 152-156).

Should we not say this passage, written a half of century ago, is an excellent criticism of the individualistic framework and is, indeed, an empirical argument for the relationalistic paradigm?

3. Instantiation

Is a relational paradigm possible? It seems we could use the Chinese ethical tradition as an example. In traditional Chinese ethics, a person is relation-based. This means that it is the ethical relations that make a person what he is. The essence of person lies in how that person can group with other persons (Xun Zi: *"Ren neng qun,"* see Lee, 1993, pp. 180-205). Goodness (benevolence), a key concept of Chinese ethics, means loving men (Confucius: *"Ren ze ai ren,"* see Shi Jun, 1988, p. 67). Each person is a product of these moral relations which are summed up as the Five Moral Relations (*wu lun*), namely the relations between parents and children, the monarch and subjects, husband and wife, the elder and the younger, and friend and friend.

Although they are not meant to be comprehensive, they do represent, in the eyes of the Confucianist at least, basic human relations. Since their claim to universality is based on commonly experienced modes of human-relatedness, they signify no more than five ordinary manners of human interaction. Despite their commonality, however, each of them is a world in itself. So far as interconnectedness among them is concerned, we can neither generalize, in terms of one particular form of relationship, nor specify which among them really occupies the most prominent position... Furthermore, the five relationships are governed by five carefully selected moral principles, each representing an important dimension of human community (Tu, 1989, Ch. 4).

These are called respectively as filial affection, rightness, differentiation, order and credit, etc. A real man is a man who lives in the context of these ethical relations and a genuine man is a man who follows these moral principles.

There are also different schools in the traditional Chinese ethics, and with them different understandings of relational paradigm in ethics. Regarding the essence of man as products of moral relations, Mencius emphasized the *a priori* aspect of human nature. He said:

> The ability possessed by men without having been acquired by learning is intuitive ability, and the knowledge possessed by them without the exercise of thought is their intuitive knowledge. Children carried in the arms all know to love their parents, and when they are grown a little, they all know to love their elder brothers. Filial affection for parents is the working of righteousness. There is no other reason for those feelings; – they belong to all under Heaven (Mencius 7:1; also see Shi Jun, 1988, p. 113).

That is a theory of "Xingshan" or "goodness by nature." But such an inherent goodness is only an original one, a seed of goodness; its development depends on the role of posterior environment and education.

III. SOME QUESTIONS AND COMPARISONS

1. Some questions

We have briefly clarified and justified a relational paradigm of ethics in response to the contemporary moral and social chaos generated by an individualistic ethics. Also, it has been shown that this relational understanding of ethics, key to understanding the relationship among individual, community and society, is well developed in Confucian thought. Still, some questions arise such as why my transcendental argument seems not to accord with our daily individualistic experiences. In addition there is the question of the relation between my transcendental argument and my empirical argument. Finally, there is the question whether we should take the Chinese traditional ethics as an ideal model of relationalistic ethics.

To the first question, in search of a solid foundation of empirical knowledge, Kant defined subjective forms as "transcendental" when they are *a priori* and universally suitable to objects of knowledge. In search of the conditions for the possibility of a concept of the individual self, what we could do is to define an objective relational structure as "transcendental" which is universally applied to all our experiences of a person. Similar to Kant who used different "transcendental" modes of recognition to constitute a system of knowledge, we use different "transcendental" forms of relations to constitute the pattern of a person, the transcendental logic of a relational individual.

To the second question, the reason why we are usually inclined to an individualistic and not to a relationalistic theory in daily life, the answer is the same as what Kant said about realism, namely we could admit both the reality of phenomenon and its foundation of transcendental ideas. "The transcendental idealist is, therefore, an empirical realist" (Kant, 1929, p. 347, A371). The transcendental realist who represents all external phenomena as things by themselves, on the contrary, may be an empirical idealist. The point is that the transcendental relationalist is coherent with the empirical individualist with whom we are so familiar; we could call him a relation-based individualist (RI). Similarly, the transcendental individualist who takes the individual as *a priori* may well be an empirical relationalist, and termed an individual-based relationalist (IR). So when my transcendental argument emphasizes the transcendental properties of relations, my empirical argument aims to show we could find the relational requirements at an empirical level in a transcendental individual tradition. The difference between these two views or perspectives lies in which one, the relation or individual, is *a priori* on the transcendental level.

As to the third question, during most of the Chinese ethical tradition, interrelations between persons were overemphasized. This severely limited or set aside the freedom of some persons. How should we face this history? According to my analysis, it is similar to the transcendental individualistic tradition which has two branches: empirical individualist and empirical relationalist, both with the same base of the transcendental individualist, i.e., individual-based individualist (II) and individual-based relationalist (IR). The transcendental relationalist also has two branches, namely relation-based relationalist (RR) and relation-based individualist (RI). The above-mentioned phenomena are the results of (RR) that misused the transcendental logic of relations in the empirical area of individuals so that no free space was left for the activities of individuals under such a double-level web of relation. There is an analogous problem in a society running under the guide of individual-based individualist theory (II). Each person would have too much free space against others' space and rights, as we noted at the beginning of this paper. So, it seems that the only reasonable choices we have are what I called the relation-based individualist (RI) and individual-based relationalist (IR) accounts.

2. Comparisons of two paradigms

Now we are left with two ethical theories that may respectively be called an empirically individualistic theory grounded in a transcendental relationalism, and an empirically relational theory grounded in a transcendental individualism. The forces of these two paradigms is respectively a relation-based individual (RI) and the individual-based relation (IR). From an ethical perspective, the advantage of the former metaphor is that because the ethical relation is prior to individual, the ethical imperative becomes an *a priori* condition, an inner requirement, and a sort of self-consciousness, not imposed forcefully by outside social institutions and ethical dogmas. Meanwhile, the autonomy of individuals remains in one's daily activity field. The account can thus satisfy both requirements of the transcendental definition of person and the empirical description of autonomy, while still understanding the pursuing of a harmonious relation with others and ethical self-perfection as the individual's inner purpose. The potential shortcoming of this account is that such a relationalistic theory can be misused by both outside social forces and by the individual himself in order to discount his autonomy, i.e., be misused as a theory of relation-based relations (RR). In such a paradigm, one must be vigilant to a tendency that a community may force a person, or a relation-based person himself may be inclined, to ignore the ethical values of himself, to constrain or even sacrifice his freedom and autonomy in order to submit to the demands of the community in the pursuit of the moral perfection. This approach may also lead to over-emphasizing the harmony thus hindering the creativity of individuals. These are all practical problems. On the other hand, while it might be beneficial in its emphasis on freedom, rights and the autonomy of persons, the focus on individual-based relations, which is the core of transcendental individualism, this approach, central to the Western ethical tradition, is fraught with puzzles such as "how is a transcendentally autonomous individual possible?" and "why must a transcendentally autonomous individual be in a moral relation with others?" etc. These problems as conceptual problems are difficult to resolve in terms of this account.

From the bioethical perspective, the notion of the relation-based person is an attractive one because harmonious relations are so important between physicians and patients, and also because so many considerations concerning human interrelation must be taken into account in this area. According to this metaphor, the order of the three basic bioethical principles that were developed originally in the individualism-oriented and right-focused framework of sociopolitical philosophy should be beneficence, justice and autonomy, namely giving priority to those principles in this order concerning human interrelations. In fact, there are some doctrines in bioethics that also advocate relational theory, such as feminist care ethics. In the light of this relational approach to ethics, the cases mentioned at the beginning of this paper cease to be moral puzzles. A person could not be reduced to his genes; a person is a product of a set of relations, including both *a priori* and postnatal. The doctor should tell the girlfriend of the AIDS patient, because the principle of beneficence has priority. Puzzles in those vulnerable situations, like the fetus in abortion debate, the terminal patient in euthanasia controversy, the Alzheimer patient with mental dysfunction, etc., should be considered on the basis of two kinds of relations. One is the family relation. There should be a consultation between the patient and his family members. Only then would a deliberative decision be made. The other is a legal relation that must establish rules, procedures and prohibitions to guide in making such decisions.

From a practical perspective, a relation-based individual would be an active and autonomous agent in social affairs, free and independent in making decisions in the empirical area. Although by the transcendental properties of relation, he would be aware that there is a bottom-line to his choice and actions: doing nothing harmful to other persons; doing what might be harmonious with others and for the happiness of other people. While the individual-based relation is only an outside constraint, an ethical convention or legal rule, it would be difficult to serve as a solid base for a convincing ethical theory and a successful moral practice.

IV. CONCLUSION

In response to the contemporary moral fragmentation which brings into question the proper relationship among individual, community and society, I put forward a relational paradigm as a substitute, drawing on an analogy with physics. Taking insights from Kant's transcendental logic, Einstein's ethical reflections and Chinese moral tradition, I argued for a relational moral paradigm, thus comparing a theory of relation-based individuals and a theory of individual-based relations. The goal is to show that the former is much superior in resolving the puzzles that appear in an individualistic moral framework. As a genuine individual one should be a relation-based interactive individual, an individual who lives in and pays attention to the collectives (family, community and society), an individual who pursues the happiness of others as well as his own interests. While maintaining individuals' autonomy, the account is conducive to an increase in the harmony, integrity and stability of the group or the community. I would say that such a theory is not a completely new one, even in Western philosophy. In Confucianism, such a genuine individual, or the true self, is an open system. Here the "true self" "refers to the Confucian idea of the self in terms such as self-cultivation, in contrast to the idea of the 'private ego' in terms such as self-centeredness." "Despite its embeddedness and rootedness," the true self "draws spiritual resources from the common spring of humanity. It reaches the common spring of humanity by probing deeply into its own ground of existence. The self need not depart from its locale to find a proper niche for spiritual development. However, it is essential that the self recognizes its own face, a face common to all members of the human community, and listen to its own voice (the rhythm of the heart longing to establishing sympathetic resonance with other human beings and with nature). Without this self-awareness, it can easily degenerate into the limited and limiting structure of the private ego." So, "the true self, as an open system, is a center of relationships rather than an isolated individual" (Tu, 1989, pp. 108-110)

In Western philosophy, as I mentioned above, Kegley also put forward an ideal that she called "genuine individuals and genuine communities." She argued that "what is needed is a new philosophical view, a new paradigm, which sees individual and community relations differently – namely, by recognizing that the tasks of building authentic individuality and of developing fulfilling moral communities supportive authentic persons are tasks inextricably bound together. Genuine individuality and genuine community arise out of mutual interaction as a creative, ongoing process" (Kegley, 1998, pp. 54-60). An individual is a relation-based person who lives in a harmonious relation with others, and the community is a person-cherished group that can create a relaxed and happy environment to let each individual in it freely develop his personal interests, to give full play to his capacity and creativity. Any attempt to frame an account of the relationship of individual, community and society must proceed in this light.

Institute of Philosophy
Chinese Academy of Social Sciences, Beijing;

Research Center for Philosophy of Science and Technology
Shanxi University, Taiyuan, PRC

REFERENCES

Allison, H.E. (1983). *Kant's Transcendental Idealism.* New Haven: Yale University Press.

Brodbeck, M. (1968). Methodological individualism: Definition and reduction. In: M. Brodbeck (Ed.), *Readings in the Philosophy of the Social Sciences.* New York: The Macmillan Company.

Durkheim, E. (1968). Social Facts, In: Brodbeck (ed.), *Readings in the Philosophy of the Social Sciences.* New York: The Macmillan Company.

Einstein, A. (1954). *Ideas and Opinions.* New York: Crown Publishers, Inc.

Heap, S.H. et al. (1992). *The Theory of Choice.* Oxford: Blackwell.

Hollis, M. (1977). *Modes of Man,* Cambridge: Cambridge University Press.

Jammer, M. (1966). *The Conceptual Development of Quantum Mechanics,* New York: Mcgraw-Hill.

Kant, I. (1929). *Critique of Pure Reason.* Norman K. Smith (trans.). London: the Macmillan and Co. Limited.

Kant, I.: (1949). *Foundations of the Metaphysics of Morals.* L. W. Beck (trans.). Chicago: the University of Chicago Press.

Kegley, J.A.K. (Ed.). (1998). *Genetic Knowledge, Human Values & Responsibility,* Lexington: An International Conference of Unity of Science Book.

Lee, S-C. (1993). *Philosophical Advances in Contemporary Neo-Confucianism.* Taipei: Wenjin Publisher.

Luo & Hu. (1996). Relational realism. In: R. Cohen, R. Hilpinen & R. Qiu (Eds.), *Realism and Anti-Realism in the Philosophy of Science.* Dordecht: Kluwer Academic Publishers.

Marx, K. & Engels, F. (1976). *Collected Works,* Vol.5. Moscow: Progress Publishers.

Mencius. (1895). *The Works of Mencius.* James Legge (trans.). Oxford: Clarendon Press.

Newman, F. (1996). *Performance of a Life Time.* New York: Castillo Int. Inc.

Qiu, R-Z. (1998). Reshaping the concept of personhood: A Chinese perspective. A paper on the International Conference on the Concept of Personhood held at Baptist University of Hong Kong, Hong Kong.

Qiu & Hu. (1998). A Chinese perspective on the use of genetic knowledge. In: Kegley, J.A.K. (Ed.), *Genetic Knowledge, Human Values & Responsibility.* Lexington: An International Conference of Unity of Science Book.

Shi Jun (Ed.). (1988). *Selected Readings from Famous Chinese Philosophers*, Vol.I, Beijing: People's University of China Press.

Tu W. (1989). *Centrality and commonality: An Essay on Confucian Religiousness*. New York: State University of New York Press.

Watkins, J. (1968). Methodological individualism and social tendencies. In: Brodbeck (Ed.), *Readings in the Philosophy of the Social Sciences,* New York: the MacMillan Company.

Xia, Z-T. (1997). Man: relations, activities and developments. *Philosophical Research* (in the Chinese language), No.10, pp. 6-15.

GERHOLD K. BECKER

THE ETHICS OF PRENATAL SCREENING AND THE SEARCH FOR GLOBAL BIOETHICS

The question whether global bioethics is possible is indicative of the specific conditions of post-modernity. It implies not only the fragmentation of contemporary morality along the line that divides secular society from particular moral communities, but also uncertainty about the nature of morality and about the justifiability of its claims. Thus it raises some rather fundamental issues in moral philosophy both at the normative and the metaethical level.

I. BETWEEN PARTICULARITY AND UNIVERSALITY: THE CHALLENGE OF GLOBAL BIOETHICS

In the past, normative theory may have been able to claim some success in its effort to bring unity and coherence into moral norms by developing highly general sets of normative principles. Such success may have been due to the limited scope of its moral world in the relative homogeneity of its Judeo-Christian cultural setting. In the secular, multicultural, and borderless world of today, the task of normative theory has become more arduous and, it seems, less promising. Conflicting moral visions of particular communities within Western society and the increasing reception of various systems of Eastern ethics (Confucian, Taoist, Buddhist, etc.) have raised doubts that a relatively stable and discernible set of moral principles can be identified around which individual norms could be meaningfully organized. The ongoing dispute between liberalism and communitarianism is a case in point, as is the critique on recent attempts at the systematization of the principles of bioethics (Gert, 1997).

This suggests that the division of normative and metaethical theory is less defined than might be expected and that questions about the universality of ethics are inescapably tied up with questions about the very meaning and justifiability of ethical claims. While some deplore the lack of coherence in contemporary ethical discourse and see the language of morality today in a "state of grave disorder" (MacIntyre, 1984, 256),

Julia Tao Lai Po-wah (ed.), Cross-Cultural Perspectives on the (Im)Possibility of Global Bioethics, 105–130.
© 2002 *Kluwer Academic Publishers. Printed in Great Britain.*

others argue that it is not appropriate to search for "a single moral vocabulary and a single set of moral beliefs" that could be shared by "every human being and community everywhere" (Rorty, 1993, 265).

Although it may be tempting to give up claims to universality in ethics and settle for 'pragmatic' solutions, particularly in view of the urgency of ethical decision-making in areas of bioethics, it is a distinctive feature of morality proper to aim at objectivity and universal validity for its assessments. In spite of the fact that "we always speak the language of a time and place, the rightness and wrongness of what we say is not just for a time and a place" (Putnam, 1987, 242). Unless one subscribes to a non-cognitivist theory of morality, moral norms and assessments are not simply 'given' but can be challenged and require reasons for their justification; thus morality makes claims of objectivity and truth which in turn imply universality. The justification of moral norms can only be expected in an open, rational discourse from which nobody ought to be arbitrarily excluded. The distinction to be drawn, then, is between the factual global recognition of individual norms and principles of morality and the claim to objectivity and universality implicit in morality proper. While ethics is always contextual, embedded in and shaped by socio-cultural traditions, the very idea of morality is based on the assumption that its language must (and can) provide for standards that obtain not just for or within a particular tradition. As Aristotle put it, "it is the good" that people seek and "not simply the traditional, or the way of their ancestors" (*Politics*, 1269a3). Though this may seem a typically Western approach to morality, it is in fact not exclusively Western but has its equivalent even in moral traditions that occasionally are being invoked to critique what is seen as the rational bias of Western morality. Aristotle's approach has its parallel in the Chinese philosopher Mozi, who poignantly asked, "Why should we follow the traditional value system?" This question implies "the crucial Socratic distinction between customary mores and morality proper" (Hansen, 1992, 108). Thus Mozi opens his own culture to moral scrutiny in search of principles that are well founded and accessible to all who care for rational inquiry into the norms of ethics.

This is to deny neither the existence of conflicting accounts of substantive ethics in particular communities nor the frustration the search for moral commonalities frequently generates. It is, however, argued that in a rapidly shrinking world there is really no alternative to the search for moral principles that can be shared cross-culturally and for a moral discourse in which anyone can participate. While the goal of global ethics may seem rather elusive, we must not forget that the history of secular ethics is short (Parfit, 1984, 453-454) but has nevertheless yielded some remarkable results. It is certainly true that

we owe much of what seems most admirable in modern societies – movements for political democracy and universal suffrage, for the emancipation of slaves and women, for the elimination of racial, ethnic, and religious discrimination, for universal social provision of basic needs, and for international law and human rights – in significant measure to universalizing and generalizing pressures that have precisely gone against the grain of some entrenched (and still powerful) particularistic moral conceptions and individual or group commitments (Darwall, Gibbard and Railton, 1997, 31).

Though even these results may not yet be fully recognized everywhere, as the so-called Asian values debate illustrates, it is difficult to ignore that they are highly attractive the world over, even in countries where extant political conditions hinder or try to prevent the free flow of information and a moral discourse open to all.

Obviously, the complexity of modern ethics is mirrored in bioethics. While individual issues may seem hopelessly perplexing and entangled in conflicting moral intuitions, the very fact that they are usually explored within shared parameters of moral discourse and solutions sought that, in principle, can be endorsed by all its participants, presupposes some common ground in the midst of diversity.

In what follows, I will focus generally on some concrete moral issues in the context of prenatal screening and testing for Down's Syndrome (PND) and only obliquely reflect on the broader issue of the (im)possibility of global bioethics. The main reason for this strategy is that the discussion of such large categories tends to go easily astray if it is not closely tied to specific issues and the established requirements of moral discourse within particular bioethical settings.

PND is highly representative of what has become the dominant paradigm of medicine as it developed in the Western world. Although it did not go unchallenged by alternative approaches inside (homeopathy) and outside the Western world (the recent revival of traditional Chinese medicine is a case in point), the global success of Western medicine, as of modern science in general, is an undeniable fact. Not only is Western medicine without borders in aiming at universal coverage for the sick and injured, but also its technology and methodology are universally applicable and indeed applied. If global bioethics would mean nothing more than that modern medicine all over the world has generated the same kinds of moral issues that need to be addressed through reason and argument, and that e.g. women in Sri Lanka, China, and the USA are facing similar ethical questions when they decide to undergo PND, then global bioethics would be a reality.

The challenge for global bioethics comes, however, with the apparently conflicting moral intuitions that are at the core of the substantive moralities of individual communities or entire cultures. Although they may complicate consensual solutions of specific bioethical problems enormously, much common ground can be secured in clarifying above all the factual issues by removing misunderstandings and precisely defining the moral issue(s) at stake. Once those communities engage in the search for (consensual) solutions to moral problems, a decisive step has been taken towards global bioethics. It is argued that global bioethics should not exclusively be understood as the existence of globally shared sets of moral values and principles, but rather as the open and methodological process of inquiry into specific moral issues from which nobody is excluded who cares about participating. On this assumption global bioethics is not only possible, but confirmed in daily practice wherever ethics committees reach consensus at local or international levels.

The concrete issues arising from PND may illustrate that there is really not much of an alternative to such global bioethics if one believes that the moral problems modern medicine generates are of the same kind everywhere in the world and can only be addressed and explored in an open moral discourse. It is the imperative of such discourse that it tries as thoroughly as the subject permits to truly understand all the relevant factual issues. As the experience of those involved in PND seems to confirm, the increase in comprehensive and specific knowledge about all the relevant factors not only greatly facilitates the clarification of the moral issues, but also considerably narrows down the field of relevant moral principles and the potential conflict they might create. By identifying the moral issues individuals and society as a whole need to address, moral discourse opens up narrow individual perspectives and directs them to the larger challenge society and humankind as a whole are facing. If it is a fundamental feature of the moral point of view to step into the shoes of others and look from their perspectives at the issues at hand, moral discourse raises the hope of consensual solutions to moral problems.

II. ETHICAL IMPLICATIONS OF PRENATAL SCREENING AND TESTING

Ever since Jerome LeJeune's discovery in the late 1950's of the causal connection between the chromosomal abnormality trisomy 21 and the severe form of mental retardation of the so-called Down's Syndrome (DS), named after its first description in 1866 by the British physician John Langdon Haydon Down, great efforts have been made to detect this condition early on in pregnancy. DS is caused by a chromosomal aberration consisting of 3 free copies (trisomy) of all or a critical portion of chromosome 21. The most likely causes for this condition are errors in maternal meiosis. The chromosomal abnormality is non-treatable and there is to date no way of preventing the conception of DS fetuses.[1]

Since the late sixties, when the development of prenatal diagnostic techniques began with the culturing and analysis of amniotic fluid cells, prenatal screening and diagnosis has proliferated dramatically and a variety of procedures have been integrated into routine and standard prenatal health care services. In most cases, prenatal screening and diagnosis is performed to exclude chromosomal abnormalities (75 percent of all amniocentesis and 80 percent of all tests using chorionic villus sampling have this objective). About 50 percent of chromosomal abnormalities detected through PND in women of 35 years of age and older are trisomy 21 causing DS.

Reflecting the gradual developmental stages of the nasciturus from the zygotic (pre-embryonic) to the embryonic and fetal stages, as well as the related opportunities for testing, prenatal care has evolved into several options, which range from screening procedures that provide data for statistical risk calculation to the highly conclusive diagnosis of fetal defects. The distinction between screening and testing (or prenatal diagnosis in the strict sense) is not only justified by the difference between the predictive or probabilistic nature of the former and the considerable room it leaves for interpretation of the screening results in comparison to the latter, but also, at least at present, marked by the difference between non-invasive procedures and invasive, surgical operations.

Health care services are supposed to offer medical assistance to those in need by treating and curing disease, or at least by ameliorating suffering and providing palliative care. In the absence of a cure for DS, the following questions must be raised: what are the objectives of PND and who should be morally entitled to set them? Is the purpose of PND, above all, to minimize the number of infants born with DS? Or, alternatively, is PND primarily intended to maximize parental choice about a DS pregnancy and to enable an informed decision within the range of available options? The former question can also be put with a significantly different emphasis as follows: Is there a moral responsibility to promote the genetic health of the population (Chadwick, ten Have, Husted, Levitt, McGleenan, Shickle and Wiesing, 1998, 267), and if so, whose responsibility is it? While the two ways of asking this question address a fundamental issue of general health care policy, which involves not only the medical profession but the whole of society, the issue cannot be decided, at least not in a society based on democratic values of participation in political decision-making, without giving due recognition to the interests of those directly affected by such a policy. Yet directly affected is not only, and not even primarily, the pregnant woman, but also the human life she carries.

Many in the medical profession seem to take it for granted that the primary goal of PND is ridding society of the occurrence of DS, since DS is, after all, a severe and incurable disease. It follows from this assumption that it is imperative to increase both the efficiency of PND procedures for the accurate detection of trisomy 21 and their range of application, which should ideally extend to all pregnant women. This goal is clearly reflected in the great effort of clinicians and service providers to ensure that no incidence of DS goes undetected – and their frustration when it happens nevertheless. While the reasons for setting this ambitious goal may not be entirely clear or uncontroversial, they surely depend to a large extent on social policy preferences. Apart from being a genuine expression of the general medical ethos of healing and curing disease, it may also be an indication of the more self-serving interest at the individual level of avoiding possible litigation on the grounds of 'wrongful birth'.[2]

On the assumption that the minimization of DS incidence is the justified objective of PND, unrestricted and equal access to screening and testing is not simply a requirement of justice but, above all, a matter of social policy in the best 'eugenic' interest of society. Aside from the difficulty to define clearly what that means and what concept of health may be involved, the near universal eugenic restraint suggests at least some common ground in its ethical evaluation. It is certainly of some significance for the issue of global bioethics that moral cooperation in such areas is possible even though the underlying reasons may differ in individual cases. The short but awful history of eugenics is a strong reminder of the need for the careful evaluation of the potentially serious implications of seemingly benign health care objectives.

Yet the more imminent moral issue in this objective arises from the fact that it cannot be achieved by genuine therapeutic efforts aimed at the cure of the disease or its prevention, but only through the termination of pregnancy. The ethical ambiguity of this situation is highlighted by linguistic attempts to gloss over the perceived difficulty and to come up with a parlance that appears still anchored in the therapeutic ethos of medicine. The result is the coinage of the problematic term *therapeutic abortion.*

A solution to the problem of determining the objectives of PND will have to be judged by its ability to accommodate at least the two following fundamental moral concerns: the concern for human life and the concern for autonomy. Ideally, the objectives of PND must be defined in such a way that they are in conflict neither with the moral appreciation of human life, its value, and its quality, nor with the value of individual freedom and autonomy.

III. THE VALUE OF HUMAN LIFE

The first major ethical issue concerns the moral status of the fetus and the kind of respect it may deserve. This is a notoriously complex and highly controversial issue, which cannot adequately be explored within the constraints of this paper. In the context of the termination of DS pregnancies, this issue has been examined differently depending on the particular moral discourse within which it is being raised. Religious communities, particularly Roman Catholicism, usually take the strong view that all human life is a gift of God and therefore sacred and inviolable, requiring protection from its very beginning at conception. At the opposite end of the spectrum are those who link any right to life to a normative interpretation of human personhood and define this through a list of specific criteria that must be met to qualify. Although there is no unanimity among its proponents about the criteria of personhood, most lists would include properties such as self-consciousness, rationality, moral sense and freedom (Engelhardt, 1996, 139).

It is, however, clear that any account of personhood as the basis for moral status would disqualify human fetuses, unless the so-called potentiality argument could be successfully applied. This argument states that although the human fetus cannot be regarded as a person yet, it is nevertheless a potential person, since it would undoubtedly develop all the requisite characteristics of personhood at a later stage if its development would not be disrupted either artificially or by illness. Yet if the argument fails, as is generally assumed by bioethicists, there is no reason to attribute to the human fetus moral status that would prevent its destruction on demand at any time before birth.

Evidently, this conclusion conflicts with the rather strong and globally shared moral intuition about the prime value of human life. It is this intuition that is reflected in legal provisions, which usually stipulate certain time frames outside of which manipulations on embryos for non-therapeutic or research purposes are deemed unacceptable and the destruction of fetuses is not allowed. The most commonly agreed upon stages in embryonic development which carry moral significance are the appearance of the primitive streak and the moment when the fetus is regarded as viable. While the former has been widely used as the cut-off for wasteful embryo research and experimentation, the latter serves as a reference point for the legal protection of the fetus. This was obviously the case in the landmark ruling of *Roe vs. Wade* when the U.S. Supreme Court described viability as the "compelling point" at which the state might assert an interest in regulating abortion (Shannon, 1997, 419). It should, however, be noted that the moment of fetal viability is neither a factual nor a moral absolute. Advancing technology in pediatric neonatal care is likely to affect the time frames of fetal viability. When the U.S. Supreme Court made its ruling, fetal viability was assumed at about 28 weeks of pregnancy. In the meantime, the time frame has been considerably reduced and fetal viability may be possible much earlier, at about 21 weeks.

The legal protection of certain phases in human fetal development reflects the general moral intuition that human life commands respect and that the degree of such respect increases relative to the developmental stages of the fetus. Though there is apparently no consensus in bioethics on the moral status of the fetus prior to reaching viability, the moral significance of fetal viability itself is widely recognized. As LeRoy Walters observed, for "the period between viability and birth there has been little ethical or legal argument in favor of pregnancy termination, or early induced delivery, of fetuses with open NTD [Neural Tube Defect] or Down's syndrome" (Walters, 1989, 56).

This may suggest two things: Firstly, the usual dismissal of the potentiality argument may not be supported by the moral intuitions of the general public (or 'common morality') and therefore be premature. Various studies have pointed to the long-lasting emotional distress and the feeling of guilt experienced by women who agree to terminate their pregnancy during the second trimester (White-van Mourik, Connor and Ferguson-Smith, 1992; Rucquoi and Mahoney, 1992). The feeling of guilt suggests that at least for the women concerned there is no sharply defined moment prior to which the life growing in their womb would not be perceived in the light of what it is to become and as an object of love and affection. Secondly, the clarification of the factual matters of human development can indeed facilitate moral consensus across moral communities.

It should be noted, however, that there is no unanimity about the foundation of the moral respect the human fetus commands.[3] In the absence of a human being of moral standing, the respect cannot be derived from the fetus itself but must emanate from other goods and values. H. Tristram Engelhardt, Jr. has indeed suggested that without full-fledged personhood a fetus is simply a "non-person" and "a special form of dear property" that gains value from the interests of its producer, who has "the first claim on effectively determining its use" (Engelhardt, 1996, 255). Although he admits that this is a rather shocking perspective, he holds that it is the only one available in general secular morality. Other accounts of fetal moral standing draw on concepts of social personhood (which Engelhardt reserves to small children, 'former' persons and to "the severely and profoundly retarded and demented who never were and will never be persons in the strict sense," [Engelhardt, 1996, 150]) and suggest that the only reason why fetuses and indeed even small children may be morally considerable is social and thus indirect. The argument can be read as an extension of Kant's account of the moral status of animals to fetuses and small children. As will be recalled, Kant denied animals moral standing but nevertheless prohibited any form of cruelty against them on the grounds that such behavior would lower the moral threshold and thus, in the long term, be detrimental to human moral development.[4] Similarly, the willful destruction of human fetuses would not be allowed for the adverse affects such a practice would have for the

moral development of persons in the strict sense and their appreciation of
human life in general.

It is doubtful, however, whether this line of argument can account for
the strong emphasis moral intuition and legal stipulations place on fetal
viability. Even on reaching viability the human fetus would not qualify
for personhood and in this regard would be no different from a fetus at a
much earlier stage of development. Yet unlike in the case of animals, in
almost all countries the willful killing of a fetus in the third trimester
legally constitutes a criminal act. While it may not be possible to arrive at
a universally agreed upon position on the underlying reasons for the legal
and moral protection of the fetus, moral cooperation in legislation as well
as in clinical settings would still be possible even if the reasons for such
cooperation may differ individually.[5]

1. The life not worth living: Who is to decide?

From the moral point of view, human life and its quality can only be
assessed from the first-person perspective of those whose life it is. Since
this perspective is not yet available to DS fetuses, the question has been
shifted to what would be in their best interest. The answer to this question
is usually discussed under the heading of the quality of life and depends
as much on certain empirical facts (e.g. levels of pain and debilitation) as
on moral intuitions and values.

Examining the ethics of abortion with regard to models of parental
authority, Charles E. Harris (Harris, 1991) has offered a five-stage
schematic classification of fetal abnormalities, which ranges from fetuses
incapable of any significant degree of human development (e.g.
anencephalics) to those suffering only from minor impediments such as
extra digits. While the second stage would be comprised of fetuses
capable of some very minimal degree of development followed by early
death (e.g. trisomy 18), characteristics of fetuses in the third stage would
include considerable capability of development which would, however,
be arrested and followed by death (e.g. Tay-Sachs disease). Harris
concludes that fetuses in the first three categories could be justifiably
aborted, since they would have little chance of achieving the primary
natural and social goods[6] that are considered necessary conditions of a
meaningful life. While he leaves open whether it may be justified to abort
fetuses of the fourth category, which are characterized by normal lives

until their thirties but will then suffer premature death, as in cases of Cystic Fibrosis, abortion is ruled out for fetuses of his fifth category.

It is remarkable that this classification is unable to account for the characteristics of fetuses afflicted with DS. In view of their impairment it would certainly not be easy to group them in the first three categories, and it may even be doubtful if they would fall into category four. Yet most surveys concur that in over 90 percent of cases of DS diagnoses, pregnancies are terminated.[7] This may suggest that the reasons for the high incidence of abortion of DS fetuses has more to do with general health care policies and parental preference than with objective criteria for the assessment of the quality of life. Moral consideration about whether the life of a DS patient is worth having require objective standards for the evaluation of life's quality.

2. Down's syndrome and life's quality

There is no question that DS is a severe, debilitating illness. It is one of the most common chromosomal abnormalities in liveborn children, with a frequency ranging from 1 in 650 to 1 in 1000 live births (Hook, 1982; Motulsky and Murray, 1983). DS patients present a major challenge to any family, which is frequently even exacerbated by the adverse attitude of the general public towards the sufferers. This adverse social attitude is noticeable in Down's unfortunate "ethnic classification of idiots" and his original description of the disease as "mongolism" (Down, 1867), a term with implicit racial overtones of degradation and inferiority.[8]

One of the reasons for the complexity of the ethical issues DS presents is the fact that it is an illness that encompasses a wide range of conditions, which differ greatly in individual cases. Depending on the degree of severity, a DS patient may be totally incapacitated, or may occasionally even become an actor like Pascal Duquenne, who won the Best Actor Award at the 1996 Cannes Film Festival for his part in the film *The Eighth Day,* in which he played a DS patient. The image of DS, both as a disease and as a public perception, is complex, too, and changing. While DS patients may be slow in physical movement and intellectual development, it is proven that they are able to learn, and many reach the level of mastering the skills of everyday life if adequate support and suitable conditions are provided. Some may even become writers publishing books, with titles such as *Count Us In: Growing Up With*

Down Syndrome.[9] Individual DS sufferers also have been found in the ranks of social workers, instructors, or even ballet dancers.

Recent research and therapeutic developments seem to confirm the hope of many parents with DS children that, while a cure is not possible, the remaining abilities of DS patients may be dramatically improved. Even some clinical features of DS patients vary widely and are susceptible to improvement. The editors of the book *Down Syndrome: A Promising Future Together*, which grew out of a recent conference on Down Syndrome with a focus on clinical, educational, developmental, psychological, and vocational issues, noticed "the sense of excitement that accompanies many recent advances in the Down syndrome arena" (Hassold and Patterson, 1999, preface, x).[10] Medical intervention and improved forms of treatment have steadily increased the life expectancy of DS patients from about 9 years in 1929 and 12 years in 1947 to over 50 years in 1970 (Rasore-Quartino and Cominetti, 1995, 239). Furthermore, DS patients are often described as cheerful and even happy persons, and there is an increasing amount of literature that gives detailed accounts of the joys, and not merely the burdens, they can bring to their families and the enrichment they can offer to the lives of those who care for them. Literature reviews reveal that the life satisfaction and happiness of families whose children have mental retardation is equal to the levels of satisfaction and happiness reported by families whose children do not have mental retardation (Wertz, 1997b).

Evidently, these and similar factors should and do play a role in decisions about DS pregnancies following prenatal screening and testing. The extent to which they are known or are even part of women's personal experience is likely to influence their deliberation about whether or not to continue the pregnancy. They are clearly related to the question of the appreciation of human life and its quality. Surveys have confirmed the assumption that women without personal contact with DS patients are more likely to decide in favor of terminating a DS pregnancy than women with personal experience of children suffering from DS (Julian-Reynier, Macquart-Moulin, Moatti, Loundou, Aurran, Chabal and Ayme, 1993). It appears that a more realistic perception of the real-life situation of DS patients can diminish maternal anxiety and serve as a corrective of biased views about the quality of life of both DS patients and their families. This is also reflected in the increased incidence of adoption cases of DS children. The standard assumption that for DS patients life cannot be anything but miserable and should have been avoided at all cost does not seem supported by the facts.

3. Psycho-social concerns of handicapped people

Concern has also been raised about the psycho-social effects of prevention strategies for congenital diseases in general and for DS in particular. At issue is their effect on people with disabilities and the implications for the moral evaluation of such strategies. The argument seems to be that, given the practical impossibility of preventing all the births of children with congenital disease and handicap, impaired people will always form a considerable contingent in a population. Their social acceptance and the respect they are owed would be adversely affected by a health care policy whose declared objective was the prevention or, if that should not be realistic, the minimization of congenital disease. In the case of DS and other severe diseases this could only be achieved through a 'therapy' which, by eliminating the disease, would also eliminate the patient. It is argued that all those whose condition either went undetected during pregnancy, or whose families decided against such 'therapeutic abortion', would be stigmatized simply for being born and being alive. Such stigmatization and the many adverse effects it may have on the lives of these people would clearly exacerbate their suffering and overshadow any form of happiness they might otherwise still be able to enjoy.

The argument deserves particular consideration when it is put forward by the handicapped persons themselves or their organizational representatives. Fears have been expressed (particularly in the Netherlands and in Germany) that long-term harm will result from screening programs that may be used to ensure the optimal quality of future children (Chadwick, ten Have, Husted, Levitt, McGleenan, Shickle and Wiesing, 1998, 260). While the dangers may be exaggerated, the legal possibility of selective abortions not just for severe congenital diseases but for almost any unwanted fetal characteristics that have been diagnosed prenatally, together with the rapidly expanding range of genetic therapies, make some commentators wonder whether PND might in fact be the first step down the slippery slope that begins with negative eugenics and ends with a shopping list for designer babies.

While the argument is occasionally charged with emotions, as the so-called Singer-debate and the public response to the arrival of *Dolly the Sheep* illustrate, it is difficult to assess its merits in the absence of unambiguous empirical data. PND service providers used to point out that even a full utilization of prenatal testing[11] with subsequent selective abortions would be able to prevent only about 20-25 percent of DS incidences, and thus any impact on the general population would be minimal (Motulsky and Murray, 1983, 280). Yet, non-invasive screening available to all pregnant women has certainly changed this and the overall prevalence of DS among live births has been sharply reduced.[12]

Obviously, prenatal screening and testing has had a strong impact on DS incidences in the population, and the concern that this may translate into diminished respect for DS patients cannot be ignored or easily dismissed. The Council of Europe and other European bodies formally recognized that selective abortion on the grounds of genetic disorder might lead to decreased acceptance of those suffering from such disorders (Chadwick, ten Have, Husted, Levitt, McGleenan, Shickle and Wiesing, 1998, 266). The bioethics convention of the Council of Europe, which in 1996 was endorsed by all but three of the thirty-nine participating nations, explicitly prohibits "any form of discrimination against a person on grounds of his or her genetic heritage" (McGleenan, 1999, 11).

One must, however, be careful not to hold professional PND service providers responsible for something that is clearly outside their control and a task for society as a whole to tackle. While it is difficult to change public perception of DS and prevent discrimination against individual members in a community, moral respect for persons is globally recognized as a fundamental principle of bioethics. The problem then needs to be addressed through educational programs intended to provide more factual information and to instill understanding that may eventually lead to greater empathy and tolerance for some of the most disadvantaged members of society.

IV. MORAL DECISION-MAKING IN PRENATAL SCREENING

The question whether there is some common ground for the moral evaluation of PND and its objectives is closely related to the more fundamental concern of what can count as a life worth living and who should have the moral right to decide this.

H. Tristram Engelhardt, Jr. has argued that within the conditions of secular society and its value pluralism, fundamental questions about the value and quality of life, which figure as prominently in individual decisions in bioethics as they do in defining social policies and the provision of health care, may be unanswerable since they presuppose content-full moralities that by definition are particular and outside particular moral communities no longer available. All we can do in situations where we meet as "moral strangers," and no longer as "moral friends" who share the same moral frame of reference (Engelhardt, 1991, 3), is to retreat to those minimal conditions even moral strangers must accept if they want to live in peace with each other. The fundamental value this transcendental analysis is most likely to uncover is the value of the individual, and the only moral principle that can be shared even by moral strangers is *autonomy*. In other words, the value of human life and its quality can only be decided by the individual whose life it is, and nobody has the right to interfere with such a decision unless it would infringe on the rights of others.

One may wonder, however, if that is not too pessimistic an account of the basis of ethics in secular society, which rigidly separates it from the substantive ethics of particular communities. The borders that separate conflicting interpretations of substantive ethics are neither clearly drawn nor impenetrable to reason and argument, and particularistic moral communities are in flux and not immune to the outside world. The possibility that the realms of secular and communal ethics may overlap suggests that moral strangers may be rather the exception than the rule in secular society, since most of us do share a rich concept of autonomy that branches out into a variety of fundamental and common moral values. Being neither entirely strangers to each other nor friends, we may well be "sufficiently morally acquainted to enable fruitful dialogue across even starkly different cultural groups" (Loewy, 1997, 3). Thus the goal of global bioethics may not be entirely elusive.

In the context of prenatal screening and testing, there is unanimity among bioethicists that respect for individual autonomy should be the "predominant ethical value guiding the counseling process and its outcome" (Murray, 1995, 928). Ethicists as well as service providers concur in their view that decisions about the use of PND must ultimately rest with the woman concerned, and that it is a moral duty of clinicians and health care providers to empower her to informed and independent decision-making as to which course of action to take at each step in the screening and testing process. Clinical protocols aim to ensure the ultimate authority of the competent patient about courses of treatment. It is assumed that the woman's autonomy is best respected if the service empowers her to independent decision-making, and that the benefits of screening lie primarily in its potential of "enabling individuals to take account of the information for their own lives and empowering prospective parents to make informed choices about having children" (Nuffield Council on Bioethics, 1993, para. 8.20).

Autonomy, however, would be underdetermined if it were thought to be comprised of not much more than the right to be left alone (or negative autonomy). Rather, it is a multidimensional concept, and although differing accounts have been given of its various dimensions and structural components, it seems reasonable to distinguish between autonomy as the capability and disposition for independent and second-order preferences, and autonomy as a property of individual moral acts that are embedded in and derived from a larger moral reference system of obligations and rights.

Though these are different aspects of autonomy, they are not entirely separate from one another. A mere capability that is never exercised is indistinguishable from one that doesn't exist at all. Yet any activation takes place not in a value-free space but in a specific context where autonomous agents meet. "The principle of respect for autonomy should be viewed as establishing a stalwart right of authority to control one's personal destiny, but not as the only source of moral obligations and rights" (Beauchamp and Childress, 1994, 126). Unless "respect for individual autonomy" is merely used as another term for the fear of a stronger power that one avoids challenging in the interest of pure survival, it is not simply a content-less prerequisite of a peaceable life but the referent of a variety of interconnected values that buttress what could be called a rudimentary substantive ethics. Only on this assumption does it make good sense to distinguish individual autonomy as "sovereign authority" (Joel Feinberg) from the unfettered and arbitrary exercise of will power which is neither available for moral discourse nor for rational justification, and to inquire into the limits of individual autonomy. Thus respect for autonomy stands for a whole cluster of interrelated moral values and principles that may translate into more specific precepts.

Although decisions affect the expecting mother as much as her future child, a decision in favor of termination of pregnancy affects mother and child disproportionately. Abortion legislation usually offers wide discretionary leeway to the pregnant woman with regard to the subjective reasons for selective abortion. If autonomy were seen in isolation from and not as embedded in other moral values, some rather peculiar conclusions might be drawn. In exercising autonomy, an otherwise required step in the 'screening cascade' could be deliberately refused for the explicit reason that one does not want to know conditions of the fetus or whether or not it is affected by DS. This has been taken to imply that autonomy entails a 'right not to know', and that such a right requires the same level of respect owed to autonomy itself.[13]

It seems doubtful that the principle of autonomy can be invoked in support of *all* subjective reasons someone might put forward that fall within the scope of the law regulating the termination of pregnancy. Instead, it has to be balanced with other relevant principles such as maleficence, benevolence, and justice (to draw on Beauchamp and Childress' influential account of the fundamental principles of bioethics). It is reasonable to argue that solidarity with others is a moral duty that entails a 'responsibility to know' and thus curtails the 'right not to know' that we may otherwise be able to claim.[14] A similar conclusion may be reached if autonomy is understood as comprising more than the right to be left alone. The refusal of information necessary for autonomy thus understood in the name of autonomy seems contradictory and confused (Chadwick, 1997). Autonomy in the full sense then includes the capability for authentic self-determination within the parameters of substantive ethics.

The conceptual difficulties are exacerbated by practical problems of the counseling process that seeks to facilitate the exercise of individual autonomy by providing the necessary conditions for informed and responsible decision-making. While in the context of invasive testing procedures the principle of informed consent helps, above all, to safeguard the patient's autonomy, it also protects the physician from possible legal claims. Non-invasive screening, however, adds a new dimension of application to the principle, which emphasizes its genuine ethical nature over its function as a legal safeguard. Even without physical risk to the woman and consequently fewer legal risks for the physician, respect for autonomy requires that informed consent must be obtained prior to screening. The complex implications of screening and testing and the stress they are likely to cause require prior, comprehensive information about its various aspects and consequences.

In practice, this may pose significant difficulties for service providers. The pattern of women's knowledge frequently reflects more the practical emphases of midwives and obstetricians than adequate information about the implications of screening (Smith, Shaw and Marteau, 1994, 776).

Occasionally, women may not even be aware of the exact nature and purpose of the tests they have undergone or what the results mean, in spite of existing guidelines on prenatal testing, or service providers may be inclined to downplay the full range of implications (Press and Browner, 1995, S11). This may suggest that the requirements of 'informed consent' or even 'informed refusal' are frequently more an indication of the fear of 'wrongful birth' claims than of genuine ethical concern for individual autonomy and responsible parenthood.

Clearly, empowerment to autonomous decision-making calls for more than the explanation of the technical procedures and medical terms. The information must also include a faithful description of the likely consequences of the various courses of action available as well as an unbiased, realistic account of DS. This account should neither overemphasize nor undervalue the likely implications of having a mentally retarded, handicapped child in the family, but offer, as much as is reasonably possible, a realistic picture of the burdens and the emotional rewards the care for a troubled life may hold.

Instrumental to the goal of exercising autonomy is the provision of adequate information about the implications of moral decision-making in the setting of PND. Current practice has been based on the concept of non-directive and non-judgmental counseling. Non-directiveness has been understood to include the provision of full information and assistance to individuals or families to make their own autonomous decisions (Fraser, 1974; United States President's Commission, 1983). While this is certainly a valid and highly relevant concept, it seems neither well-defined in its implications nor always followed faithfully in clinical practice.

A multi-national survey of the attitudes of genetic counselors showed near unanimity in their resolve to refuse making decisions for clients even at their explicit request, or even giving any indication about their own set of guiding values (Wertz and Fletcher, 1989). Doubts have been raised whether daily clinical practice in fact does live up to such standards. The findings of the Royal Canadian Commission (1993) suggest that paternalistic and directive counseling is much more common than the stated attitudes of counselors would indicate. This may mean that the implications of non-directive counseling for autonomous decision-making have been underestimated and a reconsideration of the concept itself cannot be avoided.

Empowerment to autonomous decision-making in the strict sense rests on certain conditions. They include, in addition to the freedom from controlling influences and the desire to decide for oneself, deliberative competence as well as moral reflection (Miller, 1981). Although these dimensions of autonomy may be engaged to varying degrees, in the context of PND they can only be effective if information has not just been provided but is also adequately understood. Obviously, counseling clients come from rather different cultural, educational, and social backgrounds and are at different stages of emotional stress and anxiety. It has been noted that in a more traditional Chinese cultural setting women are less inclined to decide for themselves but instead act on the advice of significant others, most importantly their husbands, but also their physicians.[15]

The main aim of counseling must therefore be to maximize autonomy by enabling individuals to comprehend the complex medical facts before them as well as the alternatives for dealing with these facts. It is doubtful whether such comprehension could be expected in a single session in which all the relevant information is provided in a supposedly value-neutral environment, which lacks empathy and leaves the anxious woman entirely on her own. Sonia M. Suter has compared such a practice to a travel agent providing pictures of foreign destinations without providing any travel guidebooks (Wertz, 1997a). Counseling is, above all, a communicative process that engages both parties in the exploration of the various dimensions and implications of a complex moral dilemma without easy solutions. While the asymmetry of the physician- (or counselor-) patient relationship certainly requires safeguards in the interest of patient autonomy, doubts have been expressed about whether such safeguards necessitate the sterility of a value-neutral environment. It may well be that total value neutrality in human communication is neither attainable nor particularly attractive, and perhaps more of a salutary fiction for the benefit of considerations in the legal community than a true characteristic of the communicative process between client and counselor.

This is not to justify imposing one's own set of moral values on someone else, or unduly influencing someone's decision. It cannot condone the exacerbation of an already stressful situation through the offering of evaluative comments once a decision has been made. The fragility of decision-making under such emotionally charged circumstances prohibits praise, but more importantly, it prohibits blameful remarks, and instead requires a deep sense of empathy.

And yet, decisions of this nature do not occur in a value-free context. Moral values are part and parcel of human life; they impregnate our perception and thinking and ultimately determine who we are as individual human beings. It might therefore better serve the purpose of empowerment to autonomous decision-making if the moral frame of reference were made explicit so as to become an integral part of the communicative process instead of being artificially suppressed or ignored (Quante, 1997). Only when the moral values involved have been clearly identified and the moral implications of PND sufficiently understood, can a decision be made that is simultaneously an expression of moral autonomy and authenticity as well as of moral responsibility.

V. RUDIMENTARY GLOBAL BIOETHICS

The discussion of various aspects of PND clearly shows that while global bioethics cannot be considered a reality yet, there is, firstly, no viable alternative to it, and, secondly, at least some common ground on which to build. It has been argued that differences that divide moral communities may, at least partially, be due to a lack of factual information about complex and highly technical medical procedures as well as to misunderstandings and biased perceptions about the intentions of service providers. The first step towards global bioethics would require that all parties concerned engage in a fact-finding mission, which would clarify all the factual issues and clearly define the moral question that needs to be addressed. Such engagement would indicate the willingness of all parties concerned to enter into an open moral discourse in search of objectivity and truth in morality. This discourse holds the hope that a deeper exploration of the common human condition will disclose some common values and moral principles across moral communities and cultural traditions that could be truly regarded as a rudimentary global bioethics. That such hope may not be totally unfounded has been suggested in recent work by various ethicists and their search for moral commonalities. Even secular ethics may be richer in content and grounded in a common stock of values, which function as the indispensable safeguards of human life.

Taking as a point of departure the universal fear of death and the necessary conditions for collective survival, Sissela Bok has identified three fundamental categories of moral values that are indispensable to human coexistence and thus, she holds, universally shared.[16] These values include the positive duties of mutual care and reciprocity; the negative injunctions concerning violence, deceit, and betrayal; and the norms for certain rudimentary procedures and standards of justice. The values and the norms based on them provide what she, with Peter Strawson, calls a "minimal interpretation of morality," since they are the "condition of the existence of a society" (Bok, 1995, 16). They affect every level of personal and working life, the family and the community as well as national and international relations. Bok argues that these values are sufficiently broad to allow for cultural diversity without preventing the critique of abuses "perpetrated in the name either of more general values or of ethnic, religious, political, or other diversity" (Bok, 1995, 23).[17]

It may not be possible, or even reasonable, to arrive at a comprehensive set of universally agreed bioethical values and principles even at some point in the future development of the 'global village' (Marshal McLuhan) our world has become. Yet this must not necessarily imply that we cannot share a substantial body of moral norms and principles on the basis of which cross-cultural moral cooperation and decision-making would be possible and which could meaningfully be called a rudimentary global bioethics. After all, efforts by international bodies such as the United Nations towards advancing human rights and their global enforcement (including the establishment of international human rights courts) presupposes and confirms such a belief. While this would still provide space for particular moral communities and their substantive moralities, it is hard to envisage that they would be able to completely and successfully isolate themselves against the world outside their ideological borders. If the need for practical cooperation exerts some pressure to enter into a moral discourse so as to challenge opposing moral views and to justify their own, chances are that the body of shared rudimentary bioethics will expand further. Although it is difficult (and unnecessary) to speculate about the future of ethics, the hope that a truly global bioethics may eventually emerge is not entirely unfounded.

Center for Applied Ethics and Department of Religion and Philosophy
Hong Kong Baptist University
Kowloon, Hong Kong

NOTES

[1] The clinical features of DS include moderate to severe forms of mental retardation with the average IQ falling within a range of 50-60, a somewhat flattened skull and facial characteristics such as slanted eyes and a depressed nose bridge. Up to ninety percent of all DS patients suffer hearing impairment (Mazzoni, Ackley and Nash, 1994), and 30-40 percent have major congenital malformations of the heart; their risk of leukemia is significantly increased, and many patients are likely to develop the neuropathological hallmarks of Alzheimer's disease at an early age. It is estimated that up to 20 percent of DS patients die within two years of their birth. Most DS patients need lifelong care, albeit to varying degrees depending on the severity of their condition. Apart from the heavy (physical, psychological, social, economic) burden to the immediate family, care for DS patients comes also with a high economic price tag. The average cost of lifelong care for DS patients was estimated to amount to about 120 000 Pound Sterling at1987 levels (Robinson, 1998, p. 176). According to *Time Magazine* (19 April 1971), the cost of maintenance and treatment of DS patients in the USA alone ranged from one to two billion dollars (Gibson, 1978, p. 1).

[2] In Germany, a high court ruled in 1984 that it constitutes professional negligence if a medical practitioner fails to inform a pregnant woman in the high-risk category of the availability of testing for DS. Consequently, women delivering a DS child without having been adequately informed by their physician during routine prenatal consultation about the full range of PND services have the right to sue for damages and compensation (Nippert, 1997, p. 108). The situation is similar in other countries, certainly in the United States where various 'wrongful birth' cases have been successful in the courts (Press and Browner, 1995, S11).

[3] See e.g. Daniel Callahan's article "The Puzzle of Profound Respect" (Callahan, 1995).

[4] Kant, 1968, p. 443 (para. 17). In the same paragraph Kant even states that there is an indirect duty to reward animals for good services (see also Kant, 1997, pp. 212-213). As Ursula Wolf has pointed out, Kant's position is far from clear, since it denies animals moral status as ends and recognizes them only as means, but at the same time rejects their use as mere means. This suggests either that Kant dilutes his otherwise sharp distinction between animals and human persons, or that he acknowledges an empirical basis of morality in the feelings of sympathy and compassion, which would be in conflict with his moral philosophy (Wolf, 1990, pp. 33-38). - It seems that the concept of the "social person" is open to a similar critique.

[5] Charles Larmore has proposed to draw on the distinction between the universal content of moral duties and the reason one might have for justifying such content. "Why," he asks, "can we not affirm a set of duties binding on all without supposing they must be justifiable to all?" (Larmore, 1996, p. 57).

[6] Following John Rawls, primary goods are either social and comprise "rights and liberties, powers and opportunities, income and wealth," or natural and include "health and vigor, intelligence and imagination" (Rawls, 1972, p. 62).

[7] Between 1979 and 1987, the termination rate of DS fetuses before 24 weeks gestation was 100 percent at Boston's Brigham-Women's Hospital, and from 1988 to 1990 the rate was 91 percent (Caruso and Holmes, 1994). In Canada, 83 percent of women terminated pregnancies with trisomies 13, 18, 21 (Royal Commission, 1993, p. 802). "In Taiwan, termination is the preferred option" (Jan, Chen, Huang, Huang and Lan, 1996, p. 172).

[8] David Gibson has called Down's racial regression proposal "a scientific nightmare" (Gibson, 1978, p. 2).

[9] By Jason Kingsley and Mitchell Levitz, published by Harcourt Brace, San Diego, in 1994.

[10] Research has suggested a causal relation between hearing loss and mental capacities. Consequently, if DS patients' hearing problems are treated promptly, their learning potential may improve considerably. Similarly, in most cases a relatively simple heart operation can correct their congenital heart problems.

[11] Amniocentesis and chorionic villus sampling (CVS) are routinely offered only women at high risk, i.e. at or above 35 years of age, however, 75-80 percent of DS children are brought to term by women below the age of 35.

[12] According to a survey on the status of screening for fetal DS in the United States in 1992, screening for DS has nearly doubled in the four years since 1988 when a similar survey was conducted. A total of ninety-one percent of the 2,113,000 pregnancies covered in the survey, representing approximately one half of the annual births in the U.S., were provided with a DS interpretation (Palomaki, Knight, McCarthy, Haddow and Eckfeld, 1993, p.1560). Figures for individual states surveyed are even higher. In the state of Maine, two thirds of all pregnant women chose serum screening during the surveyed period from 1980 to 1993. Subsequently, DS among live births was reduced by 7.2 percent for the period of 1980-1985, by 22.5 percent for the period of 1986-1990, and by 46 percent for the period of 1991-1993.The authors conclude that their findings for the latter period are likely to reflect the "maximal participation that might be expected when services have become well established in a given region" (Palomaki, Haddow and Beauregard, 1996, p.1409).

[13] E.g. by the French *Comité Consultatif National d'Ethique*, 1995, 7 quoted in Chadwick, 1998, p. 267.

[14] See also Ost, 1984, pp. 301-312.

[15] In a Taiwanese survey, over 70 percent of the women stated they underwent amniocentesis because of their physicians' suggestions, not because of their own free wish (Jan, Chen, Huang, Huang and Lan, 1996, p. 177).

[16] Erich H. Loewy similarly identifies what he calls "six existential a prioris": the human "(1) drive for being or existence; (2) biological needs; (3) social needs; (4) a desire to avoid suffering; (5) a basic sense of logic; (6) a desire to live freely and to pursue our own interests" (Loewy, 1997, p. 141).

[17] Charles Larmore is also convinced that there exists a "broad consensus about a range of core duties...from promise keeping to respect for bodily integrity," which can be shared in spite of our differing interests and conceptions of the good (Larmore, 1996, p. 58). Reviewing international business practice and its moral presuppositions, Richard De George has similarly argued for the existence of common values between countries of different cultural backgrounds. He identifies three "basic moral norms" which "apply in all countries and across all borders," and which are independent of any particular ethical theory (De George, 1989, p. 21). They include injunctions "against arbitrarily killing other members of the community to which one belongs," the injunction against lying, and the respect for property (De George, 1989, pp. 19-20). Based on these norms, De George proposes a set of seven specific guidelines which provide a standard for assessing the conduct of multinational corporations (De George, 1989, pp. 46-56).

REFERENCES

Aristotle (1992). *The Politics.* T. A. Sinclair (trans.). London: Penguin Books.

Beauchamp, T., & Childress, J. (1994). *Principles of Biomedical Ethics,* 4th edition. New York: Oxford University Press.

Bok, S. (1995). *Common Values.* Columbia: University of Missouri Press.

Callahan, D. (1995). The puzzle of profound respect. *Hastings Center Report, 25(1),* 39-40.

Caruso, T. M., & Holmes, L. B. (1994). Down syndrome: Increased prenatal detection and dramatic decrease in live births: 1972-90. *Teratology, 49,* 376.

Chadwick, R. (1997). The Philosophy of the right to know and the right not to know. In: R. Chadwick, M.Levitt & D. Shickle (Eds.), *The Right to Know and the Right Not to Know* (pp. 13-22). Aldershot: Avebury.

Chadwick, R. ten Have, H., Husted, J., Levitt, M., McGleenan, T., Shickle, D., & Wiesing, U. (1998). Genetic screening and ethics: European perspectives. *Journal of Medicine and Philosophy, 23(3),* 255-273.

Darwall, S., Gibbard, A., & Railton, P. (1997). Toward *fin de siècle* ethics: Some trends. In: S. Darwall, A. Gibbard & P. Railton (Eds.), *Moral Discourse and Practice. Some Philosophical Approaches* (pp. 3-47). New York: Oxford University Press.

De George, R. T. (1989). *Business Ethics.* New York: MacMillan.

Down, J. L. H. (1867). Observations on an ethnic classification of idiots. *Mental Science, 13,* 121-128.

Engelhardt, H. T., Jr. (1996). *The Foundations of Bioethics,* 2nd edition. New York: Oxford University Press.

Engelhardt, H. T., Jr. (1991). *Bioethics and Secular Humanism: The Search for a Common Morality.* Philadelphia: SCM Press, London and Trinity Press International.

Fraser, F. C. (1974). Genetic counseling. *American Journal of Human Genetics, 26,* 636-659.

Gert, B. (1997). *Bioethics: A Return to Fundamentals.* New York: Oxford University Press.

Gibson, D. (1978). *Down's Syndrome: The Psychology of Mongolism.* Cambridge: Cambridge University Press.

Hansen, C. (1992). *A Daoist Theory of Chinese Thought. A Philosophical Interpretation.* New York: Oxford University Press.

Harris, C. E. (1991). Aborting abnormal fetuses: The parental perspective. *Journal of Applied Philosophy, 8(1),* 57-68.

Hassold, T., & Patterson, D. (Eds.) (1999). *Down Syndrome: A Promising Future Together.* New York: Wiley-Liss.

Hook, E. G. (1982). Epidemiology of Down Syndrome. In: S.M. Pueschel & J. E. Rynders (Eds.), *Down Syndrome. Advances in Biomedicine and the Behavioral Sciences* (pp. 11-88). Cambridge: Ware Press.

Jan, S., Chen, C., Huang, L., Huang, F., & Lan, C. (1996). Attitudes toward maternal serum screening in Chinese women with positive results. *Journal of Genetic Counseling, 5 (4),* 169-180.

Julian-Reynier, C., Macquart-Moulin, G., Moatti, J.P., Loundou, A., Aurran, Y., Chabal F., & Ayme, S. (1993). Attitudes of women of childbearing age towards prenatal diagnosis in southeastern France. *Prenatal Diagnosis, 13,* 613-627.

Kant, I. (1968). *Die Metaphysik der Sitten (1797).* In: *Kants Werks* (Akademie-Textausgabe), Bd. VI. Berlin: Walter de Gruyter.

Kant, I. (1997). *Lectures on Ethics*. In: P. Heath & J. B. Schneewind (Eds.), *Lectures on Ethics*. Cambridge: Cambridge University Press.

Larmore, C. (1996). *The Morals of Modernity*. Cambridge: Cambridge University Press.

Loewy, E. H. (1997). *Moral Strangers, Moral Acquaintances, and Moral Friends. Connectedness and its Conditions.* Albany: State University of New York Press.

MacIntyre, A. C. (1984). *After Virtue. A Study in Moral Theory.* Notre Dame: University of Notre Dame Press.

Mazzoni, D. S., Ackley, R. S., & Nash, D. J. (1994). Abnormal Pinna type and hearing loss correlation in Down's syndrome. *Journal for Intellectual Disability Research, 38*, 549-560.

McGleenan, T. (1999). Genetic testing and screening: The developing European jurisprudence. *Human Reproduction and Genetic Ethics, 5(1)*, 11-19.

Miller, B. (1981). Autonomy and the refusal of lifesaving treatment. *Hastings Center Report, 11(4)*, 22-28.

Motulsky, A. G., & Murray, J. (1983). Will prenatal diagnosis with selective abortion affect society's attitude toward the handicapped? In: K. Berg & K. E. Tranøy (Eds.), *Research Ethics* (pp. 277-291). New York: A. R. Liss.

Murray, R. F., Jr. (1995). Genetic counseling: Ethical issues. In: W. T. Reich (Ed.), *Encyclopedia of Bioethics*, vol. 2 (pp. 927-932). New York: Simon & Schuster Macmillan.

Nippert, I. (1997). Psychosoziale Folgen der Pränataldiagnostik am Beispiel der Amniozentese und Chorionzottenbiopsie. In: F. Petermann, S. Wiedebusch, & M. Quante (Eds.), *Perspektiven der Humangenetik* (pp.107-126). Paderborn: Schöningh.

Nuffield Council on Bioethics (1993). *Genetic Screening – Ethical Issues.* London: Nuffield Council on Bioethics.

Ost, D. (1984). The 'right' not to know. *Journal of Medicine and Philosophy, 9(3)*, 301-312.

Palomaki, G. E., Haddow, J. E., & Beauregard, L. J. (1996). Prenatal screening for Down's syndrome in Maine, 1980 to 1993. *New England Journal of Medicine, 334*, 1409-1410.

Palomaki, G. E., Knight G. J., McCarthy, J., Haddow, J. E., & Eckfeld, J. H. (1993). Maternal serum screening for fetal Down syndrome in the United States: A 1992 survey. *American Journal of Obstetrics and Gynecology, 169*, 1558-1562.

Parfit, D. (1984). *Reasons and Persons*. Oxford: Clarendon Press.

Press, N., & Browner, C. H. (1995). Risk, autonomy, and responsibility: Informed consent for prenatal testing. *Hastings Center Report, 25(3)*, S9-S12.

Putnam, H. (1987). Why reason can't be naturalized. In K. Baynes, J. Bohman & T. McCarthy (Eds.), *After Philosophy. End or Transformation?* (pp. 222-244). Cambridge: MIT Press.

Quante, M. (1997). Ethische Probleme mit dem Konzept der informierten Zustimmung im Kontext humangenetischer Beratung und Diagnostik. In: F. Petermann, S. Wiedebusch, & M. Quante (Eds.), *Perspektiven der Humangenetik* (pp. 209-227). Paderborn: Schöningh.

Rasore-Quartino, A. & Cominetti, M. (1995). Clinic follow-up of adolescents and adults with Down syndrome. In: L. Nadel, & D. Rosenthal (Eds.), *Down Syndrome. Living and Learning in the Community* (pp. 238-245). New York: Wiley.

Rawls, J. (1972). *A Theory of Justice*. Cambridge: Harvard University Press.

Robinson, P. (1998). Prenatal screening, sex selection and cloning. In: H. Kuhse & P. Singer (Eds.), *A Companion of Bioethics* (pp. 173-185). Oxford: Blackwell.

Rorty, R. (1993). The priority of democracy to philosophy. In: G. Outka & J. P. Reeder (Eds.), *Prospects For a Common Morality* (pp. 255-278). Princeton: Princeton University Press.

Royal Commission On New Reproductive Technology (1993). *Proceed With Care.* Ottawa: Minister of Government Service Canada.

Rucquoi, J. K., & Mahoney, M. J. (1992). A protocol to address the depressive effects of abortion for fetal abnormalities discovered prenatally via amniocentesis. In: G. Evers-Kiebooms, J.-P. Fryns, J.-J. Cassiman & H. Van den Berghe (Eds.), *Psychosocial Aspects of Genetic Counselling* (pp. 57-60). New York: Wiley-Liss.

Shannon, T. A. (1997). Fetal status: Sources and implications. *The Journal of Medicine and Philosophy, 22*, 415-422.

Smith, D. K., Shaw, R. W., & Marteau, T. M. (1994). Informed consent to undergo serum screening for Down's syndrome: The gap between policy and practice. *British Medical Journal, 309(September)*, 776.

United States President's Commission for the Study of Ethical Problems in Medicine and Biomedical and Behavioral Research (1983). *Screening and Counseling for Genetic Conditions.* Washington: Government Printing Office.

Walters, L. (1989). Ethical issues in maternal serum alpha-fetoprotein testing and screening: A reappraisal. In: M. I. Evans, A.O. Dixler, J. Fletcher & J. D. Schulman (Eds.), *Fetal Diagnosis and Therapy. Science, Ethics and the Law* (pp. 54-60). Philadelphia: J.B. Lippincott Company.

Wertz, D. (1997a). Reconsidering 'nondirectiveness' in genetic counseling. *Geneletter, 1(4) (January).* [www.geneletter.com/archives/previousissues/january1997.html]

Wertz, D. (1997b). Families of children with mental retardation and developmental disabilities: How well are they doing? *Geneletter, 1(4) (March).* [www.geneletter.com/archives/mrndd.html]

Wertz, D., & Fletcher, J.C. (1989). *Ethics and Human Genetics: A Cross-cultural Perspective.* Heidelberg: Springer Verlag.

White-van Mourik, M. A., Connor, J. M., Ferguson-Smith, M.A. (1992). The psychosocial sequelae of a second trimester termination of pregnancy for fetal abnormality over a two-year period. In: G. Evers-Kiebooms, J.-P. Fryns, J.-J. Cassiman & H. Van den Berghe (Eds.), *Psychosocial Aspects of Genetic Counseling,* (pp. 61-74). New York: Wiley-Liss.

Wolf, U. (1990). *Das Tier in der Moral.* Frankfurt: Klostermann.

IAN HOLLIDAY

GENETIC ENGINEERING AND SOCIAL JUSTICE:
TOWARDS A GLOBAL BIOETHICS?*

The early 1970s witnessed major advances in two key component disciplines of bioethics. On the 'bio' side, research teams led by Herbert Boyer and Stanley Cohen in 1973 pioneered systematic means of cloning specific deoxyribonucleic acid (DNA) fragments. The breakthrough made by recombinant DNA (rDNA) techniques, as they became known, was genetic engineering. So great were the implications that from 1974 to 1978 the international scientific community voluntarily imposed a moratorium on rDNA experiments held to be particularly risky. That a technological revolution had taken place was not in doubt in the 1970's, and is not questioned now (Wheale and McNally, 1988, p.41). On the 'ethics' side, the advance was not as revolutionary. Nevertheless, when John Rawls published *A Theory of Justice* in 1971 he was quickly acknowledged to have reshaped Western political philosophy by setting debates about justice at its center (Kymlicka, 1992, p.xi). Rawls himself held justice to be 'the first virtue of social institutions' (Rawls, 1971, p.3), and many political philosophers in the Western tradition have subsequently indicated their agreement by building on this premise to construct competing theories.

The close proximity of these parallel advances might be expected to have generated three decades of inter-disciplinary debate between molecular biologists and political philosophers keen to work out the impact of genetic engineering on Western theories of justice. Much cross-fertilization of ideas has indeed taken place. The rDNA moratorium of the late 1970s is evidence of an ethical concern among molecular biologists that remains vibrant today. Similarly, many political philosophers have sought to think through the implications of biotechnological progress (an excellent example being Harris, 1998). In the process, what some call genethics has emerged as a sub-discipline of bioethics (Suzuki and Knudtson, 1989). On the whole, however, genetic engineering has been a marginal concern for writers on justice. Some certainly have brought it within their theories, but many more have not. Even at the end of the 1990s, books and articles on justice usually had little or nothing to say on the matter.

Julia Tao Lai Po-wah (ed.), Cross-Cultural Perspectives on the (Im)Possibility of Global Bioethics, 131–147.
© 2002 *Kluwer Academic Publishers. Printed in Great Britain.*

In this context this chapter seeks to analyze the consequences of continuing advances in genetic engineering for Western theories of justice. It is not a survey of ways in which political philosophers have sought to incorporate biotechnological progress into their theories, for as has already been said this is not an activity in which they have engaged very much. Instead, it is an attempt to think through the means by which theories of justice developed in the Western tradition need to be reshaped as genetic engineering becomes an ever more central part of our lives. In making this attempt, the chapter focuses particularly on the themes of individual, community and society that structure this collection. In so doing it takes as its central interest the possibility of a global bioethics in an age of genetic engineering. The chapter begins by briefly examining advances in genetic engineering and debates about justice. It then moves to analyze some of the theoretical and practical implications of biotechnological progress.

I. GENETIC ENGINEERING

The genetic engineering that became possible in the 1970s built on many earlier advances, notably the achievement of James Watson and Francis Crick in describing, in 1953, the structure of the defining genetic material, DNA. It was their double helix, the fame of which endures to this day, that cracked the genetic code by establishing the central features of 'the thread of life' and making feasible its subsequent manipulation (Aldridge, 1996). Manipulation itself, the set of techniques for modifying and recombining genes from different organisms developed by Boyer, Cohen and their teams, in principle placed two main sets of possibilities, some distant, some not, before humankind: combating disease and re-engineering society. It is by addressing these possibilities that what has been called the genetic revolution is analyzed here. Our interest is strictly human genetic engineering; no other biotechnological advance is considered.

A point to be made at the outset is that the dividing line between what might be called the reactive and proactive, or defensive and aggressive (Ryan, 1999, p.129), dimensions of human genetic engineering is so insecure as to be almost unsustainable. These apparently distinct dimensions are continuous in important respects, chiefly because what counts as 'disease' is a function of what is statistically 'normal'. As progress is made in eliminating the genetic basis for various life-threatening and severely disabling diseases, the statistical norm necessarily 'rises', and comes to include (the absence of) complaints previously regarded as marginal or trivial. Whilst, then, we might want to say that some measures count unambiguously as disease eradication (a cure for cancer is an example) and others as social engineering (the Nazi eugenic program is an instance), we cannot rely on the distinction often made here. The 'undiseased' baseline will itself progress over time, because it is a social construct. The continuity of the reactive and proactive dimensions of human genetic engineering is a critical shaping point for the rest of this discussion.

Looking at where we stand at the turn of the millennium, we find that many diseases have already been combated. By the early 1980s more than 500 diseases resulting from recessive mutations in single genes were known, including sickle-cell anemia, muscular dystrophy and cystic fibrosis (Watson *et al.*, 1983, p.211). Much progress has continued to be made in this sphere, and gene replacement therapy has developed alongside abortion of an afflicted fetus as an important medical response. When the human genome project produced a first draft of the complete human genome sequence in 2001 a further key step was taken (Sudbery, 1998). Social engineering is also very much on the contemporary agenda, not least as a result of the cloning techniques that produced lamb number 6LL3, better known as Dolly, in February 1997 (Kitcher, 1997, p.327). Whereas the idea that molecular biologists possess 'a vast organic Lego kit inviting combination and continual rebuilding' (Yexon, 1983) may have seemed fanciful when it was floated in 1983, it no longer seems so today. As Harris wrote recently in a book that acknowledges it is looking perhaps 20 years into the future: "For the first time we can literally start to shape not only our own destiny in terms of what sort of world we wish to create and inhabit, but in terms of what we ourselves wish to be like" (Harris, 1998, p.2). Real social engineering is so close as to be just over the horizon.

In this chapter the full range of possibilities opened up by human genetic engineering is taken either to be real for us now at the start of the third millennium, or likely to become real within a very short period of time. The questions addressed here are the extent to which our theories of justice need recasting to cope with that range, and the distinct ways in which the claims of individual, community and society are affected by it.

II. SOCIAL JUSTICE

The theories of justice that have emerged in such quantity in the West since the early 1970s are theories of social justice. "Our topic," says Rawls at the start of *A Theory of Justice*, "... is that of social justice." He has long and short ways of stating what he means by this. Tending towards long is his very next sentence: "For us the primary subject of justice is the basic structure of society, or more exactly, the way in which the major social institutions distribute fundamental rights and duties and determine the division of advantages from social cooperation" (Rawls, 1971, p.7). But perhaps an easier way of capturing the subject of debate lies in Rawls's notion of a "well-ordered society" (Rawls, 1971, p.8). Among the many disagreements that have characterized post-Rawlsian debate, theorists of justice have all followed him in seeking to describe the well-ordered society. This is the Holy Grail of contemporary political philosophy in the West.

There is no point in trying to describe here the full array of contributions to the debate. Utilitarians, liberal egalitarians, libertarians, Marxists, communitarians and feminists have all had their say, and have done so in ever more subtle ways (Kymlicka, 1992). Instead, our focus is on the three major categories of individual, community and society that pervade this collection. How appropriate is it for communities to have particular moral visions that contrast with those endorsed by the society as a whole? What implications do answers to this question have for individuals? How feasible, in consequence, is a global bioethics?

The issues raised by human genetic engineering affect these sorts of question, and the debates about justice that stand behind answers to them, at every step. If we look at ways in which theorists approach the topic of justice – by analyzing what they say about its scope, its structure and the place of universals therein – we find that the biotechnological advances of the past 30 years have an impact on each and every aspect. If we then move to consider issues of substance – the permissible extent of human genetic engineering and the framing of public policy in this new era – we again find that scientific progress affects debate at every turn. Each of these issues needs to be analyzed before we can develop answers to our central question concerning the possibility of a global bioethics.

III. THEORIZING JUSTICE IN AN AGE OF HUMAN GENETIC ENGINEERING

Disagreement between theorists of justice goes all the way down, so to speak. They have distinct ideas about the way in which the topic should be approached, about how you should 'do' a theory of justice. The impact of human genetic engineering on three elements of methodological dispute is analyzed here as a means of setting up debates between the competing claims of individual, community and society.

1. The scope of justice

Biotechnological advance means that we now have the capacity to alter the genetic structure of human beings. This clearly affects the scope of justice, for it raises issues about access to, or protection from, possibilities that simply did not exist before. As we enter the new millennium we are confronted with questions about the just distribution of genetic advantage and disadvantage. That, certainly, is a new situation.

One impact on debates about justice is straightforward. The list of matters that Rawls argued 'seem arbitrary from a moral point of view' (Rawls, 1971, p.15) is now a little shorter than it used to be. Talents, at least in the sense of basic genetic make-up, are currently in the process of being shifted from the category of nature's givens to the category of things over which humans can exercise a degree of choice. In one sense this impact is fairly minor, for it could be said simply to constitute a mechanical reshaping in response to scientific advance. But there is a more substantial way in which debates about justice are affected. On the one hand, the kinds of imaginative leap in theorizing that have long been made by liberals now need to be made by all theorists of justice, for human genetic engineering means that we can shape the individuals who are to populate the just society in quite fundamental ways. On the other, the 'whole person' approach to justice that is characteristic of communitarianism becomes essential, for very many aspects of who people are now fall within the scope of justice.

As it happens, these sorts of moves have been made by a number of theorists in the past 20 years. As liberals have become increasingly dissatisfied with the Rawlsian approach, they have sought to take more account of whole persons in developing theories of justice as impartiality (Barry, 1995). Disembodied liberalism has rather few takers today.

Similarly, communitarians have never sought merely to describe the norms of justice that characterize contemporary societies, but have been keen to make imaginative leaps of their own. In looking at ways in which these two approaches need changing in the wake of biotechnological advance, the central point may simply be this: the scope of justice has been enlarged in recent years, and those theorists who fail to notice it are missing an important trick. Human genetic engineering enables the genetic structure of an individual to be altered either pre-natally or post-natally. The scope of justice is increased by the extent to which this is the case at any given moment in time. Or, if justice takes on an inter-generational dimension, it is increased by the extent to which it is likely to be the case in the plausibly foreseeable future. Very few contemporary theorists of justice take all this on board.

2. The structure of a theory of justice

This expansion of the scope of a theory of justice has consequences for its structure. In a revealing article published in 1983, Ronald Dworkin argued that the distinct theories of justice developed to that date in fact had quite a lot in common, in that they were similarly structured. The linking thread, he said, was the shared premise of the 'egalitarian thesis', which held that "the interests of the members of the community matter, and matter equally" (Dworkin, 1983, p.24). Differences, Dworkin maintained, arose from distinct answers given to two questions. What are people's interests? What follows from supposing that those interests matter equally? It has to be remembered that in 1983 the communitarian critique of liberal egalitarians like Dworkin was in its very early stages. MacIntyre's *After Virtue* was published in 1981, Sandel's *Liberalism and the Limits of Justice* in 1982, Walzer's *Spheres of Justice* in 1983. Nevertheless, there is a clear liberal slant to this description. In particular, it overlooks a question that communitarians would want to give priority over both Dworkin's questions. Who are the members of the community?

In Dworkin's defense, it could be said that it was just about possible in 1983 to capture the essence of this additional question in his own first question. That is, in focusing on people's interests it is in fact possible to say something about who they are – not much, communitarians would no doubt say, and on the whole we would have to agree with them, but something. In an age of human genetic engineering this is not in the least bit possible, for the aspects of who people are that molecular biologists have brought within the realm of human choice go way beyond anything that could ever be identified as interests. The very fundamentals of an individual's being can now be changed, and this means that we simply have to take the communitarian question – Who are the members of the community? – as our starting point. This is an important structural issue which, again, has not been picked up by many contemporary theorists.

3. The place of universals

The place of universals in a theory of justice has always been contentious. For, say, liberals the whole point of theorizing is to develop universal principles of justice. For communitarians, by contrast, no such principles can ever be formulated, as the particular contexts from which we all come to debate justice mean that we end up at different points when the debate is over and it is time to get on with living.

Here human genetic engineering would seem to work to the advantage of liberals, for its effect is to make a degree of non-particularism – or, to put it the other way round, a degree of universalism – an inevitable component of all theories of justice. We are very nearly capable of creating an entire society of clones and setting them to debate justice. We already have the ability to make significant changes to individuals' genetic make-up. In these circumstances, the world for which theorists are trying to prescribe just social arrangements is itself moving towards at least an element of universalism. The degree of thought experimentation undertaken by liberals, and often strenuously criticized by others, must now feature in any serious theory of justice.

This is not to say that foundationalist liberals are wholly vindicated, for if we return to the three questions developed in the last section the impact of human genetic engineering is only felt with respect to the first. 'Who are the members of the community?' is a question to which elements of a universal answer can now be given. But this is not true of questions to do with interests. Nevertheless, this impact is important, for in recent years much of the running in this particular debate has been made by communitarians (Barry, 1995, p.3).

4. Theorizing justice in the age of biological control

In the age of biological control (Wilmut, 1999), the range of things we can seek to be just about has been substantially expanded. Kymlicka notes that *A Theory of Justice* had this sort of impact by shifting the focus of debate from that which happens within existing social practices to the practices themselves. Since Rawls, he writes, "questions of justice arise, potentially at least, wherever social institutions and practices affect the distribution of life-chances" (Kymlicka, 1992, p.12). Human genetic engineering has a similarly scope-expanding effect. It also means that our first question in theorizing concerns the nature of the individuals who are to populate the just community. We now have the ability to decide many aspects of their make-up, and cannot simply say that on Dworkin's egalitarian plateau their interests are our only concern. In addition biotechnological advance means that some element of universalism must feature in all theories of justice. Each of these changes must be taken into account as we seek to determine the possibility of a global bioethics.

IV. INDIVIDUAL, COMMUNITY AND SOCIETY IN AN AGE OF HUMAN GENETIC ENGINEERING

"Our discussion," Plato famously wrote in Book 1 of *The Republic*, "is about no ordinary matter, but on the right way to conduct our lives." Well, how should men and women conduct their lives in an age of human genetic engineering? And how should the just society be ordered in this new age?

If we take a look at what is at stake here we find that the various techniques that are now or soon will be available generate both costs and benefits. We need to determine how the costs should be borne and the benefits enjoyed. Of course, the framing of the issue can be disputed. Some hold that human genetic engineering comes close to generating nothing but costs, for the dangers of interfering with 'nature' are so great as to take the human race to the brink of the apocalypse (Ho, 1998). Others maintain that it comes close to generating nothing but benefits, for the possibilities of extending human control over 'nature' are so substantial that the actual costs incurred pale into insignificance (Silver, 1998). But it is best to take a more balanced view and say that costs and benefits are both present in significant measure. How should those costs and benefits be distributed in the just society? This question can be considered through analysis of two linked questions. Who should have access to human genetic engineering? Who should be protected from subjection to it? But to get to these questions we need to consider a prior issue. Who are we talking about here and what say do they have in the matter?

1. Humans and genetic engineering

The genetic engineering that is either possible now or likely to be possible in the very near future can take place at any stage in the human cycle. While the issue of who we are talking about here can of course be debated endlessly, it can also be given a relatively straightforward answer: all living persons and all yet-to-be-born persons. (Difficulties concerning what constitutes a yet-to-be-born person [Harris, 1998, ch.2] can be overlooked for the purposes of this discussion.) From this answer it is immediately evident that the issue of what say the relevant population has in the matter is not clear-cut: yet-to-be-born persons cannot say anything.

Taking first those who can have a say, many theorists would argue that the relevant population is all living persons. Rawls places what he calls human agents in the original position, and engages in very little discussion of the matter (Rawls, 1971, p.202). Many others use this sort of foundation. But this clearly cannot be the right answer, for just as yet-to-be-born persons cannot say anything, nor can newborn babies or very young children. We therefore need a cut-off point that is rather further down the human cycle. Hillel Steiner argues that it should come at the dividing line between childhood and adulthood, for the whole point about children is that they are incapable of making the responsible decisions that are required of participants in debates about justice (Steiner, 1994, pp.245-6). The issue then becomes the age at which the dividing line should be drawn. This looks controversial, but it must be the correct answer. In any case, it is the answer used here (on the understanding that the dividing line be placed somewhat earlier than in most contemporary societies).

We have, then, two categories. Yet-to-be-born persons and children are not capable of having a say about genetic engineering. Adults must therefore take responsibility not only for the genetic engineering they seek for themselves, but also for that which is to be undertaken on yet-to-be-born persons and children. Theorists are likely to take different positions on the age of majority. For liberals the broad criterion is attainment of moral agency. They do not need to claim that this occurs at any particular age – in some tragic cases, it never occurs – and they can

probably allow to communitarians and others that part of its variability is
the result of differences in cultural factors. Whether they would want to
fault cultures that have the effect of delaying the attainment of moral
agency is an interesting question. Theorists may also disagree about the
significance of moral agency, with liberals seeking rights-bearing
individuals who can be fully responsible for their actions and
communitarians taking a more relaxed attitude towards responsibility.
But, in an age when we are able to manipulate human genetic structures
at all stages of the life cycle, it is certainly valid to ask a theorist of justice
when an individual is capable of making the relevant decisions and when
someone else must bear the burden.

2. Access to human genetic engineering

Who, then, should have access to genetic engineering? A liberal would
argue that access should not be denied to those adults who wish to have
their genetic structure re-engineered and are able to pay for it. Few
communitarians would disagree. But this probably covers only a small
proportion of the relevant population. Excluded from it are two categories
of person or potential person. First, there are those adults who seek
treatment but do not have the ability to pay. As human genetic
engineering becomes increasingly routine, costs will fall. However, the
segment of the population able to pay for genetic structure engineering
will likely remain small. Second, there are those persons or potential
persons for whom responsibility is borne by others. What is to be done
about these two categories?

It is characteristic of liberals to focus on rights. If we go to the
libertarian end of the broad liberal spectrum, rights are pretty much the
sum total of the story. At the liberal egalitarian end, some form of
redistribution also figures. A uniquely rights-based account will state that
individuals in category one – crudely, poor adults – have no access to
genetic engineering because they have no claim on the resources
currently held by others that are needed to pay for it. This account will
argue that individuals in category two – yet-to-be-born persons and
children – have access to genetic engineering to the extent that those who
bear responsibility for them are willing to sanction treatment and able to
cover the attendant costs. A more egalitarian account will agree with this
division of responsibility, but will find ways of generating the resources
needed to cover treatments for many of the individuals in both categories.

There is, however, a further dimension to the liberal position, as those who exercise responsibility on behalf of others can often be held to account for actions they have taken. In particular, we are now becoming familiar with suits brought against adults for, say, abuse during childhood, or a deficient education. Why, then, should a case for 'wrongful life' not be sustainable? Steiner, for one, argues that it could be in circumstances in which an adult is able to prove that during the pre-adult phase of her or his life identifiable persons failed to make the decisions that would have given her or him an acceptable genetic endowment (Steiner, 1999). Here, then, a rights-based account generates grounds for holding that access to human genetic engineering should be quite extensive.

Communitarian accounts respect rights only to the extent that they are respected by the community in question. They will therefore recognize that liberal societies are likely to reach liberal conclusions. In other societies, however, very different conclusions may be reached. Indeed, it must be possible for a communitarian account to hold that issues concerning human genetic engineering should be decided strictly on the basis of community values. To put it at its most extreme, individuals should have access to genetic engineering to the extent that they do or do not conform to the genetic attributes generally valued by a particular society.

3. Protection from genetic engineering

Who should be protected from subjection to genetic engineering? In many ways this is the flip side of the question concerning access. In the liberal universe, rights-bearers are thus fully protected, for whatever else they trump (and the accounts of course differ), rights certainly trump attempts to reshape an individual's genetic structure. The doctrine of self-ownership found in the writings of Locke, Kant and others does all the necessary work here. Non-rights-bearers do not have this protection: they can be subjected to genetic engineering by the individuals who exercise responsibility for them. However, it is important to note that on Steiner's account some non-rights-bearers will eventually become rights-bearers, as children mature into adults. At this point they are able to seek redress for any maltreatment they may have suffered, including subjection to gene therapy or genetic restructuring that can be shown not to have been in their interests. In this there is a clear element of protection. For communitarians the issue is determined locally. This could generate both more and less protection, depending on particular circumstances.

4. Questions of substance in a theory of justice

Returning to the book that launched a thousand theories, we find that *A Theory of Justice* was written chiefly to contest the conclusions reached by a non-rights-based approach to justice that Rawls felt was incapable of respecting the separateness of or distinction between persons. That approach was what Rawls called classical utilitarianism. It held society to be just, he wrote, "when its major institutions are arranged so as to achieve the greatest net balance of satisfaction summed over all the individuals belonging to it" (Rawls, 1971, p.22). In the familiar phrase, the greatest happiness of the greatest number equals justice. There is a basic egalitarianism here, for in calculating total human happiness no person is to count for more or less than one. But there is not much protection for the integrity of the person, and this was what concerned Rawls.

Responses to the Rawlsian critique of utilitarianism quickly made it clear that there was more to be said on the matter than had in fact been said in *A Theory of Justice*. In particular, it was argued that rule utilitarianism could be distinguished from act utilitarianism, and that this made the theory a lot more plausible and acceptable. None of this need concern us here, for the key point remains that which was made by Rawls. Any approach to justice that fails to "take rights seriously," as Dworkin (1977) subsequently put it, fails to protect the integrity of persons. In the days before human genetic engineering had become a real possibility, this issue was held to relate most clearly to property entitlements. It was to redistribution that Nozick (1974) objected. But now that human genetic engineering is very much on the agenda, the issue also reaches deep into the person.

There would seem to be an important lesson in this, and it is particularly relevant for communitarians, socialists, feminists and all those who fail to place rights at the heart of their theories of justice. It is simply hard to see how, in this day and age, a non-rights-based account of the just society will suffice.

V. TOWARDS A GLOBAL BIOETHICS?

If only a rights-based account will do, what are the prospects for a global bioethics? It has already been noted that guidelines for the treatment of rights-bearers are unlikely to be hard to generate. But, as has also been noted, rights-bearers are by no means all of the relevant population. The particularly critical question is what to do about those who can have no say in the genetic engineering to which they may or may not be subjected. That is, how are we to regulate treatment of yet-to-be-born persons and children?

Steiner's argument makes it clear that we cannot re-engineer the genetic structures of such persons and potential persons at will, for when they attain the age of majority and assume their rights they can issue claims for maltreatment against those who have hitherto exercised responsibility for them. What should be the metric or standard by which those claims are judged? Here we get to the heart of the debate about individual, community and society. If the distinction floated earlier between the reactive disease-combating and the proactive social-engineering capacities opened up by human genetic engineering could be made to stick, we might find a solution in restricting interventions to the former and wholly outlawing the latter. But as medical reality is socially constructed, the distinction cannot be sustained and cannot therefore provide a way out of the moral dilemma. In short, in seeking to determine the contents of a rights-based account, the issue of values has to be addressed at this point.

One obvious way forward is to act at the level of communities by taking local standards as the basis for judgments of this kind. This would generate a series of answers in the many different communities that can be identified around the world. The liberal critique of communitarianism's failure to respect the separateness of persons of course raises doubts about this approach. If rights become infused with a series of local understandings, then the protections offered by them are entirely lost. To caricature the communitarian position, though not by very much, in one community a claim for wrongful life may be upheld on the grounds that blond hair and blue eyes are a necessary condition for making anything of a life. In another community such a claim would be laughed out of court. However, if we look at what counts as injury in civil law, we find significant variation from one society to another. Moreover, liberalism is not particularly well placed to object to this. It would be

quite well placed were the world to embody the traditional liberal value
of free movement, for then it would be harder to sustain the claim that
persons' vital interests are determined by reference to the norms of their
local cultures. In such a world, persons would be less tied to their
localities and, having the possibility of exit, would be less able to claim
that some characteristic which is only locally disvalued is disabling *per
se*. However, in a 'second-best' world that is illiberal to the extent of
preventing free mobility, the liberal's criterion of harm probably does
need to embrace some local variability. It is this kind of consideration that
makes issues like female circumcision so troubling for liberals.

The alternative way forward would seem to lie in an approach that is as
international or cosmopolitan as possible. Cosmopolitanism usually holds
that moral obligations extend in principle across the entire globe. It is
associated in international relations theory with Kant's project for a
perpetual peace, and with such heroic political failures as the League of
Nations. In some writings it proposes forms of world government. None
of this need be debated here, for what we seek is a means by which
individual claims for or against genetic engineering might be judged.
Cosmopolitanism seems the correct way forward for two main reasons.
First, in an era in which advances are being made not simply in
biotechnology but also in information technology, there is a sense in
which a global community with at least some common values is
emerging. In part those values can be captured in the language of rights,
with the 1948 Universal Declaration of Human Rights being a prime
example. In part, however, they must be captured at the level of shared
understandings, for it is simply very difficult to develop an understanding
of rights that encompasses, say, the unborn child. Second, the sorts of
shared understandings that would emerge from a cosmopolitan approach
would clearly be of the lowest common denominator variety. In a world
in which human genetic engineering still raises many concerns, this is
entirely desirable.

It has already been conceded that the liberal finds it hard to sustain an
objection to some form of communitarian input to this debate. Yet in an
age in which universal understandings and treatments are increasingly
becoming the norm, a fusion of these two approaches that tends more to
liberalism than to communitarianism seems most desirable. To be a bit
more specific, what we seem to need to identify is a fairly narrow zone,
fuzzy at the edges for communitarian reasons, of treatments that might be
said to be akin to the cosmopolitan birthright of those whose genetic

endowments do not conform to the global standard. This is a zone both of expectation and of limitation. That is, individuals whose genetic endowment does not reach the standard that marks out the zone have a legitimate expectation that action will be taken to ensure that they are brought up to it. Individuals whose genetic endowment is already at or above the standard have a legitimate expectation that their genetic structure will not be tampered with. Developed in a predominantly cosmopolitan way, the zone would be defined in a very conservative manner. In effect, it would be little more than a baseline. Corrective treatments for widely acknowledged congenital malformations and disorders known to be attributable to genetic make-up would feature. Anything else would not, for it would simply be very hard to generate the necessary cosmopolitan consensus. Genetic engineering as a developed form of social engineering would certainly be hard to get off the ground. There will be some variation across communities, but the argument made here is that it should not be too great. There will also be variation through time, as the baseline effectively 'rises'.

Three additional issues are worth mentioning in passing. The first is that this sort of approach could have substantial redistributive consequences. For a libertarian, the legitimate expectations of individuals concerning the pre-adult phases of their lives will have to be balanced against the resources held by those who exercised responsibility for them in those phases. An inability to pay would clearly be a legitimate defense, and the extent of redistribution would consequently be limited. For a liberal egalitarian, however, the resource claim could easily be made against society as a whole, and then the redistributive consequences would be substantial.

A second issue concerns perversion of the cosmopolitan ideal. In the real world of politics, multinational business and power structures, it would be wrong to place naïve faith in an approach that is likely to become infused with the conflicting claims of distinct brands of identity politics. Indeed, there is a danger that cosmopolitanism becomes a screen behind which the values of a dominant sub-set are imposed on the rest of humanity. In the present world, those values would very probably be the values of corporate America, to the discomfort of much of the rest of global society. Anyone who states that Walt Disney plus genetic engineering is a potentially dangerous mix makes a very fair point. This, however, is one of the main reasons why cosmopolitanism is promoted here. The cosmopolitan approach has precisely the advantage that it is likely to make it impossible to generate a global standard that moves towards a developed form of social engineering.

The third issue links to this, and concerns the way in which the necessary standard might be developed. Distinct approaches are already visible, in the largely *ad hoc* procedures adopted in the English-speaking world and in the national ethics committee formed in France (Harris, 1998, p.4). Neither is ideal, for what is truly needed here is public debate and open, democratic procedures. Genetic engineering raises fundamental questions about human identity. It should be discussed as widely as possible, through exploitation of the advances made by the communications revolution over much the same period as that on which this chapter focuses. Again, it would be naïve to place excessive faith in the possibilities of global democracy, but something close to an international forum is needed.

VI. CONCLUSION

It is hard not to mention *Brave New World* at some stage in a discussion of this kind. Indeed, it appears in the titles of a number of recent books (Ho, 1998; Silver, 1998). There is certainly something of a Huxleyan quality to the debate, and, as has already been said, it often provokes real worries (Ho, 1998). Those worries, though often ill-informed and exaggerated, are real. But there is also a positive side to the debate, for advances in human genetic engineering could make a substantial change for the better to many people's lives. The disease-eradication potential genuinely is enormous and raises the possibility of real change for the better.

The approach taken here chiefly embodies a kind of dual Kantianism in a rights-based cosmopolitanism that seeks to put the necessary access and boundaries in place for both rights-bearers and non-rights-bearers. On this basis an argument for some form of global bioethics has been made. This solution will not be perfect, and may easily strike some as too cautious at the same time as it strikes others as too ambitious. But in seeking to balance practical possibilities for alleviating human suffering and genuine fears about social engineering, it is one plausible way forward.

Department of Public and Social Administration
City University of Hong Kong, Kowloon, Hong Kong

NOTE

* I thank Hillel Steiner for very helpful comments on a draft of this chapter. I also thank participants in the conference on which this collection is based, notably H. Tristram Engelhardt, Jr, who was a discussant for my paper. The usual disclaimers apply.

REFERENCES

Aldridge, S. (1996). *The Thread of Life: The Story of Genes and Genetic Engineering.* Cambridge: Cambridge University Press.
Barry, B. (1995). *Justice as Impartiality.* Oxford: Clarendon Press.
Dworkin, R. (1977). *Taking Rights Seriously.* London: Duckworth.
Dworkin, R. (1983). Comment on Narveson: In defense of equality. *Social Philosophy and Policy, 1,* 24-35 (reprinted in: W. Kymlicka (Ed.), *Justice in Political Philosophy Volume I: Mainstream Theories of Justice* (pp. 3-14). Aldershot: Edward Elgar).
Harris, J. (1998). *Clones, Genes, and Immortality: Ethics and the Genetic Revolution.* Oxford: Oxford University Press.
Ho, M-W. (1998). *Genetic Engineering – Dream or Nightmare? The Brave New World of Bad Science and Big Business.* Bath: Gateway.
Kitcher, P. (1997). *The Lives to Come: The Genetic Revolution and Human Possibilities.* New York: Touchstone.
Kymlicka, W. (1992). Introduction. In: W. Kymlicka (Ed.), *Justice in Political Philosophy Volume I: Mainstream Theories of Justice* (pp. xi-xxiii). Aldershot: Edward Elgar.
Nozick, R. (1974). *Anarchy, State, and Utopia.* New York: Basic Books.
Rawls, J. (1971). *A Theory of Justice.* Cambridge, MA: Belnap.
Ryan, A. (1999). Eugenics and genetic manipulation. In: J. Burley (Ed.), *The Genetic Revolution and Human Rights: The Oxford Amnesty Lectures 1998* (pp. 125-132). Oxford: Oxford University Press.
Silver, L.M. (1998). *Remaking Eden: Cloning and Beyond in a Brave New World.* London: Weidenfeld & Nicolson.
Steiner, H. (1994). *An Essay on Rights.* Oxford: Blackwell.
Steiner, H. (1999). Silver spoons and golden genes: Talent differentials and distributive justice. In: J. Burley (Ed.), *The Genetic Revolution and Human Rights: The Oxford Amnesty Lectures 1998* (pp. 133-150). Oxford: Oxford University Press.
Sudbery, P. (1998). *Human Molecular Genetics.* Harlow: Longman (with additional material posted at http://www.awl-he.com/biology).
Suzuki, D. & Knudtson, P. (1989). *Genethics: The Clash between the New Genetics and Human Values.* Cambridge: Harvard University Press.
Watson, J.D., Tooze, J. & Kurtz, D.T. (1983). *Recombinant DNA: A Short Course.* New York: Scientific American Books.
Wheale, P.R. & McNally, R.M. (1988). *Genetic Engineering: Catastrophe or Utopia?* Hemel Hempstead: Harvester Wheatsheaf.
Wilmut, I. (1999). Dolly: The age of biological control. In: J. Burley (Ed.), *The Genetic Revolution and Human Rights: The Oxford Amnesty Lectures 1998* (pp. 19-28). Oxford: Oxford University Press.
Yexon, E. (1983). *The Gene Business.* London: Pan.

CORINNA DELKESKAMP-HAYES

GLOBAL BIOMEDICINE, HUMAN DIGNITY, AND THE
MORAL JUSTIFICATION OF POLITICAL POWER*
A Kantian Approach

Contemporary bioethics is characterized by an aspiration on the part of
many to establish in international law, conventions, and policy a single,
globally guiding understanding of moral principles that can shape health
care policy across the world. This aspiration to universal validity has deep
roots in Western European moral and philosophical commitments that
framed the Enlightenment and found its particularly stringent and
influential expression in the thought of Immanuel Kant (1724-1804).
These commitments have inspired the human rights movements that took
shape after the Second World War and gave new force to the French
Revolution's endeavor to establish positive moral claims that are
supposed to be compellingly rational and therefore universally valid. This
vision of a general humanitarian morality presupposes that individuals
fully realize their human dignity only in a society that is at the same time
a moral community bound by a content-rich understanding of human
rights that should transcend national borders.[1] In terms of these
commitments not only all particular value communities, but equally all
particular political societies should be recast so as to frame one globally
unified society as a universal moral community, within which those
positive rights are realized.

In this essay, I shall criticize this universalizing model of international
bioethical cooperation. I shall distinguish between Kant-inspired
contemporary understandings of autonomy with their thick commitment
not only to human choice but also to a particular good held to be intrinsic
for the moral life on the one hand, and an understanding of human
autonomy grounded in the secular moral centrality of individual choice
exclusively. As I will show, the Kantian commitment, both in its
Christianity-enriched contemporary version and in those elements of his
philosophy that respond to his own secularizing intentions, to a view of
human dignity that is loaded with moral value implications diminishes the
political space for moral difference. It is hostile to the peacable co-
existence of diverse moral communities, especially in Asia and Africa,
whose values differ from those of the West.[2] Thus, the Enlightenment's

*Julia Tao Lai Po-wah (ed.), Cross-Cultural Perspectives on the (Im)Possibility of Global
Bioethics*, 149–177.

central project of securing respect for autonomy as the condition for peaceful interaction of individuals (as well as communities and societies) turns out to be self-defeating.

This predicament, which also hampers predominant modern human rights theories, is illustrated by the *Convention for the Protection of Human Rights and Dignity of the Human Being with regard to the Application of Biology and Medicine (Convention on Human Rights and Biomedicine)* which was issued by the *Europarat* as a legal document in 1996 and signed by almost all European countries (and some non-European ones as well). This *Convention* attempts to transcend particular nations and moral communities through fashioning an international bioethical understanding. Though framed by Europeans for Europeans, it is surely by its authors regarded as normative for bioethical cooperation on a global scale as well. In opposition to such aspirations, my position is that the human rights movement in its particular value-loaded endorsement of positive rights and proscriptions as expressed in that *Convention* compromises the fundamental forbearance rights of individual choice it professes, and thus collides with the right of peaceable moral communities to their moral integrity.

In a first step (I), I shall expose the conceptual inconsistencies which result from the *European Bioethics Convention*'s aspiration to combine respect for autonomy rights with protection of a human dignity that is conceived in a content-rich, Christianity-inspired sense. These inconsistencies destroy the distinction between teleological and deontological considerations, on which the anti-utilitarian (anti-instrumentalization) thrust of human rights advocating depends. In a second step (II) I shall trace the origin of that predicament to the Enlightenment project of harmonizing the True and the Good on the plane of a thickly moral justification for the use of political power. In the end, I shall recommend a more consistently secularized, meta-moral view of human autonomy as a base for a more thoroughly culture-difference-tolerating human rights advocation for national and international cooperation in the field of biomedicine.

I. JUSTIFICATION DILEMMAS IN HEALTH CARE POLICY
THE EUROPEAN *BIOETHICS CONVENTION*

The *Bioethics Convention* deals with ethical issues. As a legal document however it also justifies sanctions. It provides ethical justification for the use of political constraints. Since membership in a polity (unlike membership in a private community) is not a matter of choice, commitment to human rights imposes somewhat rigorous conditions on the justification of such non-chosen constraints. Consistency, as well as the universally obligatory character of the ethical principles invoked are the least one would expect. I shall first (1.) expose the conceptual inconsistencies involved in the *Convention*'s regulations, afterwards (2.) interpret these as deriving from the ambiguity of formal and material aspects of the concept of human dignity, which forms the ethical basis for those regulations.

1. Inconsistencies in the Convention

Modern public bio-medicine in Europe (a) does not deliver what is morally desired and (b) is not efficient in what is morally required.[3] In democratic polities, what is morally desired and required depend on the granting of rights and the enforcement of duties. The European *Convention* defines these rights and duties for the major areas of modern biomedicine.[4]

a) The right to health care: Delivering what is morally desired
The *Convention* (in its *Preamble* and art. 3) confirms the universal right to health care, as expressed in previous Human Rights Documents.[5] Granting social rights means promising that the public will pay for those unable to afford what they have a right to. State-sponsored insurance systems covering the majority of citizens in Europe serve as redistributive devices for implementing that promise. Advances in medical technology and the rising cost of medicine-related labor have put increasing strains on economic resources. Whether publicly acknowledged or not, European countries as a matter of fact cannot afford the health care they grant as a matter of right.[6]

Modern Western democracies rest on the pointedly anti-utilitarian moral foundation of respect for human dignity. The notion of human dignity is peculiar in that it allows for no limits on the demands which can be made in its name against a public that is morally obliged to respect. No argument for restricting these demands, and for limiting the redistribution imposed for satisfying them, is safe from the charge of violating human dignity. Respecting dignity becomes tantamount to offering an encompassingly practised societal solidarity. Moreover, from the equal dignity as humans, attributed to all members of a polity, an easy transition leads to a fairness principle which in medicine demands equal health care for everyone in the struggle for a dignified life, thus endorsing ever more encompassing equalizing policies. The granting of rights is thereby transformed into a promising of desirables.

While the motive behind such granting is obviously benevolent, the practical outcome is unfortunate: Societal resources being always limited, public health care as a matter of principle cannot secure its avowed respect of dignity in terms of what is required as being morally desired.

The *Convention* accordingly concedes that "appropriate health care" depends on "available resources" (1996, art. 3). Indeed, one should promise only what one can deliver. Yet the granting of rights generates promises of an unconditionally binding nature and requires leaving them untouched by contingent interests. This is obvious in the analogous case of freedom rights: Except where the existence of a polity is threatened (as in war), no human rights advocator would permit limiting their "appropriateness" in view of the moral (or even economic) resources of a society. Rights single out an area of deontological commitment from issues that may be subjected to utilitarian weighing of costs and benefits.[7] If the *Convention* renders "appropriateness-evaluation" subject to such weighing, it acknowledges that some humans as a matter of policy will not get the health care to which they were granted a right. On the *Convention*'s own terms, they will not have their human dignity respected.

b) Regulation of medical interventions: Efficiency in what is morally required

In its regulative articles (1996, arts. 1-22) the *Convention* accounts for the fact that in biomedicine today human dignity not only is fostered (as through the granting of public health care as such) but also comes under risk. Individuals must be protected both against medical paternalism and, especially when they belong to vulnerable minorites, against instrumentalization for the sake of the common good. For that purpose the *Convention* must – again, but now explicitly – separate the (compellingly) deontological from the (discretionary) teleological. Three provisions are introduced for securing respect of human dignity:

Patient consent is required as a side constraint upon medical interventions (1996, art. 5-9). Regardless of even professionals' judgement concerning the (true and long term) interest and wellbeing of a patient himself (*i.e.* even before societal instrumentalization comes into play), it is the latter's own preferences on which the legitimacy of medical interventions depends. Not only public spirited instrumentalization (as in research), but already benevolent paternalization is thus ruled out in the name of human dignity. Respect for human autonomy in medicine takes on the special meaning of affirming a *prima facie* right to be left alone.

Non-Discrimination among humans is affirmed for securing equal respect for the human dignity of all individuals (1996, art. 1).

Extension of rights-bearers: Throughout the text of the *Convention*, human dignity is affirmed of all members of the human species (all human beings). Thus everything from persons (legally able to consent) to the fertilized ovum and even – in articles concerning genetic interventions – members of future generations are included in the class of dignity-bearers.

This enhanced protection of health-related dignity-rights is responsible for further inconsistencies in the *Convention*. These hamper both (1) the protection of those unable to consent and (2) the respecting of those able to consent.

1) Protecting those unable to consent

The *Convention* fails appropriately to protect in the areas of (i) germ line therapy, (ii) research and tissue donation, and (iii) medically assisted procreation.

(i) Humans unable to consent require a proxy consent (1996, art. 6), which truly pursues the interest of the patient (1997 #48). Yet the *Convention* prohibits therapeutic interventions with the aim "to introduce any modification in the genome of any descendants", even if these are designed to serve the "interest and wellbeing of future humans", irrespective of parents' (proxy) willingness to consent (1996, art. 13).

This is unfortunate. First, since future humans were included among the bearers of human dignity and thus of protection-rights, and since human rights are accorded to their bearers without qualification (concerning different classes of such rights), future humans are by definition also bearers of social rights. If available therapy is denied them, their social right to health care is being disrespected. (The *Convention*'s ban on discriminating against human beings with respect to their rights is thereby violated as well.) Second: While the motive behind that denial is obviously benevolent, its consequences defeat the purpose. The authors express their concern about non-therapeutic or merely eugenic manipulations of future humans (1997, #89): Once germ line therapy were permitted, future humans would get exposed to particularly serious risks of instrumentalization. But since some future humans are thus denied a human right on the ground that granting them that right may endanger others, then the formers' "interest and wellbeing" has been sacrificed to the interest of (future) society, which already constitutes an instrumentalization.

(ii) Research on those unable to consent is permitted, provided (among other conditions) that proxy consent has been secured (1996, art.17.2.iv). Yet proxy consent – even for therapy – had been accepted only if it pursued the interest of the individual at stake (1997 #48). Hence, no proxy consent to nontherapeutic research should have been accepted by the *Convention*, regardless how small the risks and inconveniences are (1996, art. 17.2.ii). If violation of the fundamental right to be left alone is accepted on the ground of such smallness, a utilitarian weighing of costs and benefits has been permitted to compromize a fundamental right.

While the motive behind that acceptance is obviously benevolent as well as grounded in a necessary concern for rights (because without it certain minorities would suffer from the discrimination of being excluded from the fruits of medical progress), the consequences are thus again unfortunate. To be sure, the *Convention* stipulates a "solidarity feeling" in humans that links their "interest" to other humans "in the same age category or afflicted with the same disease or disorder or having the same condition"(1996, art.17.2.i). Yet on persons able to consent such a supposition is never enforced. Hence persons unable to consent are discriminated against with regard to their fundamental right to be left alone. They are instrumentalized in the interest of (their segment of) society.[8]

(iii) By implicitly permitting medically assisted procreation,[9] the *Convention* accepts the fertilization of surplus ova, which will not be chosen for implantation but might be destroyed. While the motive behind that acceptance is a benevolent concern for the suffering that comes with unfulfilled hopes for children, the consequences defeat the *Convention*'s commitment to the dignity of unborn humans. Those discarded ova's right to life is disrespected.

In all three cases the *Convention* fails to protect – on its own terms of protecting – those unable to consent.

2) Respecting those able to consent
The *Convention* fails appropriately to respect persons' autonomy in the contexts of (i) selecting the sex of future children and (ii) of using their body at will.

(i) In the context of medically assisted procreation, future parents' choosing the sex of the fertilized egg to be implanted is proscribed. While the motive behind this prohibition is a benevolent concern about such parents' future offspring's basic rights to sexual non-discrimination (1996), art. 14), that concern again defeats the relevant respect. Tolerating medically assisted procreation implies tolerating that human life, the generation of which was (as it were) undertaken with a view to exposing it to the risk of being discarded, comes to enjoy a merely conditional existence, or an existence that hinges on the future parents' choice. It means tolerating that the right to life of such generated human beings exists only under condition of the parents' consent. Discrimination, on the other hand, presupposes an actual right, not the merely conditional right at stake here. Nor does there exist any right of non-existing humans to their existence. Consequently, once the practise of medically assisted procreation has been accepted (along with the granting of conditional rights to the generated human beings), the only remaining right relevant here concerns the parents' autonomy right in wanting a boy (a girl), and that right is being disrespected.

(ii) The use of bodily parts (in their biological function) for financial gain (sale of organs, surrogate motherhood, etc.) is forbidden (1996, art. 21). Human dignity is deemed incompatible with a body (medically) invaded (in extreme cases) by commercial organ harvesters (uterus renters). While the motive behind this proscription is clearly a benevolent concern for those whose poverty exposes them to instrumentalization for the medical needs of the rich, the consequences expose the underlying conceptual muddle. Since respect for human dignity underlies respect for autonomy, a dignity that can be set against someone's using his autonomy at will renders the concept inconsistent.

As a result, just as the *Convention* failed to protect against instrumentalization, so it fails to respect autonomy. In both respects, it defeats its central purpose, and suffers from inconsistencies. In the next section, I shall interpret this failure in terms of an ambiguity in the underlying ethical principle of human dignity, which also compromises the universal validity of the ethical norms, the political enforcement of which the *Convention* undertakes to justify.

2. The ambiguity of human dignity

Human dignity is used sometimes (as when grounding the requirement of consent) in a formal sense as resting on a human autonomy that must be respected (regardless of its wise or unwise, dignified or undignified employment). At other times it is used in a material sense as imposing an ideal of human excellence, in which certain particular values are realized. In the first case respect means leaving people free to choose as they will, in the second case it means restricting them in order to make them choose as they should. By vascillating between both senses, the *Convention* mingles the deontological concern for human rights with a teleological concern for desirables and usefulnesses.[10] It becomes understandable, then, that modern public health care in Europe, and even already on national levels, repudiates its underlying moral commitment.

Correspondingly, the inability of public biomedicine in Europe to deliver what was held to be morally desired (the issue discussed under [1.a] above) can be conceived as resting on the same ambiguity of human dignity. Here as well dignity's formal sense (securing autonomy) can be seen to be mingled with its material sense (demanding some materially correct, *i.e.*, morally dignified, use of autonomy). For it is the concern about securing the economic and societal conditions that might improve the practical likelihood of rights-bearers' materially dignified use of their (formal) autonomy, which are generally believed to demand, in the name of "dignity-respect", the granting of ever more encompassing (and thus in principle unaffordable) social rights. Thus, the moral failures of inter-European bioethics legislation (just as of similar national legislations) can be traced to their implied imposition of a materially loaded ideal of human dignity, or to an encompassingly misplaced moral benevolence politically enforcing that ideal.

How could things have gone so wrong? The confusion already permeates the political self-understanding of European societies taken individually. Their constitutions or founding documents also mingle formal principles and material values. Historically speaking, the former can be traced to the Enlightenment quest for secular rationality, the latter to Europe's specifically Christian cultural roots.[11]

The Enlightenment derives human dignity from moral autonomy. As morality presupposes freedom of choice, that respect concerns dignity in a merely formal sense. It theoretically should imply tolerating even undignified or immoral uses of human freedom, or the violation of material values.[12] Christianity, on the other hand, affirms a double commitment to formal as well as one particular material understanding of human dignity: It imposes formal respect on the basis of God Himself having trusted humans with moral autonomy, but adds a heavy concern for the moral use of that autonomy, and thus for a content-filled (materially value oriented) ideal of Christian humanity. From a Christian standpoint, modern democracies are commended for respecting humans' autonomy by granting the relevant liberties, and at the same time blamed for unleashing morally deplorable ways of abusing these liberties. The material implications of human excellence are argued to require superior efforts at protection. Christians thus come to be more involved in advocating particular communitarian values, even if this places them in opposition to respect for autonomy in the formal sense.

Very schematically, we can therefore allot championing the formal side of human dignity to the Enlightenment, and championing its material side to the Christian tradition. The resulting inconsistencies in Europeans' political self-understanding, as these became visible in the *Convention*, can thus be blamed on an unfortunate mingling of two incompatible strains of thought:

1. The including of merely desirable social goals in the class of political rights arose from confusing the state as a protective agent for individuals' ability to act autonomously with the state as an agent for furthering their ability to act morally. The solidarity principle, which is invoked for justifying the necessary redistributive policies, derives from the (Judaeo-) Christian love-thy-neighbor imperative.[13]

2. The unlimited expansion of the range of human rights bearers is incompatible with the Enlightenment imperative to respect humans

because of their moral accountability. Fertilized eggs are not morally accountable. It is the religious regard for man as God's beloved creature, as encompassing all developmental stages of human life, which motivates the raising of (conditional moral) obligations to protect these stages to the level of (unconditional political) obligations to grant them the respect which is due persons. Since persons (exercising their own autonomy) simultaneously were granted rights to give proxy consent with respect to human non-persons (in the moral, or legal sense), mutually incompatible claims to respect could be derived.

3. The disrespect for persons' autonomy involved in the prohibition of any technological manipulation of future humans' nature, and of any rendering one's body an object to be disposed of at will, is motivated by the same creational understanding of humans as merely stewards, not owners, of their natures and bodies. A religious commitment to the natural course of things as somehow ordered by divine design tacitly underlies the prohibition of germ line therapy or marketing bodily organs.

As a result, human rights advocating in Europe, both on the national and on the international level of bioethical regulation, is not in a universal sense a humanitarian undertaking as its defenders would like to think. The political societies in Europe are constituted as value communites with a very particular bias. Already with respect to the large portion of non-Christians (and the very large portion of not-much committed Christians to boot) in Europe, the particular Christian elements of politically enforced human dignity constitute a violation of autonomy rights. Any attempt at global implementation would merely increase the number of individuals and communities with different value commitments who get exposed to such injustice.

Moreover, since legal regulation comes with the imposition of sanctions (*Convention,* art. 23–25) and thus with the use of force against trespassers, the peace-securing function of human rights advocating is obviated, wherever what is enforced on all responds to the merely particular value preferences of some. On a global level, European-style human rights regulation in biomedicine would provide an excuse for the employment of force against value-communities with different views of what constitutes human dignity. Such regulation would in fact be hostile to peaceful inter-culture cooperation.

Is there a way out for universal human rights advocation? Could one purge the purportedly rational Enlightenment project of morally justifying the use of political power from its merely particular Christian value residua? That project, after all, being age-hallowed, is central to Europeans' self-understanding: it goes back (ultimately) to the Platonic vision that the True and the Good are intimately related.[14] The Enlightenment emphasis on rationality had rested on the (more or less) tacit supposition that human nature, as determining the conditions of that rationality, was teleologically designed. It rested on the belief that reliable (true) knowledge not only about the world but also about moral Goodness could be attained. If the Good can truly be established in terms of such rationality, then (conceivably) it is valid for all humans and thus can also be politically enforced. Political society thus is still seen as value community, but now in secular, and because of its rational credentials truly universal terms. Kant's philosophy presents the most compelling aspiration towards such moral justification, and thus the most persuasive realization of the Platonic vision.

Kant endeavored to liberate philosophy from its unacknowledged Christian elements and to restrict his moral and political theory to what is secularly rationally knowable. He posits an unconditional obligation to respect a human dignity (1786, p. 65), which encompasses on the one hand its formal sense of autonomy (1786, pp. 76f.), along with a ban on instrumentalization (1786, p.66), and on the other hand – within that very ban – at least some regard for dignity's material sense: all moral action must envisage humanity as a purpose. Thus, Kant's practical philosophy both is pointedly anti-utilitarian (1786, p.13) in the sense required by the modern commitment to human rights, and aspires to a rationality that promises universal validity. In the second part of my essay, I shall examine the possibility of framing a truly global bioethics along the lines of his arguments.

II. THE KANTIAN PERSPECTIVE:
VICISSITUDES OF MORAL JUSTIFYING

Quite in line with Enlightenment theorizing in general, Kant derives human dignity from humans' moral autonomy (1786, p.77). Humans' moral obligation, when dealing with other humans, to respect the latter's autonomy also provides the moral foundation of political societies (or polities such as, to use Kant's example, the state): The "social contract", which ideally (1797, I, pp.168 f.) underlies such societies, represents the necessary condition for the possibility of respect among humans to be realized, and thus of interactive moral action itself (1797, I, pp.154 f.). Polities, after all, secure that "everybody's acting autonomously" remains compatible with "everybody else doing the same" (1797, I, p.33). They thus enforce the morally obligatory mutual respect. In this sense, the existence of political societies is itself morally obligatory and thus morally justified.

In particular: the use of political power for restraining individuals even against their present actual consent to being restrained is morally justified in that it incorporates the respect it enforces: Those individuals can themselves be taken to have implicitly consented to being thus unconsentingly restrained, because they are under a universally valid moral constraint to consent to such constraint (1793, pp. 234, 250, 1797, I, pp.35 f.). Only through consenting to the use of political power can they secure protection against being subjected to every other unconsented-to use of power (and thus preserve the necessary conditions for the possibility of exercising their moral autonomy).

Can this view of morally justified political power escape the value parochialism that was found to compromise and render inconsistent European human rights advocated in medicine? As a heuristic tool, let us check in what sense and to what extent Kant's approach could remedy the inconsistencies which the mingling of (particular) religious and (universal) philosophical elements introduced into health care legislation, as this was described in the case of the European Bioethics *Convention* above.

1. The problem with social rights

Kant does not attribute moral autonomy to humans as a matter of fact. His philosophy thus avoids the embarrassment this concept would otherwise present for a civilization that relies on scientific knowledge, which (as far as humans are concerned) rests on the supposition of cause-and-effect relationships and thus is incompatible with the assumption of moral freedom. Kant's transcendental method, by contrast, proceeds hypothetically: The supposition of moral autonomy (1786, pp. 101 f.) presents the necessary condition without which unconditional moral approbation, or moral approbation in the genuine sense, would be theoretically inconceivable (1786, p. VII). This supposition envisages the capacity for rationality in the sense of a self-determination of the will (1786, p.1). It coincides with humans' corresponding self-experience

(1786, pp. 107 ff.) of an – as it were – creative act of the will, whenever they in fact hold humans (including themselves) to be morally accountable (1797, I, p.22). Since the matter-of-fact acknowledgement of autonomy (and rationality, and dignity) implied in that act is also an integral part of what moral obligation unconditionally requires, and insofar as Kant posits an interest in making unconditional moral approbation concerning such obligations possible, he derives the moral imperative to respect (1786, pp. 66 f.) human autonomy. Thus the supposition of human moral freedom as the theoretical condition for genuine morality to be possible, far from being proposed as a theoretical statement of fact (1786, pp.26, 63, 1797, I, pp. 18, 25) concerning the object of the supposing (and thus being vulnerable to disqualification on scientific grounds) appears itself as a function of the practically realized (1786, p. 75) genuine morality of those engaged in the supposing.[15] This solution to the problem of autonomy-grounded human dignity is what renders Kant the most compelling of the Enlightenment philosophers. But, as we shall see at once, rational compellingness has its price.

The attribution of moral autonomy and thus the respect for human dignity that are to morally justify the use of political power as well, rest on a resolute (and sometimes heroic) decision: Humans – irrespective of any empirical evidence to the contrary – are considered capable of rising beyond those bodily or psychic needs[16], which are usually considered to causally determine their behavior. They are considered, in Kant's technical term, capable of acting as noumena, or as purely intelligent beings. This implies that no material conditions or societal goods can be argued to be relevant to respecting their human dignity. Such arguments, in resting on causalist considerations, would compromise the transcendental affirmation of autonomy.[17] Hence, no immediate dignity-related derivation of social rights, and of health care rights in particular, is conceivable within Kant's theoretical framework.[18]

But could the granting of such rights at least be considered possible? Can the redistributive measures imposed for their implementation be rendered compatible with Kant's theoretical framework?

In order for his moral justification of power to satisfy the requirements of a political theory, it must allow for the protection of property, and hence place rigorous conditions on the justifiability of redistributive policies. But how could respecting a dignity that (for Kant) rests on moral autonomy allow for protecting property in the first place? Noumenal autonomy, after all, resides in the mind, and the impossibility of deriving social rights had rested on the need to keep the noumenal clear of phenomenal issues, to which property seems to belong.

Let us concede that Kant's initial quest for unconditional goodness concerned not the beauty of souls taken by themselves, but human action. To be sure, what is unconditionally good about human action is placed in the moral will that bears witness to a moral frame of mind (the *Gesinnung*, 1786, p.78). Still, the attribute of goodness applies to the action determined by such a will (1786, pp. 15, 63). If the state is morally justified insofar as – given the fact that people cannot help interacting socially – it provides the necessary condition for the possibility of moral goodness, then it must (among other things) enable people to act in such a way as not to violate the respect they morally owe one another. It must legally define the sphere of individuals' privacy (1797, I, p.163), such that illegitimate intrusion can even be recognized and identified as disrespect. (1797, I, p. 55), This sphere of privacy encompasses all the worldly means which are necessary for human action (including human bodies as well as property, *loc. cit.*) to which a person's liberty-rights therefore extend. Intruding on some one's property thus violates the obligatory respect (1797, I, pp. 55, 63). In this sense property, even though it consists of *prima facie* phenomenal things, must be considered under the perspective of justice, or considered noumenally (1797, I, pp. 56, 62, 68f.), and Kant's political theory satisfies the requirement mentioned above.

The particularly anti-utilitarian thrust of that theory lay in the fact that polities not only must force their members to respect one another's autonomy, but must also, as the representative of all members, observe that respect themselves, when dealing with any one of them. After all, political power may be used against subjects (thus violating their autonomy) only insofar as that use protects their autonomy. Even taxation, as an intrusion on people's property, *prima facie* counts as a violation of their autonomy and requires a compelling moral justification (1797, I, p.188). It is thus no matter of discretion for a polity – say – to tax its members (or otherwise burden their enjoyment of the fruits of their labors) in order to support or medically care for the needy segment of its population. Any burden that exceeds what is necessary for the securing of every one's realizing his moral autonomy through the unencumbered-by-others'-trespasses use of resources to which he is entitled constitutes an injustice (*loc. cit.*) But then no granting of social rights which would impose such burdens is even possible on Kantian grounds.[19] As a result, if polities are to respect people's property (and there is no reason for exempting income) just as they are to respect people's autonomy, then those policies' competence must be much more restricted than the European defenders of social rights would expect.[20]

2. The problems with the extension of the class of rights bearers

Kant does not discuss which classes of human beings should count as bearers of human dignity. As his account invokes a human dignity that implies moral autonomy, he would surely reserve autonomy rights (as entitling to and necessitating political respect) to those humans who are (legally or morally) able to consent. This is also what he usually expounds (1786, p. 70). Moreover, and in keeping with this restriction, his theory of the law permits only one human right: the right to liberty (1797, I, p. 45). No right to life is even considered.[21]

To be sure, Kant calls even infants (and those below any stage of moral accountability or capacity to consent) persons. He grants them a legal right to being supported and cared for by their parents (1797, I, pp. 111 f.). They were, so he argues, forced into existence without having been able to consent to such (truly existential) violence. Therefore, it is the parents' duty to reconcile them (at least *ex post*) with their lot. Parents' support and care are legally obligatory in order, as it were, to permit children, through their later consent, to ratify the violence they had suffered. Thus, Kant considers potential persons to be persons as well.[22]

Of course, with this reasoning there is no obstacle to including any members of the human species, even below the threshold of birth, down to the level of fertilized ova and further to merely potentially force-generated future human beings. Any one of these, moreover, must be thought to have suffered (or be about to suffer) the violence of having been (or going to be) generated. But then abortion (except perhaps if the life of the mother is at stake) as well as medically assisted procreation (as involving the discharge of surplus fertilized ova) would have to be subsumed under the heading of parents' failure to provide the care and support these potential humans are entitled to. Quite in contrast to established present-day European practice and to what is implied in the *Bioethics Convention*, these actions would have to be considered unjust, and thus illegal. Therapeutic efforts affecting future generations, on the other hand, which the *Convention* forbids, would have to be considered to be at least licit (*i.e.* certainly not subject to prosecution). Concerning (nontherapeutic) research on unconsenting children (down to the fertilized ova stage), perhaps one could get away with granting grown children a right to sue their parents for care- and support-failures connected with such involvement. (However, this would exclude the severely mentally handicapped, who, as they never reach the level of moral personhood, would remain entirely unprotected.)

To make matters even more puzzling, Kant's own acknowledgement of to-be-protected personhood applies only to infants begotten in marriage. Given his view of penal justice, it is significant that he considers the (in his view inescapable) death penalty for the murder of children born out of wedlock to be unjust (1797, I, p. 205). Such children, he argues, sneaked into existence illegitimately. They thus have no claim to legal protection even with respect to the more elementary right to the inviolability of their already started life (1797, I, p. 204). Their existence is seen to involve an unconsented-to violence committed against their mothers (*loc. cit.*, whose responsibility, in an age of poorly developed contraceptive technology, could obviously be disregarded).

On the basis of these considerations, it becomes difficult to understand why no political moral duty to retaliate against such children should have been derived. After all, the mother's body has been invaded, and as a result her resources for realizing autonomy through real world action is greatly reduced. Obviously, potential persons are not considered even just morally accountable. Yet if Kant still wishes to protect them as persons, then he has dissociated personhood from the moral autonomy acknowledgement that goes with holding people accountable. The moral basis of his political philosophy is thus jeopardized.

Irrespective of that difficulty, Kantian polities would have to restrict their protection of potential persons to those begotten in marriage, that is, to cases where the attack on the mother's autonomy resources was covered through her consent. The aborting or killing of fetuses and (at least) infants, just as the instrumentalization of gametes, fetuses and (at least) infants for research purposes (as long as these resulted in death), would all have to be considered licit, as long as their generation had happened outside of marriage. If we apply this reasoning to present times, the legal significance of marriage would perhaps have to be discounted, as its link with raising children has loosened. The status of human beings up to infant-stage would have to be rendered dependent on parents' (or mothers') actual (even if merely implied) *ex post* consent concerning those beings' having come into being. The formers' fate would be a matter of legal neutrality until a positive decision was formed. Abortions could then be considered licit only at an early stage (where the mother's consent implied in letting the fetus grow cannot be supposed to have been given). With medically assisted procreation, however, people producing (or permitting the production of) surplus gametes intentionally could still not claim that the existence of those surplus gametes was somehow violently forced upon them. These practices, therefore, would still be unjust. Instrumentalization of aborted fetuses for any purpose whatsoever, however, would be a legally neutral act.

3. The problem with restricting autonomy

Contemporary human rights theorizing, particularly in biomedicine, supposes that the integrity of any one's human dignity must be protected not only against others, but also against his violating that dignity in himself.[23] Similarly, Kant holds that humans are not owners of themselves in the sense that they can dispose of themselves at will (1797, I, p. 96). Man is not his own master but is obligated with regard to the humanity in his person.

To be sure, respecting people's autonomy means granting them a right to seek their happiness as they please (1793, p. 235). It is acknowledged that they tend to seek it in irredeemably heterogeneous ways (1786, p. 47). Kant's political theory is pointedly directed against any paternalistic forcing of people into a happiness someone else designed for them (1797, I, pp. 172 f, 1793, p. 261). In order to be able to restrict people's autonomy in their perhaps unusual ways of seeking happiness (or – what amounts to the same – of fleeing unhappiness) one would have to assume (at least) that the moral law also imposes on agents some universal purpose, which, in the background of any particular purpose pursued through particular actions, must be pursued "along".

Prominent formulations of the moral law in Kant's ethical writings support just such a view. Setting humanity as a purpose is even implied in the moral law (1786, p. 66). Kant formulates the self-regarding aspect of this purpose-setting as the "primary legal duty" (1797, I, p. 165). Yet Kant himself insists on distinguishing the political from the moral: The law leaves it to everyone's discretion as to which purpose he might set for himself, as long as he does not infringe on others' purpose-setting (1797, I, p. 7). Hence no legal restriction of own-humanity disrespecting uses of autonomy seems to be derivable from Kantian principles.[24]

In addition, Kant links the humanity he morally imposes (as "to-be-pursued-along") with humans' noumenality (1797, I, p. 48). He thus links the imposed purpose with their autonomy in purpose-setting.[25] This purpose (obviously) is universally compelling for all moral persons, so that its inclusion among the legal duties does not commit Kant to the political paternalism he opposes. Accordingly, we must conclude that disrespecting humanity (in this sense) in oneself is no legally neutral act. Let us suppose that someone were to drink or drug himself (more or less permanently) out of moral autonomy. As each persons' noumenality links him with all other human persons this would be an act against humanity, insofar as that person was concerned. Once he had recuperated moral personhood, he could be legally punished for that act.

Among Kant's examples of one's using autonomy so that it violates one's own humanity, two are interesting for modern biomedicine: extra-marital sex (1797, I, pp. 109 f.) and selling oneself into slavery (1797, I, pp. 116 f., 193 f.). Someone selling himself into slavery, so Kant argues, disrespects the very possibility of future purpose setting for real world action in himself. Similarly, if we take as given Kant's view that engaging in sexual relationships amounts to reducing oneself to an object to be used by another, his view that extra-marital sex involves permitting oneself to be instrumentalized becomes understandable. Kant had in mind a concubine relationship and the special vulnerability of women who let themselves be supported by someone who can drop them at whim. While marriage introduces a mutuality of legal obligedness that "makes up" for the "reduction" implied in sexual intercourse (1797, I, pp. 107 f.), outside of marriage especially females engaging in such relations must have been thought to come under failings with regard to the first legal duty: Such engaging not only annihilates their present capacity to make legally enforceable demands (concerning continued subsistence and care), but also greatly reduces their resources for ever being able to enter into a marriage relationship later (thus closing an essential option for female realization of autonomy through real world action). Translating these considerations onto the bioethical plane, one might feel tempted to posit Kantian support for the *Bioethic Convention*'s prohibition of the sale of one's bodily organs (or surrogate motherhood and related practises). This seems especially appropriate because Kant explicitly considers not only suicide but also self-mutilation (at least) immoral (1786, p. 67).

Yet on closer scrutiny neither does Kant's treatment of his own examples support any prohibition in the sense of justifying penal measures, nor is his disapprobation in the case of these examples (and supposedly the biomedical parallel cases) even compatible with his moral and political theory. Firstly, Kant himself (in his philosophy of law) speaks only of the invalidness of the respective contracts, not of prosecution. Secondly, while admittedly, self-sold slaves and voluntary concubines just as voluntary kidney-sellers or uterus-renters have reduced their resources for real world action, they have not thereby disrespected noumenal humanity, insofar as their persons are concerned (in the sense in which suicide does, 1797, II, p. 73, or destroying one's mental powers would). The amount of such resources at any one's disposal has nothing to do with noumenality (1786, p. 3), as this implies freedom from passions and inclinations (1788, pp. 212, 287) that could conceivably be practised even within the mind, and through the design of one's mental actions. Considering such reducing illegal, or even merely holding the respective contracts to be not legally binding, would presuppose that "humanity as a purpose" implies commitment to autonomy not as to what must be respected (even if it is directed against the chances of realizing it in the world), but as something the realizability of which through real world action must be protected. Autonomy would no longer be conceived in formal, but in material terms, and its respect could be even legally limited with respect to such protection.

While obviously this is what modern human rights advocation, as exemplified in the *Bioethics Convention*, must be understood to be about, the implications of such an approach reveal its untenability. Kant's extension of the concept of disrespect regarding oneself to cases in which some one permanently diminished his (her) options for future realization of autonomy, with the concept of "diminishing" being as undetermined as that of "permanent" and "the future", opens a whole Pandora's box of disturbing analogues. Would this line of argument not force political societies to also leave legally unprotected other permanent such reductions, as these attend the social practices of spending portions of one's property (and to what extent?) on charitable organizations, expensive hobbies, gambling, risky stock, or even on the raising of children[26] and caring for aged (and notoriously longevious) relatives? Would one not further have to include the keeping of (complicated sorts of) pets and even, *pace* Kant, the engaging in marriage itself under the heading of such diminishing and hence disrespect? Clearly, this would be the end of any polity that deserved its name.[27]

Moreover (as if further proof was needed!), on the basis of such a reasoning not even altruistic organ donation to the extent that this is accepted by the *Convention*, as well as the necessary contracts involved in the medical implementation of such donations, could be acknowledged as legitimate[28].

Looking back at the implications of Kant's morally grounded political theory for the major difficulties presented by biomedical regulation in the *Bioethics Convention*, the question poses itself: How helpful has this theory turned out to be with respect to our project of devising a truly universal account of human rights, which, when politically enforced on national, international and global levels, could serve its function for securing societal respect for human dignity, tolerance for cultural differences, and peaceful co-existence among divergent societies and communities? Let us go once again through each of the difficulties.

1. With regard to the granting of positive social rights, which had rendered human rights advocation incoherent because it implied using violence against the unconsenting, no justifying reason that would repudiate the violence and remedy the incoherence could be conceived on Kantian grounds. Both on national and international levels no individual, community, or society can be forced on such grounds either to contribute to or to impose the creation of societal resources required for the provision of what goes beyond basic health care, and what others deem appropriate with respect to such care. Individuals' as well as voluntary-membership-dependent moral communities' personal or collective autonomy in determining their own measures and methods for securing internal peace and extending social beneficence (and in determining the respective "appropriateness") can be respected. In this regard Kant's theory presents a valid alternative.

2. With regard to the protection of human non-persons, Kant's moralist construal of human dignity suggests excluding them from rights-bearership. If such a decision would be adopted, a considerable number of inconsistencies in the *Bioethics Convention* could be remedied. This would imply leaving the fate of human beings that do not qualify for moral personhood to the respective families and voluntary moral communities. Biomedical technologies presently accepted in many countries, such as abortion, technology assisted procreation, pre-implantation diagnosis and prenatal diagnosis (both with their respective selection-implications) as well as (annihilative) research on human

embryos could thus be rendered compatible with acknowledgement of human rights. On the other hand, biological technologies presently not accepted in many countries (such as the cloning of humans or therapeutic as well as enhancing germ line manipulation) could at least no longer be consistently prohibited. Here as well Kant's theory presents a coherent guideline.

In what concerns, on the other hand, Kant's compromises with respect to granting rights to merely potential persons, it has turned out that their policy implications led into absurdities. If Kant's theory is not to be rendered untenable on their account, these compromises must be attributed either immediately (or mediately, via the "humanity as a purpose"-construction insofar as this also carries material value implications) to unacknowledged and unfortunate Christian prejudices. It can thus with good conscience be excluded from his theory proper.

3. With regard to restricting persons' autonomy in the name of human dignity, however, the tenable elements of Kant's theory must be protected from his unfortunate commitment to regarding autonomy not just as a principle imposing respect (with regard to others just as with regard to oneself), but as a value the practical realizeability must be furthered (even if this involves opposing others' as well as one's own autonomous decisions). With the krypto-Christian tenet that man is not master but only steward of his life and body, Kant himself endorses some material value connotation of human dignity along with its formal implication. So it is Kant's own inability to tear himself loose from the Christian elements of his moral theory which renders his reasoning in this context incoherent. His attempt to devise a purely rational and hence universally compelling moral foundation for the use of political power has failed: by linking human dignity with the bodily resources for realizing autonomy through real world (as distinct from mental) action he has permitted particular value commitments to compromise his rational formalism.

But if Kant was the "best" of what the Enlightenment had to offer, what does that tell us about the Enlightenment project of harmonizing the True and the Good in pursuit of a moral justification for political power? It suggests, so it seems to me, that that project is unrealistic. There is no Good that could be known truly as valid for all. There is not one, rationally accessible, and at the same time positively determined (value-loaded) concept of human dignity that could morally justify the use of political power for the fostering of that dignity. The most one could achieve for the securing of peaceful social interaction, both within and across political communities and societies, is respect for a purely formal human dignity, that is exhaustively defined as respect for autonomy.

That autonomy would have to be conceptually separated from any of its morally dignified or undignified employments, and even from any employment that either preserved or destroyed the resources for its continued worldly employment. Only such respect would be safe from the temptation of conceiving the state as an agent for furthering any specific morality. Such respect could, however, still secure a meta-moral justification for the political implementation of forbearance rights. Admittedly, this is much less than Western human rights advocates have been expecting. But, considering the evil side effects of their more ambitious expectations, it is the only justification we should feel justified in seeking.

European Programs
International Studies in Philosophy and Medicine
Buchbergstrasse, Freigericht, Germany

NOTES

*   I wish to thank H.T.Engelhardt, Jr., for his helpful criticism of this article.
[1]   For a more detailed account of societies as communities and communities as societies, as well as societies beyond communities, see Delkeskamp-Hayes (in press).
[2]   A good summary of reasons for this un-avowed hostility is given in *Global Summit of National Bioethics Commissions: Tokyo Communique* (1999). This cultural cleavage is attenuated by suspicions that the West might be using its headstart in devising bioethics regulations as well as its ability to withhold access to medical technology not only for the "moral colonization" (Qui, in this volume, p. 80) of the East, but also as a vehicle for promoting the (distinctly amoral) interests of Western business firms.
[3]   For a more detailed account of these criticisms, see Delkeskamp-Hayes, 2000.
[4]   The two most vexing medical activities: abortion and euthanasia, are not discussed in the *Convention*. I shall nevertheless refer to problems presented by these activities wherever the argument makes this necessary.
[5]   See, for example, (1948), art. 25. The *Explanatory Report* (1997) relates the *Convention*'s article 3 "Equitable Access to Health Care" to the *International Covenant on Economic Social and Cultural Rights* (1966) and to the *European Social Charter* (1961).
[6]   This problem is usually addressed indirectly, as presenting an issue of cost containment. The responsibility is shoved back and forth between patients, physicians, their professional as well as insurance-related representatives, insurance companies, pharmaceutical industries, and various medical institutions. The underlying problem, the granting of health care rights itself with the resulting extensive public involvement in biomedicine, is usually not acknowledged.

7   This is a strictly philosophical statement. It disregards issues of implementation and the issue of its affordability (cf. the discussion of Holmes and Sunstein (1999)). European democracies and human rights documents are justified not in view of implementability but of philosophical principles. My *ad hominem* argument is designed to meet them on that ground.

8   The same problem affects the *Convention*'s permission of tissue donation (1996, art.20.2). It concerns research on in vitro embryos as well, that is, on bearers of human dignity and human rights who do not even have a concerned proxy to secure their interest and wellbeing or to speculate about their implicit solidarity feelings.

9   (1996), art. 14 presupposes that medically assisted procreation is licit.

10   A similar vascillating can be observed in the double commitment to "wellbeing" and "respect" in the *Tokyo Communique* (1999). Perhaps such muddling is not a European prerogative.

11   Historically speaking, of course, juxtaposing the Enlightenment and Christianity is problematic. Not only can some principles that are forcefully endorsed by the Enlightenment be traced to the emphasis on natural reason, which had been cultivated by scholastic medieval Christianity, and not only is the Enlightenment affirmation of humans' moral dignity recognized as having sprouted from (Judaeo-) Christian roots, there are also, in spite of the anti-religion thrust of much Enlightenment literature, among its more prominent authors many more implicitly committed Christians than outright atheists. Underneath the resolute decision to proceed along the lines of secularly rational argument exclusively, there is usually a hidden stream of robust Christian committtment surreptitiously pulling these reasonings in non-secular directions and loading them with unacknowledged Christian prejudices. On the other hand, Christianity, especially (though not exclusively) in its Protestant tradition, has been thoroughly shaped through the Enlightenment experience. To separate both elements when analyzing present-day political thought therefore involves some venture of idealizing reconstruction.

12   This understanding of human freedom is more congenial with what the British Enlightenment (as with Adam Smith) produced, namely, the insight that for human flourishing to exist there must be a place for persons to pursue their own interests and concerns.

13   The ideal status of human dignity, which rendered claims to the provision of material conditions for the full realization of that dignity so hard to limit, received persuasive force from the religious conception of a universal human brotherhood that calls for universal sharing and self-sacrifical beneficence. The Golden Rule which in the morally reduced Christianity of today is believed to capture Christianity's essence, is taken to imply not only that one should not do to others what one would not want done to oneself, but that one should not demand more for oneself than one is willing to grant to others. This is why the equalizing implications of social rights, even where they are not broadly accepted, leave opponents with no ethically tenable position from which to defend their unbrotherly lack of generosity.

14   See, for example, Plato's understanding that true goodness is attainable through wisdom, such that there is a true way of knowing the Good (*Phaedo*, 69 b).

15   Kant's project is decidedly anti-nominalist: The possibility that our interest in being able to offer unconditional moral approbation might be ill directed is never considered. Given Kant's practical interest in finding what he looked for, the necessary conditions for the theoretical possibility of conceiving an unconditional moral duty insensibly transformed themselves into practical necessities, with an unconditional moral obligatoriness attached to them. So noumenal morality was linked to empirical phenomena by a (practical) interest, which is nevertheless, on the basis of a hidden (even though harmless) teleological assumption about

the design of our experiencing, claimed to be rationally compelling. (Of course, the whole project of making philosophical sense is impossible without some such assumption.)

[16] Human autonomy is knowable only negatively as that which motivates by excluding (phenomenal) sensual motives (1797, p.27).

[17] Kant explicitly dissociates the justification of laws from any considerations of needs (1797, p. 32).

[18] To be sure, for Kant the state has a duty to intrude upon its members' property in order to care for the subsistence of the poor: Implied in the state's morally obligatory purpose of making moral autonomy possible is making subsistence possible (1797, pp. 186 ff.). Kant cautions us, however, not to exaggerate on this , because the poor will get lazier and thus increase the unjust burden on everyone.

One may surmise that Kant would also permit public expenditures for saving the lives of those who cannot afford this themselves. It is difficult to decide, however, if he provides any grounds for limiting expenditures in that regard.

[19] There is in Kant a basis for justifying the state's intrusion on its members' property (and thus autonomy) in the interest of securing its own existence (1793, pp. 252 f). Kant is thinking here about a ban on import, but one may equally apply that reasoning to the necessity of satisfying the basic needs of the poor, if this is the only means to preserve internal peace and stability. But of course this is a purely prudential matter. While the problem of overspending, which hampers public health care today, could therefore not possibly arise, still, neither the amount of care offered on such grounds nor the manner in which it is offered (as a charity used for bribing) would be considered acceptable by modern human rights theorists.

[20] With respect for dignity excluded as a ground for defending social rights, Western human rights advocates usually invoke two other aspects of Kant's philosophy. Let us shortly show why neither is helpful.

1. Suppose that satisfying the basic needs which are taken care of by those rights is a precondition for enabling people to realize their autonomy through real world action in the same way as being relieved from others' intruding upon their autonomy is a condition for the possibility of their realizing that autonomy. People who are hungry, lack clothes, shelter, adequate health care as well as the necessary education for being able to function in today's technological society are in various ways unable to realize their (even trivial, let alone moral) autonomy through real world action. Tentatively stated in Kantian terms: if the state's moral justification consists in enabling people to realize their autonomy through real world action, the state is thereby obliged to avert what hinders such realization, and thus to satisfy the basic needs of those in need. A closer look, however, will repudiate this reasoning. The Kantian state is morally justified insofar as it provides the necessary condition for the possibility of moral action. It is justified, in other words, in legally opposing whatever (and only what) renders moral action impossible. What is it that renders moral action impossible? All worldly action lies under the constraints imposed by scarcity of means, weakness of the human body, and other external circumstances. Moral worthiness was not defined in view of any success crowning the real world actions through which it is realized (1786, p.3). It was defined in view of the will acting out of regard for the moral law. No empirical lack of means for the successful realization of autonomy could repudiate the possibility of acting out of that regard. Nor would any psychological strain attendant to the state of being in need repudiate that possibility. People are expected to take into account external constraints and available means when planning to act, and this is true independently of whether they intend to act morally or not. The – as it were – advantage of having one's moral autonomy acknowledged irrespective

of actual performance thus carries the – as it were – disadvantage on its back, that one cannot make any empirical claims on behalf of getting that performance supported.

(I acknowledge that not all would agree with my interpretation of Kant. First among those in disagreement is the late Klaus Hartmann (1991), who attempts to ground within Kant's political theory an obligation to institute a system of taxation so that welfare can supplant charity. Were I to concede any of his points, I would do so only with respect to a minimal level of support not at all encompassing welfare systems, which, in addition, I believe is compatible with the general thrust of Hartmann's work.)

One might still ask: if the Kantian state's legal enabling function is limited to securing the possibility of moral action, with one aspect of this function being the definition of legally relevant spheres of privacy, why, on the reasoning just offered, should the state also enforce their non-intrusion? In what sense is this enforcement necessary for moral action to be possible? After all, moral autonomy being something in the mind (and regarding persons only noumenally), it should be impossible to hinder it through (unconsented-to) intrusion of privacy, that is, through external violence. If one supposes that violence or instrumentalization means that the person suffering from such disrespect will not be able to act as he intended to, and if one supposes further that this *a fortiori* renders him unable to act in a morally autonomous way, then indeed external violence could be understood to hinder moral autonomy. But then – so one might conclude – the natural constraints of the world would have to be charged with same injustice: In crossing people's intentions they *a fortiori* hinder moral autonomy.

Yet these suppositions are wrong. First, natural constraints were said not to make intention-forming and acting according to intentions impossible. If the intentions include a regard for the moral law, regardless how modest the resulting action, it will be a moral action. Second, the mere being constrained as such, whether by natural circumstances or through human agency in general, does not exclude the possibility of moral autonomy in the way in which being instrumentalized does. With respect to human agency, there are cases in which constraint is compatible with moral autonomy, such as when the state (through human agency) coerces a person according to its morally grounded laws, which that person is obliged to obey anyway. Due to the constraint, he will not be able to realize his intentions. He is forced to comply. But he can comply either because of the force effecting the constraint, or out of respect for the moral law behind the force. Hence moral autonomy – as this is realized in the latter case – remains possible. Being violated through instrumentalization, on the other hand, by definition excludes the possibility of acting (or forbearing) out of a regard for any moral law behind the constraint. The only option one has in this case is complying because of the force. In this case indeed moral complying, and thus moral action, is impossible.

As a result: The state in the context of Kant's philosophy makes possible moral action precisely insofar as it renders compatible the multiple autonomies of its members, or as it keeps every member from intruding upon the liberty of every other member.

For the issue of health care one could specify the difference thus: Humans' bodies undoubtedly belong to that sphere of means for realizing one's autonomy that must be protected by the state. Also, humans' bodies cannot serve their purpose if they are sick. Yet protection in the Kantian context is exhausted with discouraging all humans from unconsented-to intrusion in (or culpable infection of) others' bodies. The unconsented-to intrusion of a human body by viruses or bacteria falls outside of the Kantian state's sphere of responsibility. That state, in other words, has a duty and hence a right to intrude upon its members' property-rights (through financing police forces and courts of law) in order to avert intrusions in the former, not (at least on the reasoning developed here) in the latter sense.

2. If social claims cannot be established on the basis of a morally obligatory respect, and thus as a matter of right, there still remains beneficence. Kant's moral philosophy does indeed acknowledge a moral obligation to beneficence, as passing the test of universalizability. Moral action is distinguished as being determined by a rational regard for the moral law. The autonomy realized through moral action comes from the fact that reason, in being thus determined, is determined by itself as the rational law-giver. The content of this self-direction is captured by the well known categorial imperative, which enjoins us to act in such a way that the maxim determining our will can serve as a general law (1786, p. 52). This is how the welcome universalizability of the maxim followed must be understood as a criterion for morality, and this is how beneficence qualifies.

Yet Kant's theory of the law excludes the very possibility of that moral obligation's attaining any legal significance. The law deals with external actions exclusively (1797, pp. 15, 17, 32). A state's moral justification in enforcing the law can, as a matter of principle, not derive from that law's contributing to the realization of any moral Good (1797, pp. 13 f.). Moral goodness, in Kant's theory, resides in a frame of mind (*Gesinnung*) which expresses itself in letting one's will be determined autonomously, *i.e.*, by a regard for the moral law. Since legal rules govern only external actions, frames of mind and determinations of the will are beyond its reach. Any attempt to overcome that borderline amounts to moral despotism (1793, p. 261, 1797, pp. 172 f.). (This also forestalls any notion of a solidarity duty attaching to property, or of democracy as constituting a solidarity community, as is customary in Europe today.)

To be sure, Kant speaks of legal acts as externally agreeing with morality. One might therefore suppose, if beneficence was morally required, that external actions agreeing with the demands of beneficence could be enforced by the law as well. Societal satisfying of the needies' needs could still be defended.

There are two problems with this view. The first comes from an equivocation in the notion of universalizability. While the possibility of (welcomingly) universalizing a maxim is sufficient for establishing its private obligatoriness, only a necessity of universalizing thus (or the impossibility of universalizing its negation) warrants a moral obligation which can (at least in principle) be politically enforced. If this restriction didn't hold, it would be possible to justify a state in forcing individuals into an obligation which they (stubbornly but not irrationally bent on universalizing to the contrary) happened not to acknowledge, and thus to disrespect their autonomy. The problem with beneficence is that Kant himself believed that there was no alternative, *i.e.* that no one could consistently want to live in a universe where no one acted beneficently (1786, pp. 56 f). Kant derives this impossibility from the fact that every one must reckon with situations in which he would need others' beneficence. Yet this move justifies too much: With respect to making dishonest promises (1786, pp. 54 f., one of Kant's other prime examples) one could argue as well that everyone must reckon with situations in which he would need to resort to such stratagems. In order to escape this analogy, one would have to resort to a general consensus that while it is better to die than to be dishonest, it is not the case that it is better to die than to have accepted someone's beneficence. Yet in this case, Kant's account of morality would have presupposed a moral distinction (between unconditional honesty and unconditional self reliance) which it should have accounted for.

The second problem comes from the fact that people rank values differently. Even if the inescapability of willingly universalizing everyone's acting beneficently could be proven (which it cannot), this would involve Kant's political theory in grave difficulties. People, so Kant admits, seek their happiness in radically differing ways (1786, p. 47). If the state were allowed to enforce external compliance with any particular understanding of beneficence, then some people would be forced into others' views of (either their own or of their benefactors') happiness.

As a result, beneficence is not something to which external compliance can be legitimately enforced. Or, as we may summarize, the law, while justly enforcing moral respect for autonomy in a formal sense, cannot justly enforce any particularly content-filled morality. Or the law may serve the cause of moral goodness not in a positive sense, as through realizing moral goodness (as for example through beneficence). It may serve that cause only in a negative sense, by guarding against what violates the basic condition of the possibility of moral goodness: the autonomy needed for permitting one's maxims to be determined by respect for the moral law.

[21] As mentioned above (note 18), the Kantian state ought to support the poor, because people must be thought to have entered (ideally) into a political body (or social contract) in order to secure their preservation (1797, pp. 186 ff). Yet in the context of his other claims one must conclude that preservation is conceived – like bodily integrity – as one of the necessary conditions for the exercise of autonomy. But then again, only persons are intended.

[22] This supposition involves a distinction between respect for exercisable autonomy and protection of (minimal) conditions for such exercising within specified relationships of legal dependency. It makes sense in the context of Kant's political theory as a whole, where humans' most elementary autonomy rights are separated from political participation rights (1797, p. 166). For Kant, only those members of the state who are economically independent and able to speak for themselves in legal transactions are entitled to vote on societal matters. Infants, along with servants and wives, have only passive political rights (to protection for their claims), not active rights (to participation).

[23] Of course only social action is considered by the law; damage done to self privately is considered only insofar as emergency life saving may be required.

[24] Kant, after all, affirms the general legal principle *volenti non fit iniuria* (1797, p. 165).

[25] This is a pointedly minimalist interpretation of "humanity as a purpose", which however, so it seems to me, is the only defensible one.

[26] The obvious escape route of referring to natural biological function is closed by Kant's restricting moral issues and thus the justification of the state to the strictly noumenal perspective. (Otherwise, diverse other natural biological functions – just think of sexual desire – would have to be accepted as justifying other inter-person violations of autonomy-respect and thus upset the whole framework.) Nor would it help to invoke the state's legitimate concern about its own existence. While surely a state exists only if there are people to be protected from one another, the state's legitimate concern with its existence presupposes the existence of such people. It does not extend towards making sure such people exist.

[27] Nor could one escape this conclusion by invoking the danger that people in difficult situations might come under psychological pressure to permit an intrusion in their bodily integrity, which would reduce their resources for realizing their autonomy. Respecting people's autonomy, after all, means accepting the possibility of deplorable mistakes made under unfortunate conditions.

[28] The attempt to avoid this conclusion by having altruistic motivations repudiate the supposition of dignity-violation is also not convincing. For in this case, internal actions would be permitted to make a legal difference. Taking into account such internal acts would disrupt the restriction of the state's legitimate use of unconsented-to force against its members to what can be empirically asserted. As a consequence, in the case discussed here, it would have to be rendered a matter of psychological evaluation whether the selling of organs for financial gain was perhaps still motivated altruistically. Moreover, nothing could hinder the further consequence that even violations of human dignity in others (as being – except for lacking consent – on the same footing with violations in oneself) could be justified altruistically, as in cases of medical paternalism. Such reasoning would overturn the whole endeavor of legally protecting human autonomy.

REFERENCES

Council of Europe. (1996). *Convention for the Protection of Human Rights and Dignity of the Human Being with Regard to the Application of Biology and Medicine: Convention on Human Rights and Biomedicine*, Adopted by the Committee of Ministers on 19 November 1996. Directorate of Legal Affairs, Strasbourg.

Council of Europe. (1997). *Explanatory Report to the Convention for the Protection of Human Rights and Dignity of the Human Being with Regard to the Application of Biology and Medicine: Convention on Human Rights and Biomedicine*, Directorate of Legal Affairs, Strasbourg.

Delkeskamp-Hayes, C. (2000). Respecting, Protecting Persons, Humans, and Conceptual Muddles in the *Bioethics Convention, The Journal of Philosophy and Medicine*, Vol. 25, No. 2, pp. 147-180.

Delkeskamp-Hayes, C. (in press). *The Moral Justification of Political Power*, Kluwer Academic Publishers, Dordrecht.

Global Summit of National Bioethics Commissions. (1999). Tokyo Communique. *Eubios Journal of Asian and International Bioethics* 9, 3-4.

Hartmann, K. (1991). "The Profit Motive in Kant and Hegel", in *Rights to Health Care*, eds. T.J. Bole and W.B. Bondeson, Kluwer Academic Publishers, Dordrecht.

Holmes, S., and Sunstein, C.R. (1999). *The Cost of Rights. Why Liberty Depends on Taxes.* W.W. Norton and Company, New York.

Kant, I. (1786). *Grundlegung zur Metaphysik der Sitten*, 2nd edition, Johann Friedrich Hartknoch, Riga.

Kant, I. (1788). *Critik der practischen Vernunft,* Johann Friedrich Hartknoch, Riga.

Kant, I. (1793). "Über den Gemeinspruch: Das mag in der Theorie richtig sein, taugt aber nicht für die Praxis", *Berlinische Monatsschrift,* September, pp. 201-284

Kant, I. (1797, I). *Die Metaphysik der Sitten in zwey Theilen, Metaphysische Anfangsgründe der Rechtslehre*, Friedrich Nicolovius, Königsberg.

Kant, I. (1797, II). *Die Metaphysik der Sitten in zwey Theilen, Metaphysische Anfangsgründe der Tugendlehre*, Friedrich Nicolovius, Königsberg.

Plato. (1971). *The Collected Dialogues of Plato*, eds. E. Hamilton and H. Cairns, Princeton University Press, Princeton, N.J.

Qiu, R. (2001). "Tension between Biomedical Technology and Confucian Values", in this vol. pp. 71-88.

Sakamoto, H. (2001). "A New Possibility of Global Bioethics as an Intercultural Social Tuning Technology", in this volume, pp. 359-368.

SHUI CHUEN LEE

THE REAPPRAISAL OF THE FOUNDATIONS OF
BIOETHICS: A CONFUCIAN PERSPECTIVE

The theoretical development of bioethics has been towards the establishment of a globally applicable theory both practically useful and universally justifiable for decisions, actions and policies regarding bioethical matters. However, more and more diversity and disagreement have appeared among theoreticians of bioethics as well as practitioners. As part of the outgrowth of the enlightenment project, bioethics seems unlikely to escape the fate of the failure of the project itself. There have been numerous efforts and labors in theoretical constructions, critical responses and counter-responses to combat this disturbing failure. Some contrive to build bioethics upon commonly accepted and acceptable middle principles, such as the so-called principlism by Tom Beauchamp and James Childress.[1] Some have noted the reality of diversity and difference, especially cross-culturally, and have turned realistically to face these divisive differences and the problems they pose. Tristram Engelhardt,[2] by making the distinction between moral friends and moral strangers, has in fact made a bold theoretical endeavor to take moral differences seriously and has elaborated a procedural foundation for bioethics on the basis of a default requirement – a principle of permission (1996). However, radical critics may still regard such a meager bioethics to be, nevertheless, a culturally biased notion rooted in autonomy or rationality.

Foundational issues in bioethics have recently been critically addressed by a new group of professionals, self-styled "bioethnographers." They adopted the critique of the traditional *autonomy* liberalism by the new *difference* liberalism in the field of political theories, and then directed the same critique against the universal claims of bioethical theories. Their critique centers on the themes of difference as incommensurability, difference as justice, and difference as self-identity (Jennings, 1998). I shall elaborate how this critique affects current mainstream bioethics and how a Confucian approach can respond to this critique and defend a minimal common bioethics.

Julia Tao Lai Po-wah (ed.), Cross-Cultural Perspectives on the (Im)Possibility of Global Bioethics, 179–193.
© 2002 *Kluwer Academic Publishers. Printed in Great Britain.*

I. THE BIOETHNOGRAPHERS' CRITIQUE OF BIOETHICS

Bioethics is a field that grew out of the development of the modernization process of the enlightenment project as a response to the pressing moral issues engendered by the rapid development and applications of biomedical technologies. Traditional teachings need reinterpretations to enable practitioners to deal with such issues arising from their daily work. It is natural that bioethicists seek to construct universal theories for understanding and analyzing these issues so as to provide moral guidelines for professionals. One of the most influential attempts is the principlism of Beauchamp and Childress. However, it suffers the accusation affiliated with the failure of the enlightenment project to provide a universal rational human enterprise: it pays insufficient attention to the reality of difference across persons and cultures. In particular, Beachamp and Children are attacked as leaning too heavily on the notion of autonomy in their bioethics. This point needs some clarification from both a socio-historic and a philosophical perspective.

Historically, bioethics was set off with the emergence of biotechnology after World War II and, in particular, with the Nuremberg trial, when new moral issues in medicine and the maltreatment of experimental subjects were exposed. Up to the early 1970's, the discipline of bioethics was quite a mixture of various forums of religious, legal, as well as philosophical and ethical discussions until Beauchamp and Childress published their first edition of *Principles of Biomedical Ethics* in 1977. It provided a systematic presentation of the basic principles needed for practitioners to deal with their practical biomedical problems and became the standard approach to bioethical issues. Though it has been under severe attack throughout the years and has been significantly revised by its authors, the so-called "principlism" embodied in it remains the mainstream of bioethics (Wolpe, 1998, pp. 38-59). In their presentation, Beauchamp and Childress purposely give no hierarchy of the four cardinal bioethical principles: autonomy, beneficence, non-maleficence, and justice. However, the principle of autonomy gains prominence in bioethical explorations. Bioethics is thus heralded as the triumph of autonomy during this first stage of its development (Wolpe, pp. 38-59). Of course, philosophically, principlism does not propose the principle of autonomy as the most basic principle. However, it is listed as the first of the four principles, and the principle of informed consent, the most operative principle in bioethical matters, is subsumed under it. Together with recent trend in American politics, law and policy, autonomy, codified in the language of rights, has been no doubt the cardinal principle of bioethics up to now.

Broadly speaking, the bioethnographers' critique of bioethics centers on problems related to the notion of autonomy as a universal, colorless, and rational principle applicable to human being as such, regardless of their ethnic, cultural, religious, and ethical differences. It is the notion of difference that is brought to the fore in this critique of the mainstream bioethics of autonomy. Against traditional liberalism in which autonomy is integral to a colorless humanity that all human beings qua human beings possess, *difference* liberalism stresses that we are human precisely because we are embedded within different cultures and are entitled to respect because of our difference. Bruce Jennings elaborates three kinds of difference: difference as incommensurability, difference as justice, and difference as self-identity. He points out, first, that the ideal of autonomy cannot be easily reconciled with other moral ideals associated with social and moral pluralism, such as respect for difference, recognizing and validating the identity of others, and tolerance. Second, mainstream bioethics makes the moral mistake in normalizing and universalizing a particular set of cultural assumptions from the western version of the enlightenment project and thus fails to respect persons because it erases the particularity of persons which culturally constitutes their personal identity. Third, mainstream bioethics should be more sensitive to the culture-bound assumptions that lie at the heart of its conceptual and philosophical framework (Jennings, 1998, pp. 261-266). Put into the form of a philosophical argument, this critique of autonomy resolves into two tiers. In the form of difference as incommensurability, it amounts to a version of moral relativism holding that different persons have incommensurable values in important bioethical matters. Practically it is an attack on the availability of a universal foundation for bioethics. In the form of difference as justice and as self-identity, it is more or less an attack on the applicability of the mainstream bioethics to concrete bioethical cases. I shall address the first issue in the next section and respond to the second in subsequent sections.

II. DIFFERENCE AS INCOMMENSURABILITY AND MORAL
RELATIVISM

The thesis of difference as incommensurability derives from epistemological skepticism and relativism. Jennings may be referring to the thesis running from Thomas Kuhn to present day anti-foundationalism in epistemological studies. There are defenders of incommensurability theses both epistemological and moral. Regarding epistemological incommensurability, Donald Davidson argues that the extreme type of incommensurability between two theories or conceptual schemes in the sense of non-translatability does not exit (Davidson, 1974). Epistemologically speaking, if we could understand that two theories as truly incommensurable, we do *know* on what point they are incommensurable with each other.[3] This means that some of the concepts between the two theories must be overlapping. Hence, there is nothing that is totally incommensurable for a person who knows two competing theories or between the understandings of the speakers and translators unless, contrary to Wittgenstein, each of us does have a private language of our own. Two competing theories must be at least partially translatable and, if the overlapping is good enough, allow solutions to conflicts between them. The problem is not that one cannot understand two competing theories at the same time. It is rather that there seems to be no common standard to resolve their difference. However, the development of modern science provides a good example of how a common standard can be acquired. For Davidson, it seems that competing theories in the history of modern science did provide enough common vocabulary to justify the triumph of the accepted theories. In the same light, Alasdair MacIntyre provides a rationale in terms of the historical process of modern science. Newtonian physics, MacIntyre argues, is rationally superior to its rivals in that "it was able to transcend their limitations by solving problems in areas in which those predecessors and rivals could by their own standards of scientific progress make no progress" (1984, p. 268). Accordingly, MacIntyre holds that justification can only be secured historically and that there is no colorless way to an absolute standard for justifying scientific theories.

As regards the incommensurability thesis in moral philosophy, the problem seems to be insoluble. We have indeed two kinds of incommensurability, one in linguistics and the other in justification.[4] The linguistic one amounts to the claim that different moral theories are imprecated with different concepts of morality and thus could not be compared with each other when making different moral judgments concerning the same states of affairs. To this claim, we could make an even stronger Davidsonian rebuttal in that moral experience is the common ground of our moral concepts. In moral matters, we know on which aspects we are making judgments and we seem to have overwhelming overlapping moral experiences. We know what suicide is and what a benevolent act means, though we may make diverse and sometimes opposite judgments on them. Opposite judgments do not preclude our mutual understanding, though the rivals may never reach agreement on the final judgment. Much more often, we have similar distress across cultures and ethnics at some moral situations (such as cruelty to a child) and certain cases of moral conflicts (such as the baby Doe case). We could not help feel morally disturbed by watching an innocent little baby withering before our eyes.

It is a fact that moral experience is different from other types of experiences such as cognitive or emotional ones. To be true to our moral experience without falling into the dead end of reductionism, we need to give a positive and distinct contour to morality. In his derivation of the categorical imperative from common moral rationality, Kant has in fact formulated the basic condition of our moral experience. He disclosed the basic grammar of our moral languages.[5] The freedom of the will and its autonomy are the necessary condition for the possibility of morality.[6] This formulation has an important bearing upon our response to the incommensurability problem. However, since Kant maintains that the freedom of the will is but a postulate, it becomes a will that cannot go all the way in to our moral experience and cannot itself be realized within our daily experience. As long as the free will remains merely a back drop, our moral experience becomes problematic and autonomy as the necessary condition of morality is empty.[7] For the Confucian, the autonomy of the will must be part and parcel of our moral experience. The Mencian mind that cannot bear another's suffering manifests itself in our moral consciousness as a moral condition. Not only does it morally prescribe the direction of our action, it also constitutes a moral situation, that is, a situation for moral action. This moral mind is equivalent to practical reason. Without it, there will be no moral consciousness or experience. Our moral vocabulary is but a conception of such experience and is thus available across cultures and languages. It can prevail over

different accounts of morality and is thus immune to attack by
incommensurablists or relativists.[8]

The incommensurability thesis that most stubbornly remains as a
problem in moral conflicts takes this shape: people assign different moral
values to the same phenomenon and give different moral priorities to
different moral principles. Translated into our present concern, this means
that the competing values held by different persons cannot be resolved
when they come into conflict. There is no objective neutral standard for
rivals to refer to or weigh different values. It seems that principlism
cannot easily escape this charge because it aims to provide the common
core of moral principles supposedly embedded within our daily moral
judgments and actions. These principles themselves need justification and
there is no available method or procedure to decide regarding their
theoretical or practical ordering. Furthermore, it is precisely the attempt
to provide a common framework of bioethical principles that leads
principlism to ignore the difference between different practitioners and
patients in different cults and ethnicity. As for Engelhardt's bioethics, the
accusation is not justified since his theory takes difference seriously and
does allow personal and cultural differences in the notion of moral
strangers. In fact, his principle of permission allows moral strangers to
hold incommensurable values and asks only for each party to be bound by
the principle of permission so as to avoid the use of force to solve moral
conflicts (Engelhardt, 1996). Since he frames "permission" in the form of
a side constraint condition for a peaceful settlement of bioethical
conflicts, the principle loses much of its Kantian transcendental
significance. However, difference may remain a residue of his theory and
be relegated to the care of individual cults or communities, where again
difference is neglected between members of the community of moral
friends.

On the other hand, Engelhardt does offer an interesting solution to the
problem of moral conflicts by making the distinction between moral
friends and moral strangers. For moral friends, conflicts can be solved by
a common moral authority or a common moral standard. Among moral
strangers, solutions must abide by the principle of permission which is
only formal, not substantive. Conflicts can be resolved in a form of
compromise as long as both sides deem it necessary to remain in a
peaceful community. However, the principle affirms the autonomy of
liberalism. For the relativist, no such principle is to be upheld as the final
judge of any moral judgment, nor is a peaceful community of any special

value that must be upheld as necessary. In MacIntyre's formulation, it involves the illegal scheme/content or formal/substantive dichotomy, and thus is necessarily doomed with the Enlightenment project.

MacIntyre regards incommensurability in modern moral theories as inevitable given the Enlightenment project of accommodating diverse, incompatible types of theories and various experiences as equally competent and competitive. The project is bound to become morally relativistic. He asserts that each theory has its own rule of justification borne in its own terms. Each translates its rivals into its own language but in a neutral, detached, de-contextual fashion. On the other hand, MacIntyre does think that relativism can be transcended. He offers a different solution to such relativism (1985, p. 405). Apparently drawing on the paradigm of modern science, he offers science as a model for ethics as follows:

> Nonetheless if some particular moral scheme has successfully transcended the limitations of its predecessors and in so doing provided the best means available for understanding those predecessors to date *and* has then confronted successive challenges from a number of rival points of view, but in each case has been able to modify itself in the ways required to incorporate the strengths of those points of view while avoiding their weaknesses and limitations *and* has provided the best explanation so far of those weaknesses and limitations, then we have the best possible reason to have confidence that future challenges will also be met successfully, that the principles that define the core of a moral scheme are enduring principles (1984, p. 270).

In his *After Virtue*, MacIntyre ascribes this strategy to Aristotle's fundamental moral scheme.[9] However, I shall argue that Confucianism can better realize MacIntyre's conditions for a justified moral theory for human beings.

Let me begin with Kant's analysis of our moral experience. Kant has put into clear contour the autonomous and universalizable nature of morality. Looking more closely, Kant's ethics is not sheerly formal. His formal part lies in the universal requirement of the categorical imperative, while his substantial part is secured in his requirement of consistence of the will.[10] That is why Kant could employ false promises and suicide as illustrations of the governance of his categorical imperative. However, Kant seems to think that any substantial elements in our moral experience are rooted in heteronomy and thus relegating them to the special

conditions of particular situations. Confucianism goes further than Kant
on two points. First, for Confucianism, the moral mind is not only self-
legislative and practically rational, it is also a consciousness expressed as
our moral conscience and a motivational force for itself. It defines a
definite direction for its own actions, thus accounting for moral laws in an
autonomous fashion. As regards the direction, it is more or less formal.
Though it prescribes what is moral and what is not, it leaves the particular
action or decision to the situation that arouses our moral consciousness.
Secondly, the moral mind is an expression of our unbearable concern
with other's sufferings.[11] Here lies the substantial content of our moral
experience. It is our concern with others' lives that makes sense of
morality. Without such concern, morality is totally empty or degenerates
into a rigid form of social coercion. This concern can give a rationale to
Kant's illustrations, for they are ultimately expressions of our concern for
one's own and other's sufferings or happiness. This concern fortifies
Engelhardt's emphasis on the making of a peaceful community as well as
why we want it. It embraces the concern of care ethics as we care for the
death of an innocent child falling into a deep well as well as all sorts of
human and non-human suffering.

According to Confucianism, our concern for others spreads out in
circles as is required of substantial care. We care first for members of our
families, then our friends, then other fellows of our race and then for all
people in the world. In fact, this concern goes even further: we care for
everything with life, for our land, for Heaven and Earth, or for the whole
universe. Here we can fully recognize Hume's observation that we have
limited benevolence towards others, which is quite correct as an empirical
observation, though we have to reject his claim that we do not care more
for the death of Indians or the trembling of the whole world than for a
scratch on our finger.[12] Each of us is as distinct and non-replaceable as
every other, but we are also members of the moral community of all
human beings. Individualism is not self-sufficient, nor is it the whole
picture of our human situations and relations. We have moral concern and
care for other members of the community, which constitutes our moral
obligations to all human beings, whether we are acquainted with them or
not. Furthermore, as Mencius pointed out, this original unbearable
consciousness of another's suffering differentiates into the four moral
principles of *jen* (benevolence), *yi* (righteousness), *li* (ritual or prudence)
and *chi* (consciousness of right and wrong). These are comparable with
the four cardinal principles of principlism.[13] When these principles are

realized in the daily moral practice of a person, they become the virtues of a person, and the person becomes a virtuous person or a person with practical wisdom, to use Aristotle's terminology. The one who incorporates and manifests such virtues and moral autonomy fully in all aspects of life is a sage. However, the principles and virtues are not primordial, and Confucianism would not recognize the pure intellectual meditation of western classical philosophies as the apex of moral achievements.

To recapitulate, Confucianism regards our unbearable consciousness of another's suffering as both a free and creative will. Morality is not a thing out there to be found but is constituted by the manifestation of the self-conscious mind. It is self-legislating, law-abiding, and thus prescribes a moral direction and a series of moral principles which have concern for the sufferings of others as its content. Furthermore, Confucians realize that this unbearable consciousness of the suffering of others, though embedding the humanity of human being, is concrete and situation-sensitive. What becomes codified as rites and prudence for a certain epoch or ethnic group could be reshaped in new contexts. It seems that Confucianism fits with MacIntyre's characterization of a theory capable of giving a justifiable response to the challenges of relativism and the incommensurability thesis.

III. INDIVIDUALITY AND DIFFERENCE AS SELF-IDENTITY

I shall deal with Jennings' third type of difference as self-identity now and address the difference as justice in the next section, as this is a more natural way of presenting Confucian response to these two types of difference. By making the element of autonomy supreme, mainstream bioethicists are charged with asking a person to shed his or her self-identity as a person. Under the idea of difference as self-identity, Jennings argues that human self-realization comes in and through living in culturally meaningful practices, not by abstraction or escape from those practices. That is, a person must wear certain cultural heritage and be accepted in a culturally thick way. Thus, eliminating the self-identity element of individual difference amounts to a denial of the humanity of the person (Jennings, p. 263). Each of us has built into our personality elements of our mother culture as our second nature. Nobody is a colorless human being on Earth. This is the source of our self-identification. Each of us is unmistakably a person with difference! We are persons performing morally acceptable actions which somehow bear our individual marks if each moral act is creative and flows from our freedom of will. However, Jennings errs here, for the difference as self-identity composes not only something uniquely different from all others, it is also composed of something common to human being, called "humanity." Self-identity is a complex of our past moral as well as other activities. Our acts are located in particular space-time zones and thus bear their particular imprints; nonetheless, as long as they are the manifestations of our subjectivity or rationality, our humanity remains what is common to all human being. We all have moral experience, which is understandable across persons and cultures, though we may have different moral judgments concerning the same cases. We have identity and difference across cultures and within cultures.

From a postmodern point of view, the problem of difference as a quest for rationalization of self-identity is a challenge to the enlightenment project and it does raise a legitimate question regarding the universalizability of rationality in all aspects of a person's life. However, overarching rationality is not only a necessary dimension of the enlightenment project, it is also a condition for the respect for persons and the transcendental condition for human moral experience. On the other hand, it would become oppressive for the individual if rationality is overly emphasized and abstracted from the actual life of the individual. To prevent the subjection of the individual as mere a tool for the realization of the claims of rationality, the importance of the uniqueness of the individual must be upheld. According to the Confucian conception of morality, the individual is not only the unique way in which rationality can be actualized (as Confucius states, "it is human being who can enlarge *tao*, not the reverse"), but the individual also, to a large extend, creates his identity through his personally unique actions. I would regard this tension between rationality or subjectivity and individuality as the core problem in the hot debate between the defenders of the enlightenment project such as Habermas and the postmodernist such as Lyotard, and would venture to establish a Confucian model of the double-coding of subjectivity and individuality for the postmodern development (Lee, 1994, 1995). Individuality is given full recognition in this Confucian theory and is, in fact, regarded as indispensable for the realization of *tao*. Uniqueness with non-replaceable worth is what makes an individual a non-replaceable person with self-identity.

The legitimacy of individuality naturally generates the legitimacy of difference. Furthermore, it makes *difference* essential from a personal point of view. Difference is what makes up the uniqueness of each person and her self-identity. Any creative act, moral or not, must bear the agent's individual characteristics. This is particularly true in moral matters, since each individual has to make her decisions existentially and creatively. Though moral actions are regarded as flowing instantaneously from the unbearable consciousness of the suffering of others when moral action is called forth, one has to attend to all kinds of relevant and different conditions framing a moral situation, and must act creatively to meet the requirement of being moral. All precedents are for reference only.[14] However, the different act must bear at the same time some recognizable universal elements of being a moral act. Its legitimate claim to be considered as part of the personal identity of difference means that the universal element of humanity is embedded in one's personal identity. Hence, in a full-blown bioethics that Confucians would accept, difference as self-identity must be an inseparable ingredient of its foundation.

IV. CONFUCIANISM WITH DIFFERENCE AS JUSTICE

Under difference as justice, Jennings points out that we respect each other not because we are the same, but that we request that we treat each other equally with equal dignity and we respect precisely because we are different (Jennings, pp. 262-263). He thinks that it is an admirable and just mode of relationship if we respect the rights and freedom of those whom we find to be strangers. He finds this challenging, unsettling and perhaps threatening! In a sense, any two persons, even identical twins, could be said to be the same in some aspects and different in others. It is true that as a person, each differs from every other and demands our respect of his or her dignity. However, when Jennings requests that we treat others equally just because we are different, he has misconstrued our respect and bumped against our common conception of justice: to treat equals equally and unequals unequally. The respect we pay for each other is precisely because we regard the other as a person of humanity like us. It is because this sort of respect that we restrain from interfering with his autonomous decisions, though his decisions could be quite different from ours. Difference between persons is inevitable and difference makes persons unique and the human community more colorful and lively. This is what makes persons unique and difference valuable. However, person qua person, it is their equal and non-replaceable worth that deserves respect and equal respect. This discharges the thesis of difference as justice.

For Engelhardt, his moral friends may have received equal respect from other moral friends under the same moral, religious or cultural umbrella. So too his moral strangers, if his principle of permission signifies the due respect that we have to pay for the autonomy of the person, would receive a kind of equal respect for persons holding different and diverse moral and metaphysical values. To treat others with regard to their permission is to respect others as persons with difference. However, a moral stranger is a kind of abstract person! We are without exception the products of the interactions of our human potentials within our cultural and natural environment. We are moral friends in a basic sense before we can be moral strangers in a secondary sense. With our constitutions, we cannot help have moral experience and are thus bound to be members of the moral community. In other words, each person is in fact a moral friend not just to a small group of humankind, but to all human beings. Of course, we have different kinds of friends. Some are of closer ethnic or cultural heritage, some are farther apart. But, as the Confucian realizes, we are all friends of humanity, individually and as a whole. We each bear our own ethnic and cultural identity: Chinese act in more Chinese ways, Texans in more Texan ways, and so on. However, although we hold our cultural or metaphysical values and beliefs, we are at least concerned with the suffering of others. In fact, we are first moved by the suffering of others without inquiring whether they are of the same moral beliefs as ours. In other words, we are morally bound by the common humanity within us. For the Confucian, a person is first of all a member of the common moral community; that is, a moral friend. In response to the command of difference as justice, the Confucian stands ready to accept as just equal moral treatment of all persons of difference, and this is exactly what justice means: to treat each as a unique non-replaceable human being.

I may sum up the Confucian response to the theses of difference as follows. Our unbearable moral feeling and concern for the suffering of others cannot but compel us to actions. It prescribes the direction: to prevent, remove, and do no harm. What is contrary to this direction constitutes immorality. How to put our moral commitment into practice and action is left to the individual. By acting as an autonomous agent, personal difference manifests itself and constitutes not only the act but also the individual as a moral person. There seems enough space for each person to carry out the moral quest in different kinds of actions. Each act enriches the human world rather than causing chaos. Grey areas are waiting for our morally creative acts to carve out a new moral realm. The bottom line is to remain moral whatever happens. Immoral acts will be refuted both from our rationality and individuality.

Graduate Institute of Philosophy
National Central University
Chungli, Taiwan

NOTES

1 "Principlism" identifies theories of bioethics such as that of Tom Beauchamp and James Childress' seminar work published first in 1977. They have revised it through the years and its fourth edition occurred in 1994.

2 Principlism is the primary target of the bioethnographers. Although Engelhardt's theory is mentioned as one of the group of bioethicists who are concerned for deeper autonomy (Wolpe, 1998, p.43), this is primarily a misunderstanding of Engelhardt's term "autonomy." In fact, he uses it as a side constraint, rather than a value as his critics often mistake. This is certainly much clearer with his second edition of *The Foundations of Bioethics* where he uses "permission" instead of "autonomy" (Engelhardt, 1996). Nevertheless, I think, first, Engelhardt's theory is basically founded upon individualism and autonomy; secondly, it also has a feature that can be understood in terms of my elucidation of the Kantian and Confucian response to this kind of bioethnographers' critique.

3 Dwight Furrow gives a more comprehensive account of incommensurability in his 1995, Chapter 1, pp.28-36.

4 In his *Against Theory*, Furrow proposes three kinds of incommensurability. However, his elaboration of the Lyotardian type of incommensurability concerns more the performative aspects of language rather than the problem of justification. I shall bypass this kind of incommensurability and focus more on the problem of translation and justification problem. See Furrow, 1995, pp. 28-36.

5 This may be part of what Engelhardt says, "By examining the foundation of morals, Kant offered what could be termed the grammar of a major dimension of human thought" (1996, p. 137).

6 Kant put it into his famous saying in a footnote of his *Critique of Practical Reason*, that "freedom is certainly the *ratio essendi* of the moral law, the latter is the *ratio cognoscendi* of freedom. Moral law is not only a prescriptive principle in respect of our actions, it is what constitutes the experience that we call moral thus morality" (Kant, 1956, p. 4).

7 For a discussion of this aspect of Kant's theory, please refer to my paper 1993.

8 Incidentally, I regard this as an adequate response to the kind of meta-ethical relativism sometimes thought to be implicit in meta-ethics. See, e.g., David Wong, 1991. What he calls

normative relativism falls into the usual type of moral relativism that is treated subsequently in the text.

[9] Interestingly, in his 1985 which is delivered at about the same time or just a little latter than his *After Virtue*, MacIntyre argues much the same way against modern ethical theories without attributing the same hope to Aristotelian thought. Instead, he says, "[we are] to welcome, a possible future defeat of the forms of theory and practice in which it has up till now been taken to be embodied within our own tradition, at the hands of some alien and perhaps even as yet largely unintelligible tradition of thought and practice; and this is an acknowledgment of which the traditions that we inherit have too seldom been capable" (p. 385). It is interesting to see if Confucianism is the tradition of thought and practice that can make the proposal in this arena.

[10] Engelhardt is keen to detect that Kant smuggled content into his conclusions though they are in fact part of the content of the categorical imperative. See his 1996, p.105 ff.

[11] Haberrmas once remarked that he and Marcuse at his deathbed realized that the foundation of critical theory lies in our compassion towards the suffering of the other. See his 1985, p.77. Confucianism readily approves this compassion as the ultimate ground of practical concerns of any theoretical endeavor and for critical theory as a theory of social action. I have in fact elaborated this idea of compassion together with Lyotard's concern with liberating individuals from the suppressive power of overarching reason into a Confucian idea of postmodernization, which I later dubbed as a "double-coding theory" of subjectivity (rationality) and individuality. For fuller discussion, please refer to my articles 1994, 1995.

[12] I have elaborated more on the limitation of Hume's ethics of moral passion with respect to Confucianism in the concluding section of my 1993.

[13] I have elaborated this synthesis of principlism in my 1999.

[14] This is an often quoted saying by one of the great Neo-Confucians in Sung Dynasty, Luk Hsiang-shan (1139-1193).

REFERENCES

Beauchamp, T. and Childress J. (1977). *Principles of Biomedical Ethics. 1ˢᵗ edition*. New York: Oxford University Press.

Davidson, D. (1974). On the very idea of a conceptual scheme. *Proceedings and Addresses of the APA*, V. 47.

Engelhardt, Jr., H. T. (1996). *The Foundations of Bioethics*, 2ⁿᵈ edition. New York: Oxford University Press.

Furrow, D. (1995). *Against Theory*. New York: Routledge.

Haberrmas, J. (1985). Psyche Thermidor and the rebirth of rebellious subjectivity. In: Richard J, Bernstein (ed.), *Habermas and Modernity*, Cambridge: Polity Press.

Jennings B. (1998). Autonomy and difference: The travails of liberalism in bioethics. In: Raymond DeVries and Janardan (Eds.), *Bioethics and Society: Constructing the Ethical Enterprise* (pp. 258-269). Upper Saddle River: Prentice-Hall, Inc..

Kant, I. (1956). *Critique of Practical Reason*. Lewis White Beck (Trans.). New York: the Bobbs-Merril Co., Inc..

Lee, S. C. (1993). *Hume* (in Chinese). Taipei: San-min Book Co.

Lee, S. C. (1994). On modernity and postmodernism: A contemporary Neo-Confucianism (in Chinese). In: *Philosophical Explorations of Contemporary Neo-Confucianism* (pp. 291-309). Taipei: Wenjin Press.

Lee, S. C. (1995). Contemporary Neo-Confucianism and Postmodern Theory (in Chinese), In Shu-hsien Liu (Ed.), *Essays of Contemporary Confucianism* (pp. 51-75), Taipei: Sinica Academia.

Lee, S. C. (1999). *Confucian Bioethics* (in Chinese). Taipei: Legion Monthly Pub. Co.

MacIntyre, A. (1984). Postscript to the second edition. *After Virtue*, 2nd edition, Notre Dame: University of Notre Dame Press.

MacIntyre, A. (1985). Relativism, power, and philosophy. In: *Proceedings and Addresses of the American Philosophical Association (pp. 5-22)*. Newwark: APA.

Wolpe, P. R. (1998). The triumph of autonomy in American bioethics: A sociological view. In: Raymond DeVries and Janardan (Eds.), *Bioethics and Society: Constructing the Ethical Enterprise* (pp. 38-59). Upper Saddle River: Prentice-Hall, Inc.

Wong, D. (1991). Relativism. In Peter Singer (Ed.), *A Companion to Ethics* (pp. 422-450). Oxford: Basil Blackwell Ltd.

PART III

MORAL DILEMMAS IN HEALTH CARE

YU KAM POR

SELF-OWNERSHIP AND ITS IMPLICATIONS FOR
BIOETHICS

I. INTRODUCTION

Self-ownership is a modern idea. It has a special place in modern ethical
discourse, which is highly dominated by the language of rights. It affirms
the individualistic foundation of political morality and the possessive
nature of the individual. Consequently, it guarantees the rights and
liberties of the individual and it is congenial to the capitalistic outlook.

Can such an individualist, capitalistic and rights-based ethical outlook
be presented as a candidate for global bioethics, at least for all the
individualist and capitalist societies? The rejection of this outlook by no
means implies that global bioethics is impossible, but it does imply that
we have to look further in order to find a tenable global bioethics.

In this paper, I shall first explain the idea of self-ownership, and then
consider why people in modern society might find such an idea attractive
and acceptable. I shall then look into the implications of the self-
ownership thesis in bioethics. On the one hand, the self-ownership thesis
provides an easy way to settle disputes in bioethics. On the other hand, it
has implications that we may find difficult to accept. Such unacceptable
implications can serve as a basis for evaluating the self-ownership thesis.
This paper ends with a discussion on how to retain some sense of the self-
ownership thesis while avoiding its unacceptable implications.

II. THE IDEA OF SELF-OWNERSHIP

The view that individuals have rights is an important and powerful view
in modern ethical discourse in general and bioethics in particular. But
why do individuals have rights? One influential answer is that each
individual human being is his own master. He owns himself, so he is
entitled to decide for himself how to live his life. He is not some sort of
resource that may be used to fulfil some goals higher than what he has set
for himself. All the specific human rights such as freedom of speech,

*Julia Tao Lai Po-wah (ed.), Cross-Cultural Perspectives on the (Im)Possibility of Global
Bioethics,* 197–208.
© 2002 *Kluwer Academic Publishers. Printed in Great Britain.*

freedom of religion, freedom of work, can be derived from this super right of self-ownership.

Such a view implies that human beings are proprietors and human rights are a kind of property right. A violation of human rights amounts to a violation of private property. It is wrong to do something to an individual because it amounts to taking away from him what properly belongs to him. C. B. MacPherson regards self-ownership as the basic tenet (which he feels also creates a major difficulty) of classical liberalism:

> [T]he difficulties of modern liberal-democratic theory lie deeper than had been thought. . . . [T]he original seventeenth-century individualism contained the central difficulty, which lay in its possessive quality. Its possessive quality is found in its conception of the individual as essentially the proprietor of his own person or capacities, owing nothing to society for them. The individual was seen neither as a moral whole, nor as part of a larger social whole, but as an owner of himself (MacPherson, 1962, p. 3).

The classical statement of self-ownership can be found in Locke's *Two Treatises of Government*, but the idea can be traced back to Hobbes and even to Grotius.[1] The following quotations are from Locke:

> [E]very Man has a *Property* in his own *Person* (Locke, 1988, p. 287 [II.27]); By *Property* I must be understood here, as in other places, to mean that Property which Men have in their Persons as well as Goods (Locke 1988, p. 383 [II.173]).

Presumably it is because of this Lockean conception of human beings as possessive individuals that Marx regards Locke as the philosopher of capitalism. Marx remarks:

> Locke's view is all the more important because it was the classical expression of bourgeois society's idea of right as against feudal society, and moreover, his philosophy served as the basis for all the ideas of the whole of subsequent English political economy (Marx, 1951).

The thesis of self-ownership is still living well today. More often it is presupposed implicitly. However, there are also some philosophers who advocate it explicitly. Contemporary philosophers such as Robert Nozick (1974, pp. 28-35), Judith Jarvis Thomson (1990), and Hillel Steiner

(1994) all subscribe to the thesis of self-ownership. Cohen describes Nozick's position in this way: "[E]ach person is the morally rightful owner of himself. He possesses over himself, as a matter of moral right, all those rights that a slaveholder has over a complete chattel slave ..." (Cohen 1986, p. 109).

Self-ownership is the idea that each human being belongs to himself. He is not some sort of resource that may be used by other people. Only the person himself has the right to determine how to make use of his own life, his liberties and his capacities. This explains why it is wrong for others to take away his life, to injure him or to imprison him. All these amount to violations of his private property – taking away from him what should properly belong to him.

It is from this basic idea of a person being his own master that the three basic human rights enumerated by Locke (the right to life, the right to liberty, and the right to property) can be derived (Locke, 1988, pp. 285-288 [II.25, II.27]). Since I am the master of myself, there is at least one thing that I own, namely, I myself. No one may kill or cripple me. Hence my right to preservation. I own my body. So I may go where I like. Others are not justified to imprison me, to restrict my speech or behaviour, so long as I do not infringe upon the rights of others. Hence the right to liberty. Since I own my body, what I create by my own labour should also be mine. Hence the right to property. These three basic rights are based on the idea that a person is his own master. His life, his liberty, and his belongings are all his property. It is no wonder that Locke sometimes puts all three of these things under one general name: "property."[2]

Such self-ownership does not legitimize what I may do to others, but it does legitimize what I may do to myself. Each person is entitled to self-determination, and he is free to decide for himself how to live his life and what to do with himself. The rights a person has against others are basically negative in nature. Infringement of rights consists of doing what one should not rather than failing to do what one should. The duties corresponding to self-ownership are basically negative ones, which may be called "duties of restraints" (Ingram, 1994, p. 219).

If we accept the account that ownership is "a correlation between individual names and particular objects, such that the decision of the named individual about what should be done with an object is taken as socially conclusive" (Waldron, 1988, p. 56), then we can say that self-ownership implies that the decision of the individual should be taken by his society as final in any dispute concerning what should be done with himself.

III. THE ATTRACTIVENESS OF THE IDEA OF SELF-OWNERSHIP

The self-ownership thesis conceives human beings as proprietors. Human beings are given a unique status. Resources in nature are something to be owned. Human beings are owners. Some people own more resources than others, but each person is equal in that each person owns himself, and he cannot be owned by other people. The owner determines what may be done to the property. Human beings are their own proprietors. This implies that they have the ultimate say in determining what may be done to themselves. Such a conception of human beings provides a foundation for individual rights and is congenial to the individualist and capitalistic outlook.

In addition to its congeniality to the individualist and capitalistic outlook, the idea of self-ownership has several merits, which make it an appealing idea.

Firstly, the claim that each human being is his own master is intuitively appealing. The claim that people are their own masters corresponds with (though is not equivalent to) our moral judgments that human beings should not be regarded as resources to be used by others, that they should not be treated as means, and that there is an intrinsic value in letting people choose their own way of life.

Secondly, the claim has minimal metaphysical underpinnings. Talk of rights becomes simply talk of property. Whether we agree with the claim of self-ownership, we certainly do not find it difficult to understand. Given a self-ownership account of human rights, the nature of human rights is not metaphysical. It is simply a kind of property right.

Thirdly, this claim provides a clear way to distribute power. "To say that each individual is a self-owner is to say that the power to control her and her activities is distributed to that individual and not to anyone else" (Ingram, 1994, p. 26). Every object in the world, whether it is a person or a thing, has an owner to have the final say over its disposal. This is a convenient way of settling disputes.

IV. THE IMPLICATIONS OF SELF-OWNERSHIP IN BIOETHICS

If the principle of self-ownership is held, then a lot of problems in bioethics can have a clear and quick solution. The key to solving the problems is to find out who the owner is and what the owner prefers.

Let us now consider how the principle of self-ownership can be used to settle the following three kinds of cases in bioethics: (1) cases related to the ending of life; (2) cases related to what is going to happen in one's body; (3) cases related to what may be done to one's body parts.

Some bioethical problems are related to the ending of life, such as euthanasia and physician assisted suicide. Since it is the patient's life it should be up to the patient to decide. If the patient is an adult and is capable of making sound judgments, then the patient is in a position to exercise his rights. It is his life and it is his right to decide.

Some bioethical problems are related to what is going to happen in one's body, such as abortion and surrogate motherhood. If it is not a person's own wish to get pregnant, then she has the right to discontinue any change occurring inside her body, so long as it is within her ability to do so. Any coercion on her to do otherwise will be an inappropriate use of her body. Moreover, with the assumption that it is not wrong to give birth to a baby genetically unrelated to herself, to become a surrogate mother just involves a woman's unobjectionable decision to use her own womb. Since it is her body, it is her right to decide.

Some bioethical issues are related to what may be done to a person's body parts. Should blood be made a commodity that is ownable and saleable? May someone take away my surgically removed spleen for scientific research without my consent? If I desperately need a kidney, am I wrong to buy one at a high price if someone is more than willing to sell it to me at such a high price? If self-ownership is regarded as a first principle, then it is really up to the owner and the buyer to negotiate. Since a person owns his organs, it is up to the person to decide for himself whether it is worthwhile to sell his organs. It is the individual who has the right to decide what to do with his organs.

Self-ownership entails (among other things) a right to one's own body (Locke, 1988, p. 287 [II.27]; Thomson, 1990, p. 225). If each person has the absolute right to determine how to make use of his own body, then it seems whether the selling of organs is justified depends on whether or not the body-owner has exercised his right as an owner.

V. THE EXCESSIVENESS OF SELF-OWNERSHIP

We see from the above that the principle of self-ownership is able to provide some clear and quick solutions to a number of bioethical issues. But we have to pay a high price for those solutions. If we make use of the principle of self-ownership in solving the above problems, then we have to accept the unpalatable implications of the principle as well.

The principle of self-ownership allows not only the selling of cadaveric organs, it also allows the selling of one's organs while one is alive. Not only can organs be sold for the purpose of organ transplantation, they can actually be sold for whatever purpose, including food or collection. If self-ownership is accepted as an ultimate principle, then it is arbitrary to disallow the selling of non-cadaveric organs or the selling of organs for purposes other than organ transplantation.

Similarly, the principle of self-ownership can be invoked to justify euthanasia, but it can also be used to justify almost all kinds of assisted suicide. For a patient suffering from a terminal disease, it is up to him to decide whether his life is worth prolonging and whether something should be done to end his life and hence his suffering. But if the justification is that it is after all his life, then all forms of assisted suicide would also be justified. If it is the patient's life and it is up to him to decide whether his life is worth keeping, then why should such a right be exercised only by patients with terminal illnesses? It would be the owner's right to give up what he owns even though it may be foolish in the eyes of other people. A person who owns a broken vase has the right to throw it away, but so does a person who owns a perfectly good-looking vase. If the owner has the right to decide, then the right to choose death should also be extended to people without any serious illness.

As a result, the thesis of self-ownership seems to bring in more problems than it solves. We have to choose either to accept the thesis of self-ownership together with its implications or to reject it if we cannot digest its practical implications.

VI. THE INADEQUACY OF SELF-OWNERSHIP

In the above I argue that self-ownership is an excessive claim – it ascribes too many rights to the individual. In the below I shall go on to show that it is at the same time an inadequate claim – it ascribes too few rights to the individual.

Self-ownership is not an adequate principle in political morality, because the application of the principle gives rise to injustice. The society is unjust if some people have to sell themselves to others in order to survive. By selling themselves, the principle of self-ownership is not violated, but the principle of justice is. This implies that the principle of self-ownership cannot be an adequate principle of justice (Ingram, 1994, p. 38).

Moreover, self-ownership is not very appealing unless a person's survival needs are met and he is in a position to exercise and develop his powers. This is impossible without access to the material means of human life. So the principle of self-ownership is of little use until it is joined with a principle of rights over material resources (such as a right to health care).

Finally, the idea of being in charge of one's life is not adequately rendered by the idea of self-ownership. We may be attracted to the idea of self-ownership by taking it to be self-mastery. But merely having self-ownership is not enough for self-mastery. A patient who has no access to medical care is still a self-owner, but he is not in control of his life. This implies that self-ownership is not an adequate conception of self-mastery.

Much of the attractiveness of the idea of self-ownership comes from our association of it with the concept of self-mastery or autonomy, which we regard as attractive. But to be master of one's own life is the idea of autonomy, not self-ownership. A person who is simply owner of himself may not control or be the master of his life.

Self-ownership implies only that we should be left alone from external interference. It does not imply that we can be the author of our own lives. I can still be said to be owner of myself if my survival needs are not met, my potentials are not developed, and I lack the relevant information and the mental capacities to choose for myself. In such cases I cannot be said to be leading a life of self-mastery. But if self-ownership does not mean self-mastery, then self-ownership does not turn out to be a very appealing concept. What is the use of being the owner of myself if I cannot claim rights to anything except myself? The idea of being in charge of one's own life is not adequately rendered by the idea of self-ownership (Ingram, 1994, section 2.6).

Unlike the idea of self-ownership, autonomy entails positive as well as negative rights. The realization of autonomy requires much more than being left alone (Raz, 1986, pp. 373, 408; Nino, 1991, pp. 145-147; Ingram, 1994, pp. 158-159). Certain conditions are required for autonomy to be realized.

Firstly, in order to live an autonomous life, to be the author of one's life, it is necessary to have access to some material means. If a person lacks the basic means of subsistence, he will not really be able to choose for himself, if his concern is to survive. He would have to do things that he would otherwise be unwilling to do in order to survive. Rousseau's principle can be regarded as a prerequisite for autonomy: "In respect of riches, no citizen shall ever be wealthy enough to buy another, and none poor enough to be forced to sell himself" (Rousseau, 1968, p. 96 [Book II, Chapter 11]).

Secondly, if a person is to lead an autonomous life, he has to have certain mental and physical abilities. If he does not have any basic education, he may be easily misled by others. Since he has to form plans for himself and to carry them out in order to lead an autonomous life, he must have such abilities. So the respect and promotion of autonomy would also justify recognition of rights such as the right to education and the right to health care.

Thirdly, the promotion of autonomy also justifies a pluralistic society. If there is only one culture, one religion, or one newspaper in the society, the individual would not have any opportunity to choose. For this reason, the government is justified to intervene in order to promote an adequate range of options, such as forbidding monopoly of the news business. It should be noticed that here government intervention is justified not with the promotion of particular options, but only with the promotion of an adequate range of options.

We find the idea of self-ownership attractive because we think that it implies that we can be masters of ourselves. This attractive feature of self-ownership is captured by the idea of autonomy as well. The important thing we want is to be able to control our lives. In order to control our lives we need more than simply being owners of ourselves. On the other hand, it is not always necessary to be the sole owner of oneself in order to be able to control one's life. This line of thinking invites us to give up the ideal of self-ownership and replace it with the ideal of autonomy.

VII. ALTERNATIVES TO THE SELF-OWNERSHIP ACCOUNT OF INDIVIDUAL RIGHTS

If we do not accept the account of self-ownership as it is stated above, I think there are at least three ways out: (1) we can water down the claim of self-ownership such that some sense of self-ownership is retained but the unpalatable implications are avoided; (2) we can reject the view that individuals are proprietors and ethical issues are to be settled by referring to the rights of the proprietors, but retain the idea that individuals have a sphere of autonomy and inviolable rights; (3) we can abandon the rights approach to ethics altogether.

I regard the third alternative as too radical and the second alternative as most promising. But first let me consider the first option.

It is possible to retain the self-ownership thesis but join it with some kind of limiting condition. If we accept Kant's view that people have duties not only to others, but also duties to themselves (Kant, 1948; Kant, 1980, pp. 116-125; Chadwick, 1991), then we can argue that people do not have the right of self-ownership to the extent that they may do whatever they like with their lives and bodies. But the existence of duties to oneself is not less controversial than the thesis of self-ownership. I shall not pursue this line of reasoning further, but just take note that this is one way of arguing against the claim of self-ownership.

It is also possible to reinterpret the claim of self-ownership to make it more acceptable. Although the principle of self-ownership has unpalatable implications, we may still want to retain one version of this principle as we may still find it intuitively plausible and practically useful. The question is: how is it possible to accept the thesis of self-ownership without accepting the claim that people have a right to arbitrary disposal of themselves?

As a matter of fact, it is exactly Locke's position to hold that people own themselves but they do not have a right to arbitrary disposal of themselves. So it seems beneficial to see how Locke uses the term "ownership" or "property." In what sense can it be said that people own themselves if they do not have a right to arbitrary disposal of themselves?

Locke holds that human beings are owners of themselves. But at the same time he holds that it is not at a person's disposal to do whatever he likes to himself. He argues, for example, that it is wrong for a person to commit suicide or to sell himself as a slave. The question is: if a person may not do whatever he likes to himself, in what sense can it be said that he owns himself? If I own a book, doesn't it mean that I am free to sell or give it to someone else? Even if I throw it away or let my cat play with it as a toy, am I not just exercising my right as an owner? Even if other people do not approve of what I am doing, they have no right to intervene.

Locke's answer is this: he defines "property" as "the nature whereof is, that *without a Man's own consent it cannot be taken from him*" (Locke, 1988, p. 395 [II.193]). For Locke, the crux of ownership is a right to *exclude*, it is not necessarily a right to *arbitrary use*.

Ownership does not necessarily entail a right to arbitrary disposal. The ownership of a historical building or a masterpiece in art may not entail a right to destroy the object. Your right to exclude other people to enter the building or to transfer the painting is part of the meaning of your ownership of them. Hence ownership can be taken as a right to exclude instead of a right to arbitrary use.

Likewise, self-ownership can also mean a person's right to exclude others from dominating one's body rather than one's right to the arbitrary use of one's body. So even if the claim of self-ownership is accepted, it does not follow that one has a right to terminate one's life or a right to sell one's organs. In this way, a sense of self-ownership can be retained but the implications that assisted suicide and organ trade are justified can be avoided.

In brief, it does not seem to be impossible to rescue the self-ownership thesis to make it more acceptable. However, a more radical move is also available, and that is to get rid of the idea of self-ownership altogether. This is a move taken by contemporary liberalism.

Contemporary liberalism does not share the self-ownership thesis as classical liberalism does. Attempts have been made to derive individual rights from other conceptions such as equality or autonomy instead of self-ownership. Regarding individuals as equals who should be given equal consideration – equal concern and respect (Dworkin 1978; Dworkin 1977) – provides a foundation for ascribing rights to individuals without presupposing that individuals are proprietors who have rights in their capacity as owners. Another attempt is to start with the claim that individuals should be autonomous and derive rights from the conception of autonomy (Raz, 1986, Chapter 14).

The self-ownership thesis has its attractiveness, and an attractive thesis we do not give up lightly, unless we have something to replace it. It does seem that we have something to replace it.

General Education Center
The Hong Kong Polytechnic University
Hong Kong

NOTES

[1]   Hobbes has also spoken of a person as owning his life and body: "Of things held in propriety those that are dearest to a man are his own life, and limbs" (Hobbes, 1981, pp. 382-383 [Chapter 30]). For a discussion of Grotius' assertion that a man owns his life, body, and limbs, see Tully, 1980, pp. 105 ff.

[2]   For John Locke, the term "property" is sometimes used in the narrow sense to mean one of the three rights. But sometimes he uses it in the broad sense to cover all the three rights. Property in the narrow sense is called "estate" in the latter case. See *Two Treatises of Government*, Book II, Section 123: "their lives, liberties, estates, which I call by the general name – property." See also Book II, Section 85.

REFERENCES

Chadwick, R. F. (1991). The market for bodily parts: Kant and the duties to oneself. In: B. Almond & D. Hill (Eds.), *Applied Philosophy: Morals and Metaphysics in Contemporary Debate* (pp. 288-298). London: Routledge.

Cohen, G. A. (1986). Self-ownership, world-ownership, and equality. In: F. S. Lucash (Ed.), *Justice and Equality Here and Now* (pp. 108-135). Ithaca: Cornell University Press.

Cohen, G. A. (1995). *Self-ownership, Freedom, and Equality*, Cambridge: Cambridge University Press.

Dworkin, R. (1977). *Taking Rights Seriously*, London: Duckworth.

Dworkin, R. (1978). Liberalism. In: S. Hampshire (Ed.), *Public and Private Morality* (pp. 113-143). Cambridge: Cambridge University Press.

Hobbes, T. (1997). *Leviathan*. C. B. Macpherson (Ed.). Harmondsworth: Penguin Books.

Hyde, A. (1997). *Bodies of Law*. Princeton: Princeton University Press.

Ingram, A. (1994). *A Political Theory of Rights*, Oxford: Clarendon Press.

Kant, I. (1948). *Groundwork of the Metaphysic of Morals (The Moral Law)*. H. J. Paton (Tr.). London: Hutchinson.

Kant, I. (1980). *Lectures on Ethics*. L. Infeld (Tr.). Indianapolis: Hackett Publishing Company.

Kimbrell, A. (1993). *The Human Body Shop: The Engineering and Marketing of Life*. New York: HarperCollins.

Locke, J. (1988). *Two Treatises of Government*. P. Laslett (Ed.). Cambridge: Cambridge University Press.

Macpherson, C. B. (1962). *The Political Theory of Possessive Individualism: Hobbes to Locke*. Oxford: Oxford University Press.

Marx, K. (1951). *Theories of Surplus Value*. London: Lawrence and Wishart.

Nino, C. S. (1993). *The Ethics of Human Rights*. Oxford: Clarendon Press.

Nozick, R. (1974). *Anarchy, State, and Utopia*. Oxford: Basil Blackwell.

Raz, J. (1986). *The Morality of Freedom*. Oxford: Clarendon Press.
Rousseau, J. (1968). *The Social Contract*. M. Cranston (Tr.). Harmondsworth: Penguin Books.
Steiner, H. (1994). *An Essay on Rights*. Oxford: Blackwell.
Thomson, J. J. (1990). *The Realms of Rights*. Cambridge: Harvard University Press.
Tully, J. (1980). *A Discourse on Property: John Locke and his Adversaries*. Cambridge: Cambridge University Press.
Waldron, J. (1988). *The Right to Private Property*. Oxford: Clarendon Press.

SARAH MARCHAND AND DANIEL WIKLER*

HEALTH INEQUALITIES AND JUSTICE

In this paper we examine some issues of distributive justice in relation to the distribution of health in a population. Our focus is on socioeconomic inequalities in health within a society. Research suggests that socioeconomic status and level of education are strongly correlated with level of health such that those with lower status in a society are relatively sicker than their counterparts who have higher status. Importantly, the correlation we are concerned with is not the obvious correlation between poor health and absolute material deprivation, or poor health and bad health habits. In nearly all countries, those at the bottom of the socioeconomic scale suffer anywhere from 2-4 times the mortality and morbidity rates of those at the top. This startling statistic apparently holds true regardless of differences in the material position of the worst-off. Moreover, even in the wealthiest countries the groups who have the highest status enjoy better health and longer life than those who have the next-highest status, and so on down the socioeconomic scale. Our question is whether these social inequalities in health are unjust in and of themselves.

There are many important questions about this research that we will not pursue. First in importance must be the nature of the correlation between SES (socioeconomic status) and health. Our paper assumes what much of the research claims: that the correlation is due in the main to the impact of SES on health, and not to the impact of health on SES. Although poor health often does lead to an economic spiral downward, the research seems to show a significant socioeconomic pressure on health. The definitions and delineations of socioeconomic 'status' and 'class' are a complex matter which we also leave to one side. Suffice it to say that studies that delineate classes in different ways (including occupational status, level of education, and income) have uncovered a significant socioeconomic gradient to health. Our paper assumes all of this to be the case. If it is untrue, our thoughts here will have a short life.

Second among the issues we do not discuss are the inequalities in health between countries. This is undoubtedly a more morally important issue than our issue, which is inequalities in health within one country, based on social class. A case in which the poor in one country are

Julia Tao Lai Po-wah (ed.), Cross-Cultural Perspectives on the (Im)Possibility of Global Bioethics, 209–221.
© 2002 *Kluwer Academic Publishers. Printed in Great Britain.*

healthier than nearly all of the population in another country raises issues of justice in an international context which we will not attempt to address here.

We also omit, for the most part, discussion of the philosophical and moral issues involved in the construction and use of measures of the level of health of populations or groups of people. Inter-populational measures of levels of health depend upon a host of complex, normative judgments. Not only must a measure permit us to compare the average overall level of health of the near-poor, say, to the very poor, but ideally we might want various measures of the extent or degree of health inequality within a society much like the measures now offered for the extent or degree of income inequality within a society. There appear to be rather complex relations between degrees of inequality in social status in a society and the absolute level of health of those who have the lowest status. It is less clear, as far as we can tell, precisely what the relation is between relative degrees of inequality in social status and relative degrees of health inequality within a society. At any rate, an ordinal ranking of the levels of health of different socioeconomic classes turns out to be uncontroversial. The SES health gradient appears for most kinds of disease and disability as well as for mortality. The relatively worse off have relatively worse health in nearly every respect.

In addition, consideration of the just distribution of health states involves a choice of the unit whose distribution is at issue. In its World Health Report for the year 2000, the World Health Organization compared countries on the degree of equality of the distribution of "health expectancies," the prospects at birth for each cohort for long life and good health, reflected in a summary measure, the Disability-Adjusted Life Year (DALY) that combines these dimensions (WHO, 2000).

Having put aside these important issues, we can turn to a few preliminary observations about distributive justice and the SES gradient in health. After these preliminaries we will note the assumptions found in the empirical literature on the proper goal of health policy. The rest of the paper is concerned with various possible egalitarian perspectives on SES and health.

I. PRELIMINARIES

The SES correlation with health requires, we believe, adjustments in our approach to health and distributive justice. The most obvious adjustment is shifting our thinking from health care to health. Some have already made this shift as part of a general re-thinking and overhaul of the relevant 'index of well-being' for use in interpersonal comparisons of justice. Sen is certainly an early pioneer in this regard. This shift, from health care to health, receives new impetus from the research on the SES gradient in health. As it turns out, this gradient persists even in countries with universal access to care, where health care resources approach a fair distribution. Moreover, the effect of health care on health looks to be relatively minor compared to other socioeconomic determinants and social policies. This is not to say, of course, that expanding access to health care is any less important than we have always thought: a just health care system is a critical part of any just society. But justice may require more. It may require a just distribution of health as well. The broader concern with a society's distribution of health may even have fresh implications for health care allocation. The distribution of health care must now be placed within a larger framework, one focused on the production and distribution of risk and needs by the basic social and economic institutions of a society.

A shift from health care to health has often been rejected in the past. An influential argument has been that we cannot owe people health because health needs are often 'insatiable' or satiable at too exorbitant a cost. Because of the insatiability of health needs, so the argument goes, and because 'ought' implies 'can', or 'can within reason' anyway, we must speak in terms of owing people some fair share of health care instead of owing them health.

The research on SES and health provides a reply to this argument. First, it reminds us that whether or not we can meet a particular individual's health needs, we can increase the level of need-satisfaction in a population. If we focus, for example, on those in the lowest socioeconomic group, we can meet more needs of more people to a greater extent than is currently the case, even though various individuals within that group are beyond our help, or can be helped only at what is ultimately judged to be too high a cost. So one response to the 'bottomless pit' argument is simply to point out that it assumes obligations only to particular needy individuals, and not obligations of justice concerning the basic structure of society.

The second reply to the argument is closely related to the first. Once we shift to a 'population perspective' on health, we see that the specter of high-cost medicine is less to the point than we once thought. If the research is correct, high-cost medicine doesn't appear to promise as much by way of improving health as we might expect from the adoption of certain socioeconomic arrangements, such as narrowing social inequalities in a society.

It is easy to miss how much is at stake in this shift from individual health care to population health. We move from arguments over whether we can ever be said to owe any particular individual some minimal level of health, to arguments over whether our obligations could be as strong as 'maximizing' the health of the economically worst-off. We should make clear that this is not to deny Arrow's point made long ago in his review of Rawls's *A Theory of Justice*. Arrow noted that so long as we do not "abstract" away individuals' health needs, but include health among the social primary goods, a maximin principle of justice might require nearly unlimited health care resources directed to the worst-off, draining the economy and yielding relatively small benefits. Can justice, Arrow asked, plausibly require huge sacrifices of everyone but the worst-off in order to realize what might be very small benefits for that group? Arrow was right: we can't owe the worst-off – or any group – unlimited health care resources. But one lesson of the SES gradient in health is that this point simply does not imply much about our obligations or lack of obligations to the worst-off. Perhaps we are obligated to aim at maximizing the health of the poorest class through socioeconomic policies. Taking into account the pervasive effect of socioeconomic inequalities on health, and the much smaller effect, on the average, of such expensive health care interventions as organ transplants, it may be the case that even the strongest of distributive principles, such as equality or maximin, has at least a *prima facie* plausibility in the area of health.

Our last preliminary remarks have to do with ideal theory. Many of us share the belief that in the real world, the distribution of income and wealth in most if not all countries is unjust. This belief is a sufficient indictment of the consequences for health that follow from such distributions. But we may also claim that health inequalities are themselves an indictment of any society whose differences in social status among groups in its population would produce such inequalities. Our concern is with this latter claim: whether just social arrangements could be compatible with socioeconomic inequalities in health and if so, under what conditions. Thus our topic belongs strictly to the realm of ideal theory.

II. ASSUMPTIONS IN THE LITERATURE AND THREE EGALITARIAN PERSPECTIVES

In the literature documenting the SES gradient in health, it is widely assumed that a society's foremost concern should be with maximizing the sum total of health of its citizens. We should say here at the start, that the difficulties in determining just what a population's highest sum total of health is, is not acknowledged in this literature. The maximizing assumption is shared by cost-effectiveness analyses of health care allocation and is, of course, a local application of general utilitarian principles. Each person's interests, in this case their health, counts just as much as and no more than anyone else's. From this perspective an improvement in health for the well-off is just as valuable and carries the same moral weight as an improvement in health for the worst-off. Health benefits count equally no matter where they fall.

There are people who feel strongly that levels of health should be more equal between socioeconomic classes. Richard Wilkinson's book, *Unhealthy Societies*, is a powerful indictment of health inequalities, as is the influential Black Report. Even in these cases, however, equality is presented as a means for achieving better overall health totals. In the Black Report, for example, what is emphasized is "that eliminating social inequalities in health offers the greatest opportunity for achieving overall improvement in the nation's health" (p. 200). The focus of the literature, then, is on lowering mortality and morbidity rates wherever they can be lowered. Given reasonable assumptions, the better health enjoyed by the upper classes seems to be powerful evidence that those in lower positions suffer premature deaths and excess morbidity. Socioeconomic inequalities in health are taken to be significant because they reveal that a society is less healthy than it could be. But there is no special moral urgency given to narrowing such inequalities.

Egalitarians will obviously reject this 'distribution-insensitive' reasoning. But what they will offer in its place, we want to suggest, is a vexing question. A difficulty for egalitarians, we believe, is in deciding to what extent the vantage point of class in a just society should be morally privileged over other vantage points – the vantage points of individuals or even other groups, for example. Some egalitarians will be drawn to identifying the sick or the needy and not the relatively poor as the appropriate vantage point from which to assess a society's distribution of health. The first two views we discuss take 'class' as the appropriate vantage point. The third view is concerned with the interests and vantage point of 'the needy'.

The first two views, which we will call 'the strict equality view' and 'the maximin view', are inspired by Rawls's theory. But both views can be altered in ways that reflect different theories of just income distributions. Their point of departure is the idea that justice is concerned with the 'life prospects' attached to various social positions, or, simplifying somewhat, to socioeconomic classes. Rawls conceives of these 'life prospects' in a fairly limited way, as 'prospects' for social primary goods. Now we discover that these life prospects include prospects for health and life-expectancy as well. While we may reject Rawls's narrow definition of 'life prospects', this discovery appears to confirm the importance of focusing on the life prospects attached to socioeconomic classes. It seems that 'class prospects', as we might call them, are an even more important unit of moral concern than we thought. And like Rawls, we may believe that inequalities in these life prospects must rank very high on the list of concerns of justice. While inequalities in some of these prospects – say, in economic prospects – may be acceptable in certain cases, can inequalities in the prospects for health and life itself ever be acceptable?

Some egalitarians might favor what we will call 'the strict equality view'. On this view, health is regarded as a special good, and levels of health should not be permitted to vary with inequalities in income and wealth. Health, it might be argued, is closer in importance to the basic liberties and opportunities than to the difference-principle goods. It is closely tied to the capacities essential for autonomy and choice. If health is more like the basic liberties or essential opportunities in being closely tied to our status as equal citizens, we might argue that social inequalities in health threaten citizens' self-respect and full moral regard for one another.

Norman Daniels (1985) has developed a similar view with regard the importance of health care. Health care, many believe, should not be distributed on the basis of ability or willingness to pay but instead should have a more secure footing in equality. That the wealthier can afford to purchase a yacht while the poorer must settle for a canoe may not necessarily offend justice. But that the wealthier can purchase heart surgery while the poorer must do without is deeply unfair. Similarly, we might argue, the fact that the better off face significantly lower prospects of ever needing heart surgery than the worst off face is deeply unjust.

The 'strict equality view' might require a constraint on the operations of the difference principle much like the constraints imposed by the first principle of justice and the opportunity principle. The constraint would work in this way: we would permit inequalities in income and wealth to maximize the prospects of the worst-off, subject to the requirement that such inequalities do not produce significant inequalities in levels of health.

We cannot offer anything close to a satisfactory assessment of this proposal here. But we can point out that for some, its appeal will depend upon its costs, especially its costs for the worst-off. If such a constraint in effect rules out income inequalities that would improve the position of the worst-off, we may believe that the constraint would be unjust. Importantly, it may be because we assign special weight to health that we would reject such a constraint. A constraint which imposes losses in income on the worst-off threatens to impose losses in health as well. If we believe that the consequences of adopting such a constraint for the worst-off affects its plausibility, then we cannot hold the 'strict equality view'.

We might adopt in its stead the view that levels of health should simply be placed under the direction of the difference principle. This is the 'maximin view'. The difference principle, in its simplest form, requires that the distribution of income and wealth be equal unless inequalities improve the absolute level of income of the worst-off. The vantage points for assessing distributions are representative persons attached to the various social positions, but the representative of the worst-off class has the 'veto'. If we depart from the presumptive default position of equality, then we do so because the worst-off are better off than they would be under strict equality. The attraction of allowing levels of health to be regulated by the difference principle is that, as with income and wealth, it is plausible to argue that the worst-off would rationally prefer a higher average level of health to a lower level, even if this meant a greater inequality in health between themselves and the other social positions. But the problem is that the difference principle, as it stands, provides no guarantee that the worst-off will have a higher level of health. And even if it did guarantee this we would want that guarantee to be written into the terms of the contract, so to speak. This familiar problem with Rawls's theory is that 'life prospects' are confined to prospects for primary goods. The worst-off do not have a complaint about their prospects for absolute levels of income; they do have a complaint about their prospects for health and longevity.

If we believe that levels of health should be included in the 'life prospects' attached to the various social positions we must reject Rawls's insistence on using levels of income and wealth as an index for well-being. We must expand 'the scope of complaints' assigned to the various social positions, in particular to the worst-off.

We believe that the SES gradient in health offers a new and powerful argument for expanding the scope of complaints against Rawls's theory. Unfortunately, we can only list these arguments here. Firstly, health is an all-purpose good, important to virtually all conceptions of the good, and thus on par with the other primary social goods. Secondly, levels of health between socioeconomic classes are measurable and observable, and thus can serve as a practicable basis for agreement. Thirdly, the concern with socioeconomic inequalities in health is consistent with a focus on the basic structure and with the vantage points of social positions. Socioeconomic inequalities in health are inequalities in the basic structure; they are produced by inequalities in income and wealth, along with differences in power, control, and opportunities. We should acknowledge, we believe, that it is in part an empirical question what inequalities are actually produced by the basic structure.

It is helpful to step back for a moment and remind ourselves that inequalities in health or health needs have played an important role in arguments against the use of the social, primary goods for purposes of interpersonal comparisons of justice. Sen has emphasized the inadequacy of the social, primary goods in dealing with cases of unequal needs. As Sen has argued, people with equal amounts of primary social goods count as equally well off on Rawls's theory, but they may not be equally well off in ways relevant to judging the fairness of their situation. Someone who has greater health needs than others, for example, should not be counted as well off as others. Sen concludes that we need a better account of 'being equally well off' – one that is sensitive to such differences and therefore can tell us how well people are actually doing with their share of primary goods.

Sen's emphasis has been on the differences in specific needs and circumstances of individuals rather than on differences in needs generated by the basic structure. But we could use some version of a 'capability index' to register inequalities in health between social positions. Rather than measuring a particular person's 'capability set' given a certain amount of resources, we could measure income in terms of capabilities. We could, in other words, determine what 'the standard capability set' is, that is produced by a certain level of income. This kind of index would not reflect differences in individuals' natural abilities, needs, or health. But it would reflect the differences in levels of health, needs and even disabilities between social positions.

Importantly, the kind of index we have in mind is intended as a measure of income shares, and not as a conception of well-being that replaces the social primary goods of income and wealth. It is an income-related index: it would reflect only what a given level of income and wealth, together with the overall distribution of wealth, actually produced in terms of all-purpose goods. Using a broader index for income and wealth simply reflects the fact that inequalities in these goods may causally produce inequalities in other all-purpose goods, such as health, and that such inequalities should be included in the 'life prospects' identified with social positions.

It will usually be true, we assume, that the difference principle applied in the traditional way to income and wealth will not yield different results than the expanded difference principle. But this does not mean that expanding the scope of complaints under the difference principle is unimportant. It is important for two reasons. Firstly, it makes clear that in distributing income and wealth we are, in effect, distributing health as well, regardless of how health care is distributed. The better-off will on average live longer and healthier lives than the worse off. The difference principle must ensure that inequalities in income and wealth raise the absolute level of all-purpose goods for the worst-off, which now includes their average level of health as well as their income share.

Secondly, it may not always be the case that the traditional difference principle would improve the position of the worst-off. Richard Wilkinson, along with many others, believes that the degree of social inequality in a society has an important effect on levels of health, especially for the worse off. If this hypothesis is correct, the difference principle applied to income could come at the cost of losses in health, especially for the worst-off. It is impossible to say what the net effect of the traditional difference principle would be on health. On the one hand, the worst-off have more purchasing power and a correspondingly higher standard of living. On the other hand, their position relative to others may be worse. It might prove to be the case that less inequality in income, and thus less income for the worst-off, improves the absolute level of health of the worst-off. Faced with the possibility of having to trade-off health against income and wealth, it is clear that applying the revised difference principle would require assigning relative weights to relative levels of health and income. The revised difference principle cannot avoid these complications.

Finally, we would like to mention a third view on socioeconomic inequalities in health. We can call this view 'the urgent needs view'. We might claim that instead of giving absolute priority to the worst-off – as the difference principle does – we should give them relative priority depending upon the urgency of their level of health. It doesn't seem plausible to give absolute priority to improving the health of the worst-off if those who are next to the worst-off are also doing badly. They too may have short life-expectancies and high morbidity rates. Applying a standard of urgency to levels of health suggests that we are concerned with the minimum position in proportion to their neediness. If others are needy as well we should show more even-handed concern for their fate as well. Conversely, it doesn't seem plausible to give absolute priority to the worst-off when they are doing very well, with a generous average life-expectancy. Intuitively, we seem to apply some standard of urgency to levels of health that is not captured by the revised difference principle. While we may agree that absolute rather than relative levels of health matter, we may also believe that they matter more when they are low and less when they are high.

Now consider the following real-world example. A decade ago, the American state of Oregon initiated a prioritization process to decide which medical services should be provided to the poor under its Medicaid health insurance program. One goal of the Oregon Plan in restructuring Medicaid was to improve the health of the economically worse off as a group, by providing more services that yielded the largest health benefits and cutting back on services that promised fewer benefits. If this proved to be successful, such a plan might well narrow the health gap between the wealthier and poorer populations in Oregon and improve the latter's absolute level of health. Yet the health gap between the very sick and the healthy promised to widen in certain particular areas. For example, certain transplants were placed on the list of services to be eliminated for Medicaid patients. Those with these transplant needs would now do worse on the whole. We should also note that the inequality in life prospects for the poor needy as opposed to the wealthy needy in this case would also increase.

It is essential to stress here that our responses to such a plan are undoubtedly shaped by the fact that the underlying distribution of socioeconomic status which produced these particular choices is deeply unjust. But the example is relevant because the opportunity to make these kinds of choices is always with us. The example illustrates what we have known all along: that the absolute level of health of the worst-off can always be improved and socioeconomic inequalities in health can always be narrowed simply by leaving the sickest behind, when treating the medically needy is expensive. It is also true, of course, that we can always raise a population's health totals simply by leaving the very sick behind.

It is unclear what conclusions egalitarians will want to draw from this fact. One conclusion would be to jettison concern for the relative or absolute levels of health of socioeconomic classes in favor of focusing on the gap in well-being between the sick and the healthy, or rather, on improving the level of health of those in real need. We may assign priority to those who are threatened with very serious harms – who have short life-expectancies and severe diseases and injuries. Perhaps they should count as the worst-off. In other words, perhaps we should single out the sick and needy rather than a socioeconomic class as the morally privileged vantage point for assessing the justice of a society's distribution of health.

Imagine a society like ours in which the relatively wealthy lived on average longer and healthier lives than the relatively poorer, but that a truly devastating disease was confined to the wealthy classes. And imagine that no individuals among the worse off suffered such severe drops in life-expectancy as these stricken souls. We could conclude that we would give priority to eliminating this terrible disease visited upon the wealthy even if doing so widened the gap in average levels of health between the best-off and the worst-off. Of course the point cannot be that we should spend unlimited resources on the severely ill – that would land us back in the 'bottomless pit' discussed by Arrow. Rather, this view would simply give relative priority to trying to help such people instead of helping the worst-off, whose average health was relatively lower than other classes but whose occupants did not suffer similarly severe diseases.

There are two very brief points we would like to make concerning this view. Firstly, this view does not necessarily avoid the problem of 'leaving the worst off behind'. So long as needs are aggregated, improving the health of the 'worst sufferers' is now compatible with leaving 'the worst' of the 'worst sufferers' behind. Secondly, the intuitive judgment that a need is 'great' or that a certain life-expectancy or level of health is 'high' or 'low', is obviously a complicated normative judgment. 'Low' or 'high' relative to what? As others have pointed out, we appeal intuitively to some notion of a 'standard life-expectancy and standard level of health' that is neither descriptive of what a given population has achieved in that area, nor an ideal standard. What is 'standard' in this context is clearly shaped by ideas about material and technological feasibility. But it also seems to be shaped by a complex of ideas about what it is reasonable to achieve. We believe that critically examining the normative assumptions behind these kinds of intuitive judgments would prove helpful.

In closing, we believe that the issues of justice raised by social inequalities in health are critical issues. They are also difficult issues. In this exploratory paper, we have only been able to offer the barest of sketches of three possible perspectives on health inequalities. The arguments have yet to be developed and we hope that philosophers will place this task high on their agenda of concerns.

World Health Organization, Geneva, Switzerland
Department of Philosophy, University of Wisconsin, Madison, USA

NOTE

*   This paper reflects the views of the authors alone and is not necessarily reflective of the views of the World Health Organization.

REFERENCES

Arneson, R. (1989). Equality and equal opportunity for welfare. *Philosophical Studies, 56*, 77-93
Arrow, K. J. (1973). Some ordinalist-utilitarian notes on Rawls's theory of justice. *Journal of Philosophy, 70(9)*, 245-262
Cohen, G. A. (1989). On the currency of egalitarian justice. *Ethics, 99(4)*, 906-944

Culyer, A. J. (1993). Health, health expenditures, and equity In: E. van Doorslaer, A. Wagstaff, & F. Rutten (Eds.), *Equity in the Finance and Delivery of Health Care.* Oxford: Oxford University Press.

Culyer, A. J., & Wagstaff, A. (1993). Equity and equality in health and health care. *Journal of Health Economics, 12,* 431-457.

Daniels, N. (1985). *Just Health Care.* Cambridge: Cambridge University Press

Dworkin, R. (1981a). What is equality? Part 1: Equality of welfare. *Philosophy and Public Affairs, 10(3),* 185-246.

Dworkin, R. (1981b). What is equality? Part 2: Equality of resources. *Philosophy and Public Affairs, 10(4),* 283-345.

Kaplan, G. A., & Lynch, J. W. (1997). Whither studies on the socioeconomic foundations of population health? *American Journal of Public Health, 87(9),* 1409-1411

LeGrand, J. (1991). *Equity and Choice.* London: HarperCollins.

Lynch, J. W., Kaplan, G. A., & Salonen, J. T. (1997). Why do poor people behave poorly? Variation in adult health behaviors and psychosocial characteristics by stages of the socioeconomic lifecourse. *Social Science and Medicine, 44(6),* 809-819

Nozick, R. (1974). *Anarchy, State, and Utopia.* New York: Basic Books.

Parfit, D. (n.d.). On Giving Priority to the Worse Off. Unpublished ms.

Rawls, J. (1971). *A Theory of Justice.* Cambridge: Harvard University Press.

Raz, J. (1986). *The Morality of Freedom.* Oxford: Oxford University Press.

Roemer, J. (1995). Equality and Responsibility. *Boston Review, April/May 1995, Vol. XX, No. 2.*

Scanlon, T. M. (1995). A Good Start. *Boston Review, April/May 1995, Vol. XX, No. 2.*

Sen, A. (1993). Capability and well-being. In: M. C. Nussbaum, & A. Sen (Eds.), *The Quality of Life* (pp. 30-53). Oxford: The Clarendon Press.

Sen, A. (1980). Equality of what? In: S.M. McMurrin (Ed.), *Tanner Lectures on Human Values* (pp. 1-51). Cambridge: Cambridge University Press.

Tobin, J. (1970). On limiting the domain of inequality. *Journal of Law and Economics, 13,* 263-278.

Townsend, P. & Davidson, N. (Eds.) 1982. *Inequalities in Health: The Black Report.* Harmondsworth: Penguin.

Walzer, M. (1983). *Spheres of Justice.* New York: Basic Books.

Wikler, D. (1978). Persuasion and Coercion for Health. *Milbank Quarterly, 56(3),* 303-338.

Wilkinson, R. (1996). *Unhealthy Societies: The Afflictions of Inequality.* London: Routledge.

World Health Organization (2000). *World Health Report 2000.* Geneva: World Health Organization.

CHAN HO MUN AND ANTHONY FUNG

MANAGING MEDICAL INFORMATION:
THE MORAL DILEMMAS IN POSTMODERN SOCIETIES

This paper examines how we can manage medical information in our
postmodern societies. It addresses the moral dilemmas involved in
dealing with all the relevant channels for information dissemination at
both individual and social levels. The paper first starts with a discussion
on the nature of moral dilemmas. We argue that moral dilemmas can only
be *managed* but not resolved, and that the way in which a moral dilemma
is handled properly is highly dependent upon the context in which the
dilemma occurs. The paper challenges the view that there are universal
procedures or principles that can enable us to resolve moral dilemmas in
all possible contexts (universalism). It is argued that practices that are
acceptable in one context may not lend themselves readily applicable in
other context. Solutions to moral dilemmas are therefore alleged to be
contextual and sensitive to the domain in which they arise. Since no
procedure or principle is universally applicable to all domains, moral
consistency (weak sense) can only be maintained within a certain context
but that consistency (strong sense) is hard to maintain across different
moral contexts. This conclusion argues against the possibility of
affirming a specific set of global rules about the relationship among the
interests of individuals, communities, and societies regarding the
management of medical information.

The paper examines the usual moral dilemmas arising from the
management of medical information and then puts the issues in the theme
of the conference "individual, community and society." Medical
information is a broad term describing all the messages related to health,
medical issues and organizations. The term information can be identified
as a measure of organizations, as a resource or as a commodity (Mosco
and Wasko, 1988, p. 19). Whether it is in the domain of organizations,
resources or commodities, it should be utilized and managed not only
efficiently and effectively but also ethically.

What is the criterion to release certain medical information? What are
the means and ways? Through which channels should the information be
conveyed? Why should certain messages be disseminated and other
messages should not? These are all issues to be resolved. This paper will

*Julia Tao Lai Po-wah (ed.), Cross-Cultural Perspectives on the (Im)Possibility of Global
Bioethics*, 223–235.

center on the dilemmas in deciding whether certain medical information should be disseminated to the individual and the public, and briefly discuss the practical difficulties and solutions of the process. We argue that information management can be regarded as a recapitulation of the communities and relationships that we have lost in moving from thickly bound communities to life in an individualistic society. The ways in which the moral issues present themselves regarding information management will be a function of how societies in the East and West respond to this transformation

I. THE NATURE OF MORAL DILEMMAS

Should medical information be always released to the relevant parties involved? Are there conditions in which it should not? In case it should, in what way and to what extent should the release be done? In order to answer these questions, people need to consider whether the public, the community or the individuals have the right to know, how serious the consequences are when the information is released, what are the impacts of the information on third parties, and so on. Yet these ethical considerations can come into conflict and moral dilemmas may then arise.

People often uphold conflicting views on moral issues among themselves, but conflicting voices can also be found within a single mind in the case of moral dilemmas. The sources of conflicts are not simply results of errors in reasoning, and we often find ourselves in dilemmas because we do not know how to resolve the conflicts. A physician may on the one hand think that he should tell the truth to his patient but on the other hand find that telling the truth may do him harm. He may find himself get caught in an irresolvable conflict between veracity and beneficence.

Moral principles are not absolute rules but are only expressions of prima facie duties (Ross, 1930; Prichard, 1949). The universal claim that fish has no lung can be falsified by a single counter example, but a prima facie principle allows non-compliance. Prima facie principles, like rules of thumb or default rules, are something that people should *normally* follow. But people could be waived from following them for good reason. So the physician, after deliberation, may decide to tell the truth. Though it may hurt the patient, one cannot say that the physician has violated the principle of beneficence or that the principle is wrong. The same applies to veracity even if the physician decides to conceal the information from the patient. The two principles are not in real conflict, and they indeed survive the conflict no matter what the physician does. This may even lead us to think that moral dilemmas are not real in the sense that they are at root only psychological but not logical.

The logical form of a prima facie principle in moral reasoning, as we will show, is structurally the same as that of default rules in daily reasoning. The logic of default reasoning has been studied extensively in non-monotonic logic, a rapidly developing area in artificial intelligence and cognitive science (for a recent overview, see Brewka et al, 1997). Some of its findings indeed have important bearings on the study of moral reasoning.

In our daily reasoning, many of the statements that we believe are *default* rules. For example, we all believe that birds fly, and if we are further told that Tweety is a bird, we will then assume that Tweety can fly too. Yet if it is later discovered that Tweety is a penguin, we will retract the conclusion that Tweety can fly. But we will not give up the belief that birds fly, because "birds fly" states the default rule that for any bird it is consistent to assume that a bird can fly. Yet information updates may well make it no longer consistent to assume that of Tweety, so the conclusion needs to be retracted accordingly. But it does not mean that "birds fly" as a default statement is wrong.

Default rules are very common in our daily reasoning. For instance, if you go to an airline company in Hong Kong and say that you want to buy a single ticket to San Francisco, people at the service counter will normally assume that you want to fly from Hong Kong to San Francisco unless you explicitly state otherwise. A default rule is of the form $\Diamond P \vdash P$ (read as: if it is consistent to assert that P, then conclude that P). Reasoning with default rules (default reasoning) is non-monotonic in the sense that a conclusion may be retracted as a result of information update because the new information may make a default rule no longer applicable.[1]

In ethics, a prima facie principle P (say, thou ought to be honest) simply says that if there is no counter reason against doing P, then you ought to do P. The logical form of the principle is in fact rather similar to that of a default rule in daily reasoning. Prima facie principles can be regarded as default rules in moral reasoning, and so findings in non-monotonic logic are of direct relevance. One of the findings is that a system of default rules is, though not intrinsically, prone to be inconsistent.

Consider the proposition ~ (A & B) (i.e. not both A and B are true) and the two default rules ◊ A ⊢ A and ◊ B ⊢ B (i.e. if it is consistent to assume that A then conclude that A, and if it is consistent to assume that B then conclude that B). If we start with the first rule, then we can assert that A. A together with ~(A & B) gives us ~B. However, if we start with the second rule, we can assert that B. The two rules together with ~(A & B) are inconsistent simply because both B and ~B are derivable (Genesereth and Nilsson, 1987).

The above formal result captures the logical structure of Sartre's famous example of a moral dilemma for one of his students during the World War II (Sartre, 1948). On the one hand the student wanted to fulfill the obligation of filial piety and look after his mother, but on the other hand he wanted to be patriotic and join the French Resistance Force. There are two prima facie duties here, but it is physically impossible for him to do both. If he undertook the duty of filial piety, he had to stay home and would not be able to join the French Resistance Force. If he chose to follow patriotism, he would join the Force but would not be able to stay home.

Although prima facie principles, as we have said earlier, can survive moral conflicts, inconsistencies are bound to occur in some contingent situations in which contradictory results can be yielded from a set of prima facie principles. In other words, moral dilemmas are context-specific, and a moral dilemma is said to occur when there is a context in which the moral requirements of doing A and ~A are derivable from a set of principles Ps. It is not that the principles in themselves are inconsistent, but that they yield inconsistent results in some contexts.

In our formal example, the two default rules are not intrinsically inconsistent. The inconsistency is created by the presence of ~(A & B). The situation is similar in Sartre's example. Had the story changed a little bit, Sartre's student might have been able to take up both duties. For instance, it might so happen that he lived right next to an important Nazi German military site and that his duty in the French Resistance Force, if he joined it, was to spy and take note of the military movement of the troops there. He could stay home very often, and therefore be able to fulfill both duties. Hence, one of the possible ways to get around a moral dilemma is to change the context (say, the student might be able to get a good friend to look after his mother). On the other hand, moral dilemmas may emerge as a result of contextual changes, which make people difficult to comply with a set of relevant principles. Many bioethical problems arise because inconsistent results are derivable from new alternatives created by the advancement of technology together with some prima facie principles. The inconsistencies may not emerge before the new technology is introduced.

As an example, many have gone through a hot debate about surrogate motherhood, an arrangement that is made possible by *in-vitro* fertilization. The logic of moral dilemma seems to capture the logic of the debate quite well. Those who adopt the principle of autonomy as their starting point are prone to conclude that surrogate mothering should be allowed. Yet for those who adopt the respect for child interest and family integrity as the starting point are inclined to think that surrogate mothering should be banned. In order to resolve the conflict, it seems that we need to make a choice between autonomy on the one hand and respect for child interest and family integrity on the other.

In a particular context, one may need to use one prima facie principle to override another one so as to resolve a certain dilemma. But will that principle be overriding in other contexts as well? In the case of organ transplant, the respect for family decision-making may override a person's prior consent to organ donation after death in an Asian context as in Hong Kong, but it may not be the case in other places. Nor does it mean that in Hong Kong family decision will always be overriding in all contexts, such as the choice of treatment. Similarly, the over population problem in China, since it is so severe, may justify some stringent measures of birth control. But these will sound like obvious violations of human rights in other societies. Thus, it does not seem true that there is a universal solution to moral dilemmas across different contexts.

Furthermore, the problem of moral dilemmas has a temporal dimension which can make the process of resolution very complex. Although information updates and changes in a context in which a set of prima facie principles is applicable can make some old dilemma go away, it can create new ones as well. For example, secularization made many non-religious practices permissible, but the advancement of biotechnology at the same time has created new dilemmas. The rapid changes of social development and technological advancement make it difficult to resolve moral dilemmas in a once for all manner, and people can only manage them as we go along.

However, some universalists still have the false hope for a once for all solution. They think that some single principles, such as the maximization of utility, should override all other considerations in all possible contexts, or that some principle should always be lexically prior to others, or that some principle should be always assigned a higher weighting. Such a position has the flaw of context-insensitive. Furthermore, our discussion has shown that when conflicts arise in a certain context, we can maintain consistency *within* that context by overriding a prima facie principle by another one (coherentism in the weak sense), but it does not imply that we can consistently resolve moral dilemmas in the same way across different contexts. A principle, say respect for autonomy, may override another principle, say beneficence, in one context, but the opposite may be true in another context. (Coherentism in the strong sense is therefore false.) In other words, although local consistency can be maintained, global inconsistency is unavoidable.

II. THE SOCIAL DIMENSION OF MEDICAL INFORMATION RELEASE

In the management of medical information, in various contexts conflicting obligations can be derived from prima facie principles, including the right to autonomy, the right to privacy, the right to confidentiality, the well-being of the patient, the interest of the third parties, public interests and respect for family decisions. The following are a few cases:

(1) Should a medical doctor follow the request of the family members of a patient not to disclose to the patient that he has cancer or Alzheimer?

(2) Should one be informed that she is a HIV carrier despite the fact that she does not want to know?

(3) Should the spouse or partner of a HIV infected person be informed of the risk of HIV infection even if the person disagrees? Is the compromise of confidentiality justified?

(4) Should mandatory HIV screening be introduced for newborns, which may reveal information that their mothers do not want to know (Etzioni, 1999)?

(5)   Is it justified to introduce the mandatory screening of all pregnant women for hepatitis B so as to make the immunization for newborns more effective?

There are no obvious answers to these questions. Different people (stakeholders) in different places may have different views on these issues. Say, (5) may sound less controversial in Hong Kong and Mainland China where the rate of hepatitis B infection rate is very high. Also, although it is common to have screening policies for several genetic diseases, policies dealing with HIV infection are bound to provoke strong reactions from some walks of life in a community. Not only are moral dilemmas not resolved in the same ways across different contexts, even within a single context, different people (stakeholders) may have different views on how a moral conflict is to be resolved. In different contexts and in different cultures, different approaches to the management of conflicts among the interests of individuals, communities, and societies will seem more or less easy and plausible.

The situation is more complex when we have to consider moral dilemmas in society in which additional parties are involved. These parties may include mass media, health organizations, insurance companies, pressure groups on patients' rights, government, and other semi-official institutions, to name but a few. Among them, there must be some conflicting controversies and priorities. Thus, moral decisions justified and resolved (though unlikely) by one party may not be justifiable for others. This is not a matter of whether certain principles are justifiable to override some others. Simply, we should recognize the differences existing in different institutions. Let's take mass media as an example. The objectives of mass media are essentially different from the objectives of medical institutions.

Mass media objectives	*Public health objectives*
To entertain, persuade, or inform	To educate the public
To make a profit	To improve public health
To reflect society	To change society
To address personal concerns	To address social concerns
To cover short-terms events	To conduct long-term campaigns
To deliver salient pieces of material information	To create understanding of complexes

These differences again cannot be resolved because of the differences in standpoints and the nature of the organizations. If medical institutions can exist independently of some public institutions such as the media, the whole situation can be resolved by merely considering the doctor-patient relationship (or doctor-community relationship). However, in practice, these disparate objectives create many problems for medical institutions. They have to utilize mass communication to release information and to address the public (Atkin and Arkin, 1990, p. 16). Besides, with an ever-increasing openness of the environment, medical institutions are faced with the problem of media access. Media institutions aggressively pursue new information and establish new stringers for information. The following case is a good illustration.

In January 1998, it was found that more than 100 patients had been injected with a medical reagent possibly contaminated by the protein which causes Creutzield-Jakob Disease (CJD), after a recall of the product in more than 40 countries ordered by the British manufacturer. It was believed that the risk for contracting the disease through the administration of the test reagent was extremely remote. Moreover, there is no established way to predict whether a person will develop CJD, and there is no known effective treatment. There was a very tricky issue in this case. On the one hand there was the moral requirement of veracity. On the other hand the patients might not want to know the information at all, because they could do nothing even if they would develop CJD and the news would only bring about anxiety. However, one could not ask them for approval before letting them know about the contamination, because the very act of asking already hinted that they had been injected by the reagent in question. So somebody was required to make a surrogate decision for these patients. The British Department of Health under the advice of the Ethics Committee therefore did not contact patients in UK who had received the reagent because they wanted to avoid putting an enormous burden on them by telling that they had a remote risk of developing the disease. However, the Hospital Authority in Hong Kong acted differently. One possible explanation for this difference is that the Hospital Authority would likely come under attack by the press for being secretive and trying to cover up blunders if it did not approach the patients. Since no mechanism, such as a well-established Ethics Committee, had the adequate moral authority to make the decision on those patients' behalf in this case plus the pressure from the press, a diffcrent choice was made. The proper doctor-patient relationship came under siege by the intrusion of the media.

As the examination of the nature of moral dilemmas has shown, a moral dilemma that is resolvable in one context may not be resolvable in the same way in another. This at least informs us of the fact that while postmodernity may be characterized by a lack of absolute moral foundations applicable to all individuals, our goal is not to search for an universal or coherent system of ethics to resolve all the issues related to medical information release. In cases like CJD, it may be perfectly all right for medical institutions in Hong Kong to adopt an approach very different from the one followed by their counterparts in UK. But the heart of the matter is whether people have gone through an appropriate process of deliberation before accepting their resolution. The decision of the Hong Kong Hospital Authority seemed to be largely driven by the pressure from the press. Cross-cultural differences may depend on the presence or absence of appropriate moral authorities.

III. TOWARD A COLLABORATIVE AND REPRESENTATIVE MODEL OF DECISION

In postmodern societies, it seems that the distinction between private circle and public space is not as clear as it was in the past. Patient information in the past might have been considered "private." At most, the patient-doctor relationship may extend to the patient's family or community. Quite contradictorily, if our society is characterized only by the fragmentation of communities and the alienation of individuals, more and more relationships should have become more individualized and privatized. However, it is obvious that "the more individualized view" is not true. What was private may now be treated as "public" as when information release involves some other third parties, such as the press, and the consequences of the release are not on private individuals. At the same time, our era manifests a new web of "relationships" for which we even do not have names. We can call it "interdependence" of interests. Such linkages immediately throw the private circle into the public space, or simply transform all "the private" into "the public." Whereas the public space is more encompassing, some scholars such as Habermas (1962) have raised the issue of the degeneration of public sphere in our modern world. While we cannot guarantee the quality of public discussion, the public sphere is now expanding, and even engulfing the private life of individuals, whether in Europe, America, or the Pacific Rim.

This is exactly the fear brought out by Arendt (1963). The web of human relationships created so thoroughly entangles humans that they seem more the helpless victims than the masters of what they have done. When this becomes apparent, humans forfeit their freedom at the very instant, and we tend to condemn various social arrangements. When people conclude that the burden of action is too great, they abandon their life in community in favor of a life of supposed self-sufficiency and self-mastery in isolation. Then we may go to another extreme. That is, we have become entirely private. We would be deprived of seeing and hearing others, of being seen and of being heard by them. They are all imprisoned in their own singular experience. Then different parties in societies can only be seen as competing for their own interests.

If the parties are competing for interests and there is no overarching principle to guide us in resolving moral dilemmas in different contexts, it is impossible for medical institutions to arrive at the best resolution on their own. Atkin and Wallack (1990) have listed out three levels of issues concerning how medical institutions communicate with other parties: (1) there is a basic incompatibility between the medical institutions and other parties; (2) there is potential for misleading messages; (3) there is a practical concern over the forming and implementation of collaboration. While (1) and (2) have been discussed, what is essential is (3). That is, we need a collaborative effort from relevant parties to arrive at a solution when information issue comes up. The idea of a postmodern society implies a society in which individuals are alienated and communities are fragmented with respect to moral foundations. Universalism and coherentism in a strong sense are therefore hardly possible. All cultural identities and diversities are simply regarded as manifestation of anachronism, isolation, massification, and in incommensurability (Pasquail, 1997). Thus, when considering different parties in the process of resolving moral dilemmas regarding the dissemination of medical information, we seldom find moral decisions that equally and squarely fit all parties. However, this does not preclude the possibility of establishing some minimal ways of collaboration and representativeness. The moral dimension of communicating messages should be examined from a multicultural perspective, one that is "multi-moral" and "multisituational" (Pasaudli, 1997, p. 25). Universalities of principles may fail to resolve all technical details, disparate situations, multiple interests and the moral dilemmas. Hong Kong's approaches may not be those of France or the United States. Without an openness of interpretations, particularistic conjecture and contextual considerations in managing medical information, we simply regard the moral codes, claims or principles of a certain society as the only way to morality. These considerations weigh against global rules for the management of individual, communal and societal interests in the use of medical information.

Collaboration in the form of open discussion with other parties has pragmatic advantages. On a technical level, there is a clear need to improve medical communication skills in medical institutions, particularly, when this involves many parties in the field. On a decision level, what is essential for the medical field is to recognize the information differences in moral positions, interests and capabilities of various parties. Thus what we argue for here is the establishment of ethics committees or other platforms for open dialogues and public deliberation in order to allow people from all walks of life, not to resolve the unresolvable moral dilemmas using universal principles, but to manage the information more appropriately, in particular, when it involves different parties.

V. CONCLUSION AND IMPLICATIONS

This paper addresses the moral dilemmas of managing medical information in postmodern societies. While we argue that we may not be able to resolve the moral dilemmas in the complexities of society and the webs of relationships in the postmodern world, we can at least manage the dilemmas by means of some procedural ethics, for example, by establishing some ethics committees or some platforms for open dialogue and public deliberation. The significance of the argument not only sheds light on the practical problems of medical information release, but also re-addresses the problems of the configuration of "individual, community and society" in different cultures in our era.

While procedural arrangements are important, the philosophical illustration is as essential. The illustration can be summarized by Arendt's argument in *Human Condition (1958)*, which is a study of the central dilemma of humans in modern age. She held that humans in our age wish to escape from the "imprisonment of the earth," and to master the "human condition." Yet in the course of escape and mastery, which is characterized by the advance of science, freedom from labor, etc, Arendt argued that humans have slowly forfeited the meaning of these activities for the sake of which such "advance" or "freedom" would deserve to be won. In the bioethical context, the release of information is often approached in a way that involves weighing certain principles over some others without serious consideration of the meaning of the information release in a specific social context. What is reflected in Arendt's argument is a fear concerning our thoughtlessness as well as our willingness to communicate in our fragmented society. What we suggest is the establishment of a committee with representatives of different views towards the end of cooperative and partnership in managing medical information. It seems that in our society, as Arnedt points out, we proclaim the "freedom" to demand for appropriate release of medical information in our interdependent society. Paradoxically, through the various activities employed by the seemingly competitive fragmented parties in developing principles for release of medical information, we are more uncertain of the consequences of the decisions and our human destines, perhaps, with the results that certain parties may dominate others. What is important here are the open dialogues among the parties in considering the cases in various contexts with an approach to a procedural judgment. The dialogues constitute a web of human relationships that forms the human community. Without this vision, people in this new millennium will be continuously deprived of human community and quickly lose their sense of their own reality as well as that of their common experienced world.

Department of Public & Social Administration
Department of English and Communication
City University of Hong Kong
Kowloon, Hong Kong

School of Journalism and Communication
The Chinese University of Hong Kong
Shatin, Hong Kong

NOTE

[1] In contrast, classical logic is monotonic because the acquisition of new information will cause no conclusion to be retracted. That is, given that

$\Gamma \vdash C$ (C follows from a set of beliefs Γ), it follows that

$\Gamma, N \vdash C$, for any new belief N.

REFERENCES

Arendt, H. (1958). *The Human Condition*. Chicago: the University of Chicago Press, Chicago.

Atkin, C. and Wallack, L. (1990). *Issues and Initiatives in Communicating Health Information to the public*. Sage: Newbury Park.

Brewka, G., Dix, J. and Konolige, K. (1997). *Nonmonotonic Reasoning: An Overview*. Stanford: Center for the Study of Language and Information Publications.

Chan, H. M. (1994). *Formalization, Complexity, and Adaptive Rationality*. Unpublished Ph.D. Thesis, University of Minnesota, USA.

Etzionzi, A. (1999). *The Limit of Privacy*. New York: Basic Books.

Genesereth, M. R., and Nilsson, N. J. (1987). *Logical Foundations of Artificial Intelligence*. Los Altos: Morgan Kaufmann.

Habermas, J. (1962). *The Structural Transformation of the Public Sphere: An Inquiry into a Category of Bourgeois Society*. Cambridge: the MTT Press.

Mosco, V. and Wasko, J. (1988). *The Political Economy of Information*. Madison: University of Wisconsin Press.

Pasquail, A. (1997). The moral dimension of communicating. In: C. Christians and M. Traber (Eds.), *Communication Ethics and Universal Values*. Sage: Thousand Oaks.

Prichard, H. A. (1949). *Moral Obligation*. Oxford: Clarendon Press.

Ross, W. D. (1930). *The Right and The Good* Oxford: Clarendon Press.

Sartre, J. P. (1948). *Existentialism and Humanism*. London: Methuen.

KURT W. SCHMIDT

STABILIZING OR CHANGING IDENTITY?
THE ETHICAL PROBLEM OF SEX REASSIGNMENT
SURGERY AS A CONFLICT AMONG THE INDIVIDUAL,
COMMUNITY, AND SOCIETY*

I. INTRODUCTION

The world is getting smaller and smaller. In Europe, borders are
disintegrating and a joint currency (EURO) is already upon us.
Globetrotters encounter the same companies, the same hotel chains, and
the same fast food nearly everywhere they go. Some say the world is fast
becoming a 'global village'. And yet this 'village' is not governed by
unanimity. Taking any sociopolitical topic at random, it is impossible to
make a non-controversial statement about it. For every opinion there is a
counter-opinion; for every attempt to solve a problem, a counter-attempt.
Our opinions are divided on all matters of import, although our
differences regarding the moral evaluation of (new) medical options are
particularly marked. We have not even been able to achieve unanimity in
individual countries (the villages within the global village), as the never-
ending arguments about the permissibility of abortion, embryo research,
germline interventions and euthanasia (to name but a few) aptly show.
The reasons for this dissent are the profound anthropological differences
and varying worldviews of the individual citizens at stake. Moral
pluralism, as it is often termed, stems from the wars of religions and the
Enlightenment, through which secular and religious communities have
become mixed in many European states. Today we have to acknowledge
the fact that there is a difference between society and (religious)
community. There is no longer a single valid social stance acceptable to
all (religious) communities, even in fundamental matters of human
coexistence, even in the most fundamental of them all: the question of
what constitutes being a man or a woman. Using the example of sex-
reassignment surgery, the following paper will show that profound
differences of opinion exist between individuals, (religious) communities
and society about our understanding of males/females, and that there is a
deep rupture between traditional and post-traditional understandings of
the sex of a human being and its reassignment. The sex-reassignment

*Julia Tao Lai Po-wah (ed.), Cross-Cultural Perspectives on the (Im)Possibility of Global
Bioethics,* 237–263.
© 2002 *Kluwer Academic Publishers. Printed in Great Britain.*

surgery option requires individuals and individual communities to take a stance, which may range from 'progressive' to 'conservative-orthodox'. The problem facing the State is how to succeed in providing a framework inside which such differing positions may be adopted side-by-side. The problem facing individual citizens is how to find orientation when forming their own opinions in a world which is becoming increasingly complex in its globality. One fixed point of orientation for many is the differentiation male/female: a parameter they believe to be both 'natural' and beyond debatable.

Our entire social life is divided into male/female. The sex of a baby is registered at birth, resulting in certain obligations for the new earthling. Once adulthood is reached, national service is compulsory for all German men, for example, whilst only women are allowed to be midwives. The sex of a person is a reality which cannot be denied. The fact that there are differences between men and women seems so obvious that no further elaboration is necessary. Not a single German law, for example, defines characteristics distinguishing a man from a woman. Our legal system simply assumes that there is such a difference and that it is clear.

However, this 'clarity' becomes rather clouded under the following circumstances: on the one hand, human beings are born whose sex cannot be precisely determined at birth, and on the other hand, there are individuals who gradually come to believe that their soul has been trapped in the 'wrong' biological body. These are phenomena that occur everywhere on the globe, to which different cultures have reacted in various ways. In the Western industrial nations, plastic surgery has made it possible to change the outer appearance and adapt it to the features of the sex desired. This is regarded as a reasonable therapeutic action by some, but is opposed by others who believe it to be an unacceptable mutilation of the human body. The argument about the justification for such an intervention does not only divide religious communities and secular groups, but also members of the medical profession. More so even than transplantation medicine (keyword: brain death) or artificial insemination, sex reassignment surgery highlights a plurality of value concepts and confronts us with one of the basic problems of modern medicine at the start of the 3rd millennium: what are the medical treatments that we must, can or may carry out when there is a considerable disagreement between individuals, community and society with respect to the permissibility of the particular treatment? How can the conflict arising in society about the permissibility of certain medical treatments be resolved? These questions are examined here using sex reassignment surgery as an example. The article describes the ethical conflicts and practical problems associated with this type of treatment and proposes that a consensus should not be the sole aim in the 3rd millennium, but rather our aim should be the recognition and establishment of structures that offer scope for dissent within society.

II. NOTHING IS DEFINITE, NOT EVEN SEX

Until the middle of the 20th century, the sex of a child only became known at birth. Since then the tools of prenatal diagnostics and other examinations during pregnancy have revealed the previously 'invisible' fetus. Although this makes it possible to ascertain at a very early stage whether the parents are expecting a boy or a girl, there are always cases in which the sex cannot be clearly determined. Some children are born with both sets of sexual organs (so-called hermaphrodites). It is not difficult to see why this can cause a multitude of problems, currently the subject of fierce debate in the U.S.A. (cf. Dreger, 1998; Elliott, 1998). The following issues are particularly in need of clarification:

- Which sex is to be registered at birth if the sex is unclear? (The registrar has to make a decision between 'male' and 'female' since a third possibility [an 'intermediate stage' between man and woman] is not foreseen, although it does occur biologically).
- Which sex is the child to have if there are discrepancies between chromosomal and genital sex? (Children suffering from the rare 5-alpha-reductase deficiency syndrome have an XY-chromosome, yet are born with ambiguous sexual characteristics [Elliott, 1998]).
- With such a lack of clarity, should children be able to decide their sex for themselves at a later date? If so, in which roles should parents bring them up until that time?
- Should a surgical intervention be carried out early on in order to adapt the child externally to a particular sex, or should such early surgical interventions be refrained from in order to avoid the danger of adapting a child to the 'wrong' sex?

None of these questions is fundamentally new. Indeed, when looking at the General Law for Prussian States dating back to 1794, it is surprising just how modern the regulations seem. In §§ 19 and 20 a plea is made to leave parents to decide the sexual role in which they would like to bring up their children, allowing the children themselves to choose freely at the age of 18 (Hattenhauer, 1996).[1] The opinion seems to have been that children have a 'sex-specific true identity', which may temporarily be hidden by external features, but which will emerge with time, revealing itself most clearly of all to the person in question. Therefore, the question of sex assignment is not merely a private matter; it is equally of public interest and has an impact on society. Although the phenomenon of individuals changing to another sexual role – resulting in a certain fame, but often enough also leading to harsh punishments (cf. Laqueur, 1990) – has been known in many communities throughout the ages, it is still, at the end of the 20th century, surrounded by great uncertainty. As the techniques of sex reassignment surgery are being perfected, society is faced with the question of whether such interventions are permissible: do they represent an immoral interference with nature or are they a necessary form of therapy to alleviate suffering?

III. TRANSSEXUALITY

Whilst the sexual identity of babies can only be judged – as shown above – within certain limits, some adults and teenagers are forced to reassess their own sexual identity when perceiving a discrepancy between their outer sexual appearance and their inner feelings. Transsexuals are persons who are born with the anatomy of one sex but suffer from an identification with the other sex. Many people with 'transsexual symptoms' are convinced that they are trapped in the 'wrong' body, with the 'wrong' name and the 'wrong' documents. Without the chance to adapt their bodies to their inner perception of themselves, using drugs, cosmetics and/or surgery, and without a legal acknowledgment of this transformation, i.e. having their name and sex changed in all documents, this situation is often deemed unbearable by those concerned.

At this juncture a brief look at terminology would be appropriate: it appears difficult to find a suitable term which is not discriminatory to the people referred to here. Many things have been suggested, only to be discarded again. Even the term 'sex change' is problematic as far as an evaluation of the treatment is concerned, since some of those concerned say that they do not need to be changed, but merely adapted to their true sex. In the English (medical) language, the technical term sex reassignment surgery (SRS) has become established.

1. Examining the ethical permissibility of sex reassignment surgery

As far as medicine and medical ethics are concerned, a request for sex reassignment surgery is a request in line with the basic situation routinely encountered by the medical profession: a person in need turns to a physician and asks for help. For those who comprehend the patient-doctor relationship as a service relationship involving a contract between autonomous individuals in which anything freely negotiated between physician and patient is fundamentally possible, the issue of sex reassignment will not pose much of a problem: from an ethical point of view, if both parties are happy, the operation may be carried out. However, this understanding of the patient-doctor relationship is certainly not shared by everyone. It is far more common to believe that the objectives of medicine are determined by society as a whole and governed by professional/ethical codes (such as the Hippocratic oath). For some, 'true' medicine is solely concerned with healing and alleviating *physical* afflictions, while 'cosmetic' interventions are excluded (Kass, 1981). The psychoanalyst C.G. Jung, for example, dismissed sex reassignment surgery as 'merely' aesthetic surgery having nothing at all to do with medicine or psychology (as defined by him). "If the patient initiates treatment, defines treatment, and defines the goals of treatment, according to Jung, this is not 'real' medicine. Jung argued that it is as bad as if someone went to a surgeon and persuaded him to amputate a fully functional finger. The only difference is that the functional penis has a symbolic status in society. For Jung, any physician who undertakes such a procedure is to be roundly condemned: there is no possibility that such 'cosmetic procedures' are medicine in any shape or form" (Gilman, 1999, p. 271).

A physician is faced with a request for treatment and has to examine whether and how this request can be met. There are many reasons why a physician may not be able to fulfill a patient's request, for example, for a sick note or an unnecessary prescription when the patient is not sick or – in an extreme case - a request for euthanasia. It is easy to see how ethical, legal or health policies can run counter to the patient's wishes, and that the reasons for this can vary considerably from country to country. Concerning the case investigated here, a first fundamental question arises.

2. Is sex reassignment surgery a 'permissible' form of medical treatment? And for whom?

Several decades ago, a wish for sex reassignment surgery was attributed to a 'pathological inner feeling' which should be 'corrected' by psychotherapy, psychoanalysis, etc. A patient's belief of being a woman in a man's body was considered a 'neurotic notion'. Bound by this view, sex reassignment surgery was understandably interpreted as 'therapeutic malpractice', an attempt at adapting reality to an illusion. Surgery was considered a step in the wrong direction and there was far more of a call for the very wish for sex reassignment to be 'treated'. A patient was deemed healed when he ceased to experience this wish.[2] Accordingly, the moral responsibility of the physician was to oppose the wishes of the patient since (a) surgical intervention was the wrong approach to finding a cure, and (b) patients could otherwise use sex reassignment surgery in order to commit 'focal suicide', i.e., they wished (autodestructively) to destroy their existing bodies, extinguishing part of their person and identity, and physicians may not assist them in such 'partial suicide'.

3. Sex-reassignment surgery as focal suicide?

According to Pfäfflin (1996), what is behind the suicidal tendency theory is the negative idea that patients are 'mutilating' their bodies; what is not seen is that they are actually gaining something essential. Sex reassignment surgery has even been deemed a life saver: advocates of sex reassignment have maintained that transsexual patients are at a high risk of committing suicide due to unbearable mental suffering, and that surgery keeps them from doing so.[3] Pfäfflin compiled the following data in 1996:
- Suicide attempts are relatively frequent in patients with transsexual symptoms. At least one in five patients has attempted suicide at least once before commencing treatment. Man-to-woman transsexuals have a higher suicide tendency.
- Successful sex reassignment treatment has a favorable effect on patients.
- Extremely bad results increase the postoperative risk of suicide in individual cases.
- Sex reassignment is not, however, to be performed simply as a means of preventing suicide. It must be embedded in longer-term

psychotherapeutic and/or psychiatric treatment. "The expectation that despair could be dispelled by a somatic intervention as a stand-alone solution is unrealistic" (Pfäfflin 1996, p. 122).

If a differentiation is made between the integrity of the body and the integrity of the overall person, then this can be an argument in favor of sex reassignment: it puts an end to the suffering and makes the integrity of an overall person possible, as well as leading to 'true identity' (correspondence between outer appearance and inner feelings).

4. Surgery: the best solution in a range of bad ones?

After it became apparent that psychotherapeutic interventions alone were often unable to reduce patient suffering to a bearable level, it became the accepted opinion that, in individual cases, sex reassignment surgery may be an appropriate method of relieving the suffering of these patients.[4] And yet, medically speaking, this procedure is unusual. In the 'classic' treatment model practiced in conventional medicine, physicians try to find and remove the cause of suffering. The causes of transsexuality have yet to be sufficiently explained (Wille, 1989, p. 1049) and have been the subject of fierce controversy for years. In addition to psychogenetic or biological causes, there is also a hypothesis that gender identity develops as a result of an interaction between the developing brain and sex hormones, and that sexual differentiation of the brains in transsexuals might not have followed the line of sexual differentiation of the body as a whole (Zhou et. al., 1995, 68). If one follows this line of thought, the question arises whether the subjective gender identity could not be 'due to nature' in the same way as the chromosome-based sex assignment. Whatever the truth may be, to date there are no generally accepted explanations for the phenomenon of transsexuality.

The causes of individual suffering play a role in cases involving responsibility for the administration of hormones with irreversible consequences for biological men after 3 months, or for the performance of a surgical intervention: Could the suffering of an individual patient not be hiding a mental disease which should primarily be treated with psychotherapeutic means?[5] All investigations to date, however, show that in cases of genuine transsexuality, attempts at psychotherapeutic treatment regularly fail. Hormone treatment or plastic surgery therefore represent reasonable ways of liberating patients from their mental suffering when psychotherapeutic means are inadequate. However, although it has been found that psychotherapeutic treatment has no influence in cases of genuine transsexuality, it would be quite wrong to deduce that such attempts should not be made and surgery offered straightaway (Bosinski et. al., 1994). The fact that hardly any transsexuals agree to accompanying psychotherapy following 'successful' (!) sex reassignment surgery can be seen as further confirmation of the theory that this complaint is not open to psychotherapy (Banaski, 1998).

Is sex reassignment therefore more a *substitution therapy?* No, since the patient does not have a "somatic" deficiency and is often "physically" healthy. On the contrary, the hormone therapy can actually cause undesirable side effects and the surgical interventions can infringe upon previous 'physical' health, especially when a patient has undergone corrective interventions.

In many cases, sex reassignment surgery is able to close the gap between the patient's feeling of gender identity and the biological sex. 'Objectively' seen, however, a complete transformation or adaptation cannot be achieved, since the sex-specific set of chromosomes is not changed by surgery. The operation can only help to adapt, i.e., bring the outer appearance more into balance with the patient's inner feeling of gender identity. The *genetic* sex cannot be changed by this type of intervention. This may be acceptable to the individual patient, but at the same time this 'indeterminate state' presents a number of difficult questions in (ecclesiastical) law, for example, with a view to subsequent marriage. Since marriage is only allowed between a man and a woman, it is important to define which factors (inner feeling, outer appearance, chromosomal disposition, etc.) should be used as criteria for determining to which sex a person belongs.

Another basic problem still remains: while sexual organs can be removed, medical science is incapable of substituting functional reproductive organs of the opposite sex. Sex reassignment surgery involves the construction of simulated sexual organs devoid of reproductive capacity (ovulation, semen formation). The crucial question then is whether or not a 'real' change of sex has indeed occurred from the different moral, religious and legal points of view.

5. Ethical positions regarding the right of disposal over one's own body

One of the strongest manifestations of the Nuremberg Code is that medical interventions in human beings require the informed consent of the person concerned. The right to freedom from physical injury is accorded a high value, and human beings are personally responsible for their own bodies. And yet with the dawn of the 21st century humanity increasingly seems to view the body and its individual parts as possessions. Anyone accepting the resulting fundamental principle that all human beings have the right to do whatever they like with their bodies, will not be able to find any objections to sex reassignment surgery. Traditional Western religious values defy this, however, as outlined briefly below.

From a Jewish/Christian point of view, the body is not a 'possession' since human beings do not acquire their bodies themselves, but receive them as a 'gift' (or, for some believers, as a kind of 'destiny') and have a certain responsibility for them. The question in this ethical or religious context is therefore: Should human beings be entitled to so much disposal over their own bodies that they can have their outer sex reassigned, and should physicians be allowed to carry out such interventions?

Although sex reassignment surgery has only become possible through modern medicine, the ethical assessment of the entire complex is not at all new. Earlier generations have already tried to find the answers to some of the basic questions, since sex reassignment surgery requires the surgical removal of sexual organs (i.e. sterilization or castration), and this is a problem with a long tradition of religious and ecclesiastical teachings. However, these teachings should always be understood in their historical context.

6. The biblical/Jewish tradition

On the basis of the explicit biblical prohibition, sex-change operations are clearly prohibited. "And that which is mauled or crushed or torn or cut you shall not offer unto the Lord; nor should you do this to your land" (Lev. 22:24). Consequently Jewish law also forbids the sterilization of women and, since sex reassignment surgery necessarily involves sterilization, this intervention is not permissible.[6] This prohibition only affects Jews, however, so an Orthodox Jewish physician would not perform such an intervention. If, however, neither the patient nor the physician is Jewish, then there can be no objection to this intervention from a Jewish point of view (cf. Brody, 1981). Therefore, this prohibition does not possess universal validity, although it fully applies to the members of this religious community.

Another prohibition also of significance for sex reassignment surgery is noted by Rabbi Meir Amsel: the commandment "A woman shall not wear that which pertains to a man, nor shall a man put on a woman's garment" (Deut. 22:5) is not limited to the wearing of apparel associated with the opposite sex. In this sense, sex reassignment surgery violates the very essence of this prohibition (Bleich, 1987, p. 192). David Bleich concludes: "Although Judaism does not sanction the reversal of sex by means of surgery, transsexualism is a disorder which should receive the fullest measure of medical and psychiatric treatment consistent with Halakhah. Transsexuals should be encouraged to undergo treatment to correct endocrine imbalances, where medically indicated, and to seek psychiatric guidance in order to alleviate the grave emotional problems which are frequently associated with this tragic condition" (Bleich, 1987, p. 195).

7. Christian aspects

Since sex reassignment surgery in the case of a man-to-woman transsexual involves castration, it is appropriate to look at the attitudes towards castration in the early period of Christianity. It must however be realized that this question was above all posed with a view to overcoming sexual desire.[7] A classical example here is Father Origen (185-254 A.D.). Most research today comes to the conclusion that he interpreted the biblical words "... and there be eunuchs, which have made themselves eunuchs for the kingdom of heaven's sake" as an encouragement to self-castration and that he actually went that far himself. Origenes considered the biological difference between men and women, and their sexual inclinations, as merely a transitional phase. He believed God would soon bring about a transformation of the bodies which would be so tremendous that all the existing ideas about sexual differences would be thoroughly shaken (Brown, 1988). Against this background, castration was not a serious crime against creation, but solely served to concentrate the spiritual forces. Basically one can assume that castration was not so rare at that time, but that it was also sharply condemned by those who regarded the creation of the sexes and their differences as Divine Providence.

In this sense the first Canon of the Council of Nicea I (325 A.D.) underscored the importance of maintaining one's physical integrity. Surgical intervention for maintaining health is allowed, but castration that destroys the gender identity of the person is seen as a crime against creation. The actual intention serves as the criterion for the permissibility of this type of surgery: if it is intended to destroy the gender identity of the person without any necessity on grounds of health, then the action is not permissible. However, these prohibitions were only directed at members of the clergy: the Nicene fathers were only concerned with castration as a bar to ordination! Nothing however is said of the need to curb lay enthusiasm (Chadwick, 1959, p. 111). In the 23rd of the Apostolic Canons it is laid down that a layman who mutilates himself is to be excommunicated for three years "for he conspires against his own life." Because the body is a 'gift' of the Creator, the argument against suicide is equally valid against self-mutilation. Anyone who rebels against the providential ordering of divine creation is seen as an enemy of God's creation. Man was expected to fight his passions on a spiritual level without resorting to surgery, which deprived him of the chance to prove himself as a virtuous person. There is no merit in chastity if incontinent behavior is rendered a physical impossibility by the knife (Chadwick, 1959, p. 111).

Some centuries later, Canon VIII of the Council of Constantinople (861 A.D.) expressed a similar view:

> The divine and sacred Canon of the Apostles judges those who castrate themselves to be self-murderers; accordingly, if they are priests, it deposes them from office, and if they are not, it excludes them from advancement to holy orders. Hence it makes it plain that he who castrates himself is a self-murderer, he who castrates another man is certainly a murderer. One might even deem such a person quite guilty of insulting creation itself. Wherefore the holy Council has been led to decree that if any bishop, or presbyter, or deacon, be proved guilty of castrating anyone, either with his own hand or by giving orders to anyone else to do so, he shall be subjected to the penalty of deposition from office; but if the offender is a layman, he shall be excommunicated; unless it should so happen that owing to the incidence of some affliction he should be forced to operate upon the sufferer by removing his testicles. For precisely as the first Canon of the Council held in Nicea does not punish those who have been operated upon for a disease, for having the disease, so neither do we condemn priests who order diseased men to be castrated, nor do we blame laymen either, when they perform the operation with their own hands. For we consider this to be a treatment of the disease, but not a

malicious design against the creature or an insult to creation (St. Nicodemus, 1983, p. 163).

This argumentation essentially continues in the teachings of the Catholic Church today: the operative removal of a healthy organ is only permissible if (a) the continued presence or functioning of the particular organ within the whole organism is causing serious damage or constitutes a menace to it; (b) this damage can be lessened by the mutilation in question and (c) one can be reasonably certain that the negative effects of the mutilation will be compensated by the positive effect.

> Human beings have been given this right by God. It is founded on the principle of totality, in virtue of which every particular organ is subordinated to the body as a whole, and, in case of conflict, must cede to the good of the whole. "As a result, man, who has received the use of the whole organism, has the right to sacrifice a particular organ if its continued presence or functioning causes notable harm to the whole, a harm which cannot otherwise be avoided" (Pope Pius XII, 1953, p. 320).

But the appeal to the principle of totality is unjustified in the case where healthy fallopian tubes are removed to prevent any new conception, and the serious danger which could result there from either to the health or the life of the mother. The danger in such a case is not due to the oviducts, nor to the influence they exercise over the diseased organs (kidney, heart, lungs, etc.). The danger only arises when free sexual intercourse causes a conception. Therefore "the operation on the healthy oviducts is morally illicit" (Pope Pius XII, 1953, p. 321).

The teaching of the Catholic Church with respect to the permissibility of sterilization is quite clear: direct sterilization (which aims, either as a means or as an end in itself, to render childbearing impossible) is a grave violation of the moral law, and therefore unlawful by virtue of the natural law (Pope Pius XII, 1951, p. 305). For Catholic hospitals it is absolutely forbidden to carry out sterilizations. The question is whether this also applies to sex reassignment surgery.

From the Catholic standpoint it is also true that human beings do not 'own' their bodies, but 'mind' them and hold them 'in trust'. Changes should only be made if they serve the overall body, its integrity and health. A physician may not, for example, perform an unnecessary amputation just because a patient would like one since, according to this comprehension of the human body, 'self-mutilation' is not permissible.

When this argumentation was applied to the permissibility of sex reassignment surgery, it was found that even the attempt to change a person's sex by surgery should be regarded as morally illicit (cf. Madigan, 1956; in Bier, 1990, p. 483).[9]

Besides mentioning these critical voices it needs to be pointed out that there is no more a uniform Protestant view in these matters than a uniform Catholic one. In contrast to the Catholic view, 'nature' is not a point of orientation for Protestant ethics. Based on the mental suffering of the person involved, however, there is a strong argument in favor of the fundamental permissibility of sex reassignment which can be formulated from a Protestant point of view. Analogous to emancipatory sexual ethics, it is here too a question of people's self-designs and their concepts of a successful life. This prevents a quasi-natural comprehension of sexual normality from stipulating possible courses of action. Concepts of a successful life can only be explored with those actually affected, for it will always be necessary to check that therapeutic interventions do not infringe upon the self-determination of patients. Intervention with drugs or surgery can accordingly be performed wherever they represent help for those affected, freeing them from their mental suffering, and where this is also desired by the person in question.

IV. PRACTICAL CONSEQUENCES OF SEX REASSIGNMENT SURGERY

After examining the question of whether sex reassignment surgery is basically permissible or not, the following ethical discussion primarily focuses on whether these interventions are also permissible with regard to their effects on family and society.[10] With the increasing success of sex reassignment surgery, new applications for this type of treatment are being made, showing that the whole issue of sex reassignment ultimately touches fundamental issues of human identity.

CASE I
TRANSSEXUAL SURGERY AND MARRIAGE

In August 1981, the following case was reported in the American bioethics journal *The Hastings Center Report*, and commented upon by four ethicists from different areas (Judaism, Protestantism, Catholicism and Philosophy of Law):[11]

> *John has been married for eight years and is the father of three children. Even before his marriage he had suffered from identification with the other sex. Increasingly, however, he finds his male role unacceptable, and he now seeks hormone therapy and sex-change surgery to become a female. His wife objects, arguing that she is incapable of managing her family alone and is unwilling to live with him if he undergoes the surgery and therapy. John is unwilling to forego the therapy and surgery. (...)* (Toulmin, 1981a).

One of the questions put to the ethicists concerned the effects the sex reassignment surgery would have on the marital status of John/Joan. If we take marriage to be a contractual agreement between two adults, then this issue poses no problems. According to the Jewish point of view, however, marriage is an agreement between two people of different sexes. Is marriage then automatically annulled during the course of the operation? Does the surgery then necessarily entail one-sided divorce which the wife is powerless to prevent? These questions were completely without precedent and, to aid answering, the following parallels were put forward by some Jewish authors.

In the Jewish tradition there are reports of the prophet Elijah who did not die but went straight to Heaven and became an angel. According to the Jewish view, human beings cannot marry angels. Some pedants have asked, what then happened to Elijah's marriage. Many believe that the marriage was automatically annulled at the moment he became an angel. This could also be applied to the issue of sex reassignment: the marriage is automatically annulled through the sex reassignment surgery (cf. Brody, 1981).

1. Should John/Joan be permitted to remarrry after successful sex reassignment surgery?
The teachings of the Catholic Church do not contain any official opinions on this matter. For the Catholic understanding of marriage it is important that the marriage should constitute a community between a man and a woman (*Codex iuris canonici* 1055 § 1). However, there is no exact definition of the term 'man' and 'woman', and in particular there is no guidance as to the criteria that should be used to determine the sex of an individual.[12]

With regard to the permissibility of marriage after sex reassignment surgery, the transsexual should not be judged any differently that other persons affected by a psychosexual deviation, i.e., there is not necessarily an incapacity to perform sexual intercourse and thus consummate the marriage [*impotentia coeundi*] (Bier, 1990, p. 477).[13] Man-to-woman transsexuals can be regarded as eligible for marriage if there is an artificial vagina permitting sexual intercourse. The ability to reproduce is not laid down in ecclesiastical law as a condition of marriage, so that, for example, sterility will neither preclude a marriage nor render it invalid (provided it does not involve malicious deceit, cf. *Codex iuris canonici* 1084 § 3).

In the case of woman-to-man transsexuals, by contrast, ecclesiastical law states that they are still not eligible for marriage after 'successful' sex reassignment surgery and that they are not permitted to marry a woman according to ecclesiastical law. It is the canonical view that marriage must allow a *copula perfecta* and this demands that the man must be able to ejaculate (cf. Bier, 1990, p. 488). Even if the surgery has created a phalloplasty, ejaculation will not be possible. Consequently, woman-to-man transsexuals are *basically incapable of sexual intercourse* which, under current Catholic ecclesiastical law, precludes marriage of such a person to a woman (*Codex iuris canonici* 1084 §§ 1 and 2).

To summarize the Catholic view of marriage: While John/Joan in the case described above would basically be 'eligible for marriage' after successful sex reassignment surgery, the Catholic Church would not permit a woman-to-man transsexual to remarry.

It would go beyond the scope of this paper to discuss the exact consequences that sex reassignment has for the members of different faiths and religions today. The standpoints cited merely serve to underline the fact that this issue raises important questions about the basic comprehension of marriage. What is marriage? What constitutes marriage? When may a marriage be annulled?

2. Legal regulations to prevent moral conflicts

A further important question is: What level of responsibility does John have towards his children? How can it be measured? With what instruments can the benefits and damages in an individual case be weighed up? How can the interests of children best be taken into account?[14] Could not the marriage continue to exist *de facto* even if it no longer existed *de jure*? According to Anglo-Saxon law, a court would agree to a divorce if Susan desired one. But as long as she chose not to start divorce proceedings, her marriage would continue to exist. The courts would have no way of annulling this marriage. If, however, this were to facilitate marriage between individuals of the same sex, then no argument could be found for the continued prohibition of registry office marriages between other same-sex couples. In order to prevent this from happening in Germany, the transsexuals law, introduced for both humanitarian and social reasons in 1981, only allows unmarried persons to have their sexual status officially changed (official registration of sex from male to female or vice versa).[15] This means that applicants must either be single or must get divorced beforehand.[16] Further prerequisites for an official change of sex (the so-called 'major solution') are that the person concerned:

- can produce two experts' reports to the effect that "the chances of a change of feeling regarding his/her true sex are highly improbable" and that there is no evidence of a temporary disturbance affecting his/her feelings regarding his/her true sex;[17]
- has undergone successful sex reassignment surgery (that a significant move towards the appearance of the 'new' sex has been made);[18]
- is permanently sterile.

By officially changing sex, transsexuals are then subject to the rights and obligations existing in society, e.g.:

- a female-to-male transsexual is eligible for national service;
- a male-to-female transsexual is eligible to become a midwife;
- marriage may be entered into with a person of the 'former sex'.

This so-called 'major solution' is preceded by a 'minor solution' which enables the new sexual role to be tested in a legally recognized manner, without surgical interventions having to be performed or a marriage annulled first. A prerequisite for this 'minor solution', officially involving merely a change in name, is that the person in question has felt compelled to live according to his/her transsexual inclinations for at least three years, and that two independent experts have confirmed that "the chances of a change of feelings regarding his/her true sex are highly improbable."[19]

Whereas a first name may officially be changed without sex reassignment surgery, the official sex of a person (male/female) may only be changed following successful surgical adaptation to the 'new' sex. Practice has shown, however, that not all people who undergo sex reassignment surgery choose to have their official sexual status changed.

In Germany it is therefore possible for a married man to change his name to a female one, undergo sex reassignment surgery and then continue living with his wife 'as a woman'. At first sight this is a marriage between 'two women', yet it is legally possible since no application has been made for an official change of sex: legally it is still a marriage between a man and a woman.

3. Conflicts between the secular and the ecclesiastical interpretations of eligibility for marriage

While the secular law concerning transsexuals – both man-to-woman and woman-to-man transsexuals– allows marriage to a partner of the opposite sex after change of sexual status, it is the Catholic view, as explained above, that marriage eligibility does not exist for fundamental reasons in the case of a woman-to-man transsexual. The civil law recognition of the sex change cannot be simply accepted by the Catholic Church.[20] There are also differences with respect to the criteria used to determine the sex of an individual: civil law is guided by the outer appearance and the individual's inner perception of his/her sexual identity, while the chromosomal sex does not play a decisive role. In ecclesiastical literature, however, the criteria for the determination of the sex have so far not been clarified. Proponents of a 'materialistic' view of human beings, who consider the subjective feeling and the outer sex changes to be decisive, are opposed here by 'biologistic' advocates who believe the genetic sex to be of overriding importance. Neither of these standpoints makes due allowance for the entire human being as an entity composed of body and soul. This can lead one to the conclusion that, after sex reassignment surgery, the unity of body and soul is not given with respect to gender identity, so that a sex-specific division into male and female is precluded! In fact, the postoperative transsexual is neither a man nor a woman, but an "artificially created hermaphrodite" (Bier, 1990, p. 485), who may not remarry according to canon law because a clear sex determination, which is a precondition for marriage between a man and a woman (*Codex iuris canonici* 1055 § 1), is not possible.

However, it is doubtful whether this "theoretically clean and perhaps scientifically justified answer" can be considered satisfactory from a pragmatic point of view (Bier, 1990, p. 485). Neither civil nor ecclesiastical law acknowledges an intermediate stage, i.e. a 'third sex'. The German law concerning transsexuals – similar to the legal regulations in various other countries – wants to enable an unambiguous sex determination, i.e. it aims to create clarity in a situation that is fundamentally unclear. The Roman Catholic ecclesiastical law does not allow for a 'third sex' either. If this continues to be ignored in the future, it will only be possible to accept the civil law decision or to adopt a separate ecclesiastical position using the chromosomal characteristics as a criterion for sex determination. However, the result would be that under civil law a woman-to-man transsexual would be recognized as a man and allowed to marry a woman, whereas under ecclesiastical law this same individual would still be regarded as a woman who would not be able to be married in church, even after change of sexual status, due to the incapacity to perform sexual intercourse! Medicine does not have the means to resolve this anthropological conflict.

CASE 2
ON THE PROBLEM OF CONTINUED IDENTITY AND SOCIAL ACCEPTANCE

The question of the eligibility for marriage is closely linked with the future role in the context of society. Even if the individual feels absolutely sure about his/her identity after successful sex reassignment surgery, it is still not decided how the community will react to the new identity. There is a clear continuity in biographical terms, but at the same time the new, openly displayed life as a person of another sex, confronts the people in the immediate social environment with a 'new person'. The following case demonstrated this problem in a spectacular manner:

Norbert Lindner, who was elected honorary Mayor of the East German village of Quellendorf (in Sachsen-Anhalt) in 1996, announced two years later (1998) at a meeting of the village council that from now on he would be appearing as Ms. Manuela Lindner. A qualified engineer, he had been married for 18 years and was the father of two children. The legally acceptable 'minor solution' (see above), involving a change of his male name to a female one, officially permitted Manuela Lindner to try out life as a woman, with the option of undergoing sex reassignment surgery at a later date. Aware that this would not be easy for the population of a small village to come to terms with, a trial period of eight weeks was laid down. After just two weeks the village council submitted an application for 'her' to be voted out of office (without giving a reason). Ms. Manuela Lindner still wished to stand for election, however, since she believed that previously "a woman looking like a man" had been elected and that "nothing about this person had changed" except the "external façade." She believed the citizens should base their decision on the performance of the Mayor to date. She wished to continue as Manuela Lindner in her office as Mayor(ess) since she had not suddenly become a different person. The key issue, that of the 'true identity' of a person, was put to Ms. Lindner by the journalist Erich Boehme in a German talkshow: "To you, you have remained the same person but for others, through becoming a woman, you have become a different person. Would it not have been fairer to resign of your own accord and then to have restood for election?" Ms. Lindner repeated that the person had not changed, merely the external façade.

Has Ms. Lindner really become a different person? Assuming Norbert Lindner had bought a lottery ticket as a man and had now, after several weeks of living as Manuela Lindner, drawn the first prize, would Manuela Lindner be entitled to claim the prize won by Norbert Lindner? In the case of a (hypothetical) criminal offense committed by a 'man', the responsibility would be borne after appearing as a 'woman': the biographical continuity would be preserved with all the legal consequences. However, it is a question for debate whether there are not certain other aspects – besides the biographical continuity – that should also be considered in the case of a public office, such as the office of mayor. This could apply to the 'change of mind' that some politicians show after their election, for example, by adopting a different political position compared to their pre-election statements or even by changing to a different political party. In this context the interesting question would be: what has the individual citizen elected? A certain (political) position or a person?

CASE 3:

ETHICAL PROBLEMS OF DOCUMENTING IDENTITY

The question of the 'identity' of a person after successful sex reassignment surgery can be particularly difficult for the social environment when the person was already 'on file' before the operation. This is illustrated by the following case:

In the mid-1980s young Mr. P. trained to become a chemical laboratory assistant in a chemicals firm in Gladbeck, Germany. Ten years later (1997), Mr. P. applied for and was granted an official change of sex; Mr. P. became Jacqueline. Not content with having her personal documents altered, she also wanted to have a gap-free documentation of her career to date for future job applications, without any reference to her sex reassignment. This entailed having employers' references and records adapted to her new sex and name. Her former employer in the chemicals firm refused to alter the papers. In his opinion, this would be a violation of his obligation to tell the truth. After all, he did train a young man and not a woman.

The industrial tribunal in Hamm decreed that the previous employer had to alter all relevant employment records and documents written ten years beforehand. The employer's refusal was deemed by the chief magistrate to be a violation of his obligation not to discriminate (Appeals were permitted) (adt, 1998).

The above cases are not discussed further here. The aim was to show that sex reassignment surgery poses questions far beyond the scope of the patient-doctor relationship. Even apparently 'uninvolved' citizens, such as the former employer in the above example, are confronted with the consequences of modern medical interventions and are obliged to consider their attitudes.

V. CONCLUSION
THE GULF BETWEEN THE RIGHT AND THE GOOD LIFE

It can be summarized that sex reassignment surgery as a therapy has developed during the 20th century, while there are still strong reservations or concrete prohibitions existing in various (religious) communities with regard to this type of surgery. On the other hand, positions have emerged that are expressly in favor of sex reassignment surgery, since the patient's suffering is regarded as an illness that should also be treated surgically if this corresponds to the will and objectives of the patient.[21]

This situation is typical of today's multicultural societies in many countries of the Western world. With the onset of the 21st century, there are definite signs of a discrepancy between the citizens' concepts of what is morally permissible and the freedom of action granted by the various governments. One of the most poignant experiences in this context has been the discussion about abortion, which has revealed a deep division in society. In Germany, for example, a law was introduced in the mid-70's, which permitted pregnant women, without fear of prosecution, to have an abortion under certain circumstances and within a fixed period. This was fiercely debated, and the law has since been amended several times. At present, there is no general consensus in sight. However, this case must be regarded as epoch-making insofar as it has manifested the gulf between the right and the good life. Ever since the debate about abortion, it has become obvious that the freedom of action granted by the state can clash with what is considered morally justified within a (religious) community. In view of these separate developments it is not surprising that the German Federal Supreme Court decided in 1971 that surgical interventions for sex reassignment are not immoral, thus smoothing the way for this kind of operation.[22] The second step in connection with the formulation of health policies is always the question of cost: what should the general public (general health insurance) pay, considering that not all the citizens agree to these measures?

In the spring of 1982, for example, a 'man' underwent sex reassignment surgery in Germany. The health insurance company with which he was compulsorily insured refused to cover the costs of the surgery on the grounds that prior to the operation, an "improper physical state" had not existed requiring healing, relief or prevention from deterioration (criteria for refunding the costs). The social welfare tribunal decided later that the health insurance company had to pay for the surgery (an appeal was not permitted). The court based its decision on a concept of disease referring not only to a deviation from the model example of a healthy body or mental state, that is, whether the insured was in a position to perform normal psycho-physical functions, but also to the mental suffering existing for the individual transsexual.[23] In other words, the social welfare tribunal did not attribute disease status to every instance of transsexuality, but called for an exact analysis in each individual case (An., 1988, 1551). An attempt at psychotherapeutic treatment (an 'attempt at changing the person's mind') has to be undertaken first in order to investigate all the possibilities and determine whether only surgery can relieve the suffering. This is a prerequisite for the health insurance companies if they are to carry the costs.

Interestingly enough, Great Britain – where the health service sector has undergone extensive rationalization, which also attracted attention elsewhere in Europe – decided in the summer of 1999 that the National Health Service will pay for the costs of sex reassignment, on average amounting to some US$ 11,000 (afp, 1999).

However, these decisions do not necessarily lead to an overall consensus in society; the measure continues to be the subject of heated debate.[24] At the same time, an organization such as a hospital must ensure that individual members of staff can refuse, for ideological reasons, to take part in controversial surgical interventions. There has to be a consensus of all involved, and therefore this field of work should be especially pointed out to job applicants during the interview. No one who basically disagrees with this type of surgery should be obliged to assist (unless it is a question of saving the patient's life). Furthermore, special hospitals will exclude sex reassignment surgery—as practiced by Catholic hospitals where abortions and sterilizations are not carried out. Since the world is getting smaller and smaller, patients have the opportunity to get sex-reassignment surgery in other parts of their country and the world.

Center for Medical Ethics at the Markus-Hospital
Wilhelm-Epstein-Strasse 2
60431 Frankfurt/M., Germany

NOTES

* I am in deep debt to many who gave generously support to this article. In particular, I am especially grateful to Dr. Georg Bier, Office of the Catholic Bishop, Limburg, Germany; Prof. Dr. med. Michael Sohn, Department of Urology, Markus-Hospital, Frankfurt/M., Germany; and Prof. H.T. Engelhardt, Jr., M.D., Ph.D., Rice University, Houston, Texas. Special thanks to Sarah L. Kirkby (B.A. Hons.) and Christiane Hearne for all their work with the translation.

[1] Cf. *General Law for Prussian States* (1794):

§ 19: If hermaphrodites are born, their parents are to determine the sex in which they shall be raised.

§ 20: It is, however, to be left open to such persons to make their own decision to which sex they want to commit themselves after having reached the age of 18.

§ 21: His/her rights shall thereafter be judged according to that choice.

§ 22: If the rights of a third party depend upon the sex of the supposed hermaphrodite, then the first party may apply for an examination by specialists.

§ 23: The findings of the specialists are decisive, even against the wishes of the hermaphrodite and his parents.

[2] Examples exist in film and literature which make a direct and problematic connection between transsexuality and psychological abnormality. One of the most famous of these is *The Silence of the Lambs*. In this novel by Thomas Harris (1989), an FBI agent tracks down

Jame(s) Gumb, a mentally sick serial killer whose application for sex reassignment surgery has been refused.

[3] In a subsequent German TV hospital series more of an effort was made to highlight the suffering of transsexual patients: Following the refusal of his application for sex reassignment surgery, a man jumps to his death before the eyes of the surgeon.

[4] While transsexual surgery was performed in Europe as early as 1930, sex reassignment surgery has become prevalent in the U.S.A. only since the late 1960's. At the beginning of the 70's, the surgical interventions were initially assessed very optimistically. Later more critical voices were raised, and in 1980 sex reassignment surgery was even temporarily stopped at Johns Hopkins University (Baltimore, U.S.A.), home to the leading specialists in this field (cf. Bier, 1990, p. 476).

[5] Cf. Laszig et. al., 1995 regarding the problem of 'reconciliation with the biological sex' dominating the therapeutic relationship as a clear therapeutic goal.

[6] It can be argued that sterilization is not the envisaged goal of sex reassignment surgery, but a non-intentional side effect of the treatment. It would therefore be conceivable that a Jewish patient could have this intervention performed by a non-Jewish physician.

[7] We are particularly grateful to H.T. Engelhardt, Jr. for his help and comments on the Orthodox Christian view before his book was published (Engelhardt, 2000).

[8] Brown (1988) speaks of a 'routine operation' at that time. However, this could not be decided freely by the physician and the patient. Justinus the Martyr mentions around 150 A.D. that the physicians in Alexandria were only allowed to carry out such operations after obtaining the consent of the regional governor (Justinus Martyr, 1st Apologia c. 29).

[9] This assessment does not eliminate the question as to the actual sex of the person after sex reassignment surgery: an abortion may also be morally quite unacceptable, but afterwards, the woman is clearly no longer an 'expectant mother' (cf. Bier, 1990, p. 483, note 28).

[10] The issue of transsexuality involves not only factual arguments but also emotional aspects, which should be followed up and discussed. Precisely because many issues surrounding transsexuality represent a 'head vs. gut' conflict, any analysis should incorporate both aspects.

[11] It should be mentioned at this juncture that at the time in question, no legal regulation existed for this field in the FRG.

[12] In canon law, the sex to which a person belongs is also of significance in various other respects, for example, for ordination to the priesthood (Codex iuris canonici 1024). Since the Protestant church also permits the ordination of women, the case of a 39-year old man-to-woman transsexual pastor caused less of a problem: the woman was ordained as a female pastor five years after her sex reassignment (dpa, 1999).

[13] Bier (1990, pp. 479f.) quite rightly points out that one first needs to determine the actual sex of the postoperative transsexual before it becomes possible to assess the capacity to perform sexual intercourse. If one takes the view that a man-to-woman transsexual is still a man after sex reassignment surgery, then the assessment of the capacity to perform sexual intercourse will be negative when the penis has been amputated. However, if the same person is considered a woman, then it needs to be clarified whether the vagina created during the operation enables her to perform sexual intercourse.

[14] The special conflicts experienced by teenagers in conjunction with the sex reassignment of a parent are sensitively depicted in the Canadian film Le Sexe des Étoiles (Baillargeon, 1993).

[15] Whether future laws will change this remains to be seen (cf. protests in France about a government proposal to permit same-sex marriages [late January 1999]).

¹⁶ On the basis of legal decisions to permit changes in name and sex in accordance with the transsexuals law, the following statistics for the frequency of transsexuality can be deduced: a total of 1047 cases were granted in the period from 1981 to 1990 (former West German states only). This corresponds to a ratio of 2 transsexuals to every 100,000 adult inhabitants; the ratio of man-to-woman to woman-to-man transsexuals was given as 2.3:1.

¹⁷ Since this form of surgery has very grave consequences, which the surgeon (!) is ultimately responsible for, it is understandable that a thorough and reliable differential diagnosis by experts is required. The surgeon must always be able to reserve the right not to perform an operation. It seems remarkable that the relevant professional societies for sexual medicine did not agree to 'standards for the treatment and appraisal of transsexuals' until 1997 (!) (Banaski, 1998). The surgical situation seems to be comparable to that of pregnancy termination: the physician performing the intervention has to be able to depend upon the verdict of the investigator/counsellor.

¹⁸ The subjective contentment of the transsexual with his/her postoperative condition is recorded in some studies as significantly higher than the evaluation given by medical analysts.

¹⁹ The original age limit in Germany of 25 years for the 'major solution' has been rejected by the Federal Constitutional Court as unconstitutional. And yet experts have been keen to point out the role of 'transitory adolescent crises' in younger adults (Bosinski et al., 1994).

²⁰ In 1978 the Sacrament Congregation, responding to a question by a bishop from the U.S.A., did not permit the marriage of a woman-to-man transsexual. It is interesting in this connection that the Congregation apparently had no problem accepting the civil law recognition of this transsexual as a man (cf. Bier, 1990, pp. 480f.).

²¹ In 1979, the Johns Hopkins Hospital in Baltimore (Maryland), one of the first clinics carrying out sex reassignment surgery in the U.S.A., suddenly announced that all relevant operative measures would be stopped, since they had come to the conclusion that the problem was basically due to a psychological disorder which could not be dealt with very effectively by surgery.

²² In retrospect, legislation reacted to the phenomenon of transsexuality with incomprehension for a very long time. The courts especially were unconvinced by the irreversibility of a feeling of being trapped inside the wrong body. In a remarkable series of court cases in Germany, individuals fought as far as the Federal Constitutional Court for the right to change their names and sex.

²³ According to the verdict of the social welfare tribunal in Baden-Württemberg, transsexuality is to be interpreted by the statutory health insurance companies as a disease: "A disease is also present when the relationship between the mental and physical states of the insured does not correspond to the relationship between the mental and physical states of a healthy person" (in: Koch, 1986, p. 176).

²⁴ The situation of transplantation medicine in Germany shows that new laws do not necessarily change the citizens' moral attitudes toward a particular measure. Although the 1997 transplantation law defines brain death as the criterion for the death of a human being, there are still opponents who argue that the brain-dead person is really in an irreversible process of dying, i.e. that he or she is a dying person, but not yet dead.

STABILIZING OR CHANGING IDENTITY?

REFERENCES

adt (1998). Neues Geschlecht, neue Papiere. *Frankfurter Rundschau, 294, 18(December)*.
afp (1999). Geschlechtsumwandlung ist künftig kostenlos. *Frankfurter Rundschau, 175, 31(July)*.
An (1988). Transsexualität als Krankheit. *Neue Juristische Wochenschrift, 24*, 1550-1551.
Augstein, M.-S. (1992). Zur rechtlichen Situation Transsexueller in der Bundesrepublik Deutschland. In: F. Pfäfflin and A. Junge (Eds.), *Geschlechtsumwandlung. Abhandlungen zur Transsexualität* (pp. 103-111). Stuttgart: Schattauer.
Baillargeon, P. (1993). *Le sexe des étoiles*. Canadian movie.
Banaski, D. (1998). Geschlechtsumwandlung - Die leidige Kostenfrage. *Psycho, 24,II/4*, 84-88.
Bayertz, K. (1994). The concept of moral consensus. Philosophical reflections. In: K. Bayertz (Ed.), *The Concept of Moral Consensus* (pp. 41-57). Dordrecht: Kluwer Academic Publishers.
Becker, S.; Bosinski, H. A. G.; Clement, U.; Eicher, W.; Goerlich, T.; Hartmann, U.; Kockott, G.; Langer, D.; Preuss, W.; Schmidt, G.; Springer, A.; Wille, R. (1998). Standards der Behandlung und Begutachtung von Transsexuellen. *Psycho, 24, II/4*, 89-93. First published (1997) in *Sexuologie, 2 (4)*, 130-138.
Bier, G. (1990). *Psychosexuelle-Abweichungen und Ehenichtigkeit. Eine kirchenrechtliche Untersuchung zur Rechtsprechung der Rota Romana und zur Rechtslage nach dem Codex Iuris Canonici von 1983 im Horizont der zeitgenössischen Sexualwissenschaft*. Forschungen zur Kirchenrechtswissenschaft, Vol. 9. Würzburg: Echter Verlag.
Bleich, J. D. (1987). Transsexual surgery. In: F. Rosner and J. D. Bleich (Eds.), *Jewish Bioethics*, 4th ed. (pp. 191-196). New York: Hebrew Publishing Company. First published (1974) in: *Tradition* (Spring).
Bosinski, H. A. G.; Sohn, M.; Löffler, D.; Wille, R.; Jakse, G. (1994). Aktuelle Aspekte der Begutachtung und Operation Transsexueller. *Deutsches Ärzteblatt, 91(11)*, 726-732.
Brown, P. (1988). *The Body and Society. Men, Women and Sexual Renunciation in Early Christianity*. New York: Columbia University Press.
Brody, B. (1981). Marriage, morality, & sex-change surgery: Four traditions in case ethics. A Jewish perspective. *Hastings Center Report, 11(August)*, 8-9.
Chadwick, H. (1959). *The Sentences of Sextus. A Contribution to the History of Early Christian Ethics*. Cambridge: Cambridge University Press.
Codex iuris canonici / Codex des kanonischen Rechtes (1994). Kevelaer: Butzon and Bercker, 4th edition.
Congregation for the Doctrine of the Faith (1975). Sterilization in Catholic Hospitals (March 13). In: *Vatican Council II*, Vol. 2 (1982) (pp, 454-55). Reprinted in K. D. O'Rourke and P. Boyle (Eds.) (1993). *Medical Ethics: Sources of Catholic Teachings* (pp. 306-307). Washington D.C.: Georgetown University Press.
dpa (1999). Transsexuelle Pastorin beharrt auf Beschäftigung. *Frankfurter Rundschau, 51(March 2)*, 38.
Dreger, A. D. (1998). Ambiguous sex - or ambivalent medicine? Ethical issues in the treatment of intersexuality. *Hastings Center Report, 28(3)*, 24-35.
Elliott, C. (1998). Why can't we go on as three? *Hastings Center Report, 28(3)*, 36-39.
Engelhardt, H. T., Jr. (1994). Consensus: How much can we hope for? In: K. Bayertz (Ed.), *The Concept of Moral Consensus* (pp. 19-40). Dordrecht: Kluwer Academic Publishers.
Engelhardt, H. T., Jr. (2000). *The Foundations of Christian Bioethics*. Swets & Zeitlinger: Lisse.

Frey, C. (1988). Brauchen wir eine neue Sexualethik? *Zeitschrift für Evangelische Ethik, 32(3),* 168-171.

Gesetz über die Änderung der Vornamen und die Feststellung der Geschlechtszugehörigkeit in besonderen Fällen (Transsexuellengesetz - TSG) vom 10. September 1980. *Bundesgesetzblatt,* 1980, Part 1, pp. 1654-1658.

Gilman, S. L. (1999). *Making the Body Beautiful. A Cultural History of Aesthetic Surgery.* Princeton: Princeton University Press.

Harris, T. (1989). *The Silence of the Lambs.* London: Random House.

Hattenhauer, H. (1996). *Allgemeines Landrecht für die Preußischen Staaten von 1794,* Luchterhand: Neuwied, Kriftel, Berlin, 3rd ed.

Healy, E. F. (1956). *Medical Ethics.* Chicago: Loyola University Press.

Hirschauer, S. (1992). Hermaphroditen, Homosexuelle und Geschlechterwechsler - Transsexualität als historisches Projekt. In: F. Pfäfflin and A. Junge (Eds.), *Geschlechtsumwandlung. Abhandlungen zur Transsexualität* (pp. 55-94). Stuttgart: Schattauer.

Holl, M. (1997). *Seele im Spagat. Eine Reise zwischen den Geschlechtern.* Stuttgart: Gatzanis.

Justinus Martyr (1870). Erste Apologie für die Christen. In: Fr. X. Reithmayr (Ed.), *Bibliothek der Kirchenväter,* Vol. 25 (pp. 60-61). Kempten: Koesel.

Kamermans, J. (1992). *Mythos Geschlechtswandel. Transsexualität und Homosexualität.* Hamburg: edition hathor.

Kass, L. R. (1981). Regarding the end of medicine and the pursuit of health. In: A. C. Caplan, H. T. Engelhardt, Jr.; J. J. McCartney (Eds.), *Concepts of Health and Disease* (pp. 3-30). Reading: Addison-Wesley.

Keil, S. (1989). Sexualmedizin. 2. Ethik. In: A. Eser, M. v. Lutterotti, P. Sporken (Eds.), *Lexikon Medizin/ Ethik/ Recht* (pp. 1052-1058). Freiburg: Herder.

Koch, H.-G. (1986). Transsexualismus und Intersexualität: Rechtliche Aspekte. *Medizinrecht, No. 4,* 172176.

Laqueur, T. (1990). *Making Sex. Body and Gender from the Greeks to Freud.* Harvard University Press: Cambridge, MA.

Laszig, P.; Knauss, W.; Clement, U. (1995). Psychotherapeutische Begleitung einer transsexuellen Entwicklung. *Zeitschrift für Sexualforschung, 8,* 24-38.

Levine, S. B.; Brown, G.; Coleman, E.; Cohen-Kettenis, P.; Hage, J. J.; Van Maasdam, J.; Petersen, M.; Pfäfflin, F.; Schaefer, L. C. (1998). The standards of care for gender identity disorders: Revision by committee, Draft Nine B2, June 15. Henry Benjamin International Gender Dysphoria Association.

Lindemann, G. (1992). Zur sozialen Konstruktion der Geschlechtszugehörigkeit. In: F. Pfäfflin and A. Junge (Eds.), *Geschlechtsumwandlung. Abhandlungen zur Transsexualitä* (pp. 95-102). Stuttgart: Schattauer.

Madigan, J. J. (1956). *Intersexuality and its Moral Aspects.* Rome: Diss.

McCormick, R. A. (1981). Marriage, morality, & sex-change surgery: Four traditions in case ethics. A Catholic perspective. *Hastings Center Report, 11(August),* 10-11.

Pfäfflin, F. (1996). Zur Suizidalität bei Transsexualität. In: H. Pohlmeier, H. Schöch and U. Venzlaff (Eds.), *Suizid zwischen Medizin und Recht* (pp. 115-125). Stuttgart: G. Fischer.

Pfäfflin, F.; Junge, A. (Eds.) (1992a). *Geschlechtsumwandlung. Abhandlungen zur Transsexualität.* Stuttgart: Schattauer.

Pfäfflin, F.; Junge, A. (1992b). Nachuntersuchungen nach Geschlechtsumwandlung. Eine kommentierte Literaturübersicht 1961-1991. In: F. Pfäfflin and A. Junge (Eds.),

Geschlechtsumwandlung. Abhandlungen zur Transsexualität (pp. 149-457). Stuttgart: Schattauer.

Pope Pius XII (1951). Fundamental laws governing conjugal relations (Oct. 29). In: *The Human Body: Papal Teaching* (pp. 161-62). Reprinted in K. D. O'Rourke and P. Boyle (Eds.) (1993). *Medical Ethics: Sources of Catholic Teachings* (p. 305). Washington, D.C.: Georgetown University Press.

Pope Pius XII (1953). Removal of a healthy organ (Oct. 8). In: *The Human Body: Papal Teaching* (pp. 277-79). Reprinted in K. D. O'Rourke and P. Boyle (Eds.) (1993). *Medical Ethics: Sources of Catholic Teachings* (pp. 320-321). Washington, D.C.: Georgetown University Press.

Raab, G. (1982). Kirchenrechtliche Probleme bei Transsexuellen. *Österreichisches Archiv für Kirchenrecht, 33*, 436-465,

Smith, D. H. (1981). Marriage, morality, & sex-change surgery: Four traditions in case ethics. A Protestant perspective. *Hastings Center Report, 11(August)*, 11-12.

St. Nicodemus and St. Agapius (1983). *The Rudder of the Orthodox Catholic Church*. Chicago: Orthodox Christian Educational Society.

Tolmein, O. (1999). Intersexuell. *DIE ZEIT magazin, 5, 28, January, 12-15*.

Toulmin, S. (1981a). Marriage, morality, & sex-change surgery: Four traditions in case ethics. *Hastings Center Report, 11(August)*, 8.

Toulmin, S. (1981b). Marriage, morality, & sex-change surgery: Four traditions in case ethics. The Common Law tradition. *Hastings Center Report, 11(August)*, 12-13.

Will, M. R. (1992). ... ein Leiden mit dem Recht. Zur Namens- und Geschlechtsänderung bei transsexuellen Menschen in Europa. In: F. Pfäfflin and A. Junge (Eds.), *Geschlechtsumwandlung. Abhandlungen zur Transsexualität* (pp. 113-147). Stuttgart: Schattauer.

Wille, R. (1989). Sexualmedizin. 1. Medizin. In: A. Eser, M. v. Lutterotti, P. Sporken (Eds.), *Lexikon Medizin/ Ethik/ Recht* (pp. 1045-1052). Freiburg: Herder.

Zhou, J.-N.; Hofman, M. A.; Gooren, L. J. G.; Swaab, D. F. (1995). A sex difference in the human brain and its relation to transsexuality. *Nature, 378*, 68-70.

STEPHEN MAN-HUNG SZE

HOMOSEXUALITY AND THE USE OF REPRODUCTIVE TECHNOLOGY

I. EXPOSITION OF THE MAIN THEME

The term "reproductive technology" used here designates widely the use of donation of sperm and eggs, artificial fertilization, in vitro fertilization, fetus implantation and surrogate motherhood. These are all recent medico-technological developments, by means of which (heterosexual) couples, who were formerly unable to bear children, can give birth to children. If homosexual couples are allowed to form their own families, then they are comparable to sterile heterosexual couples in the sense that both groups have certain weaknesses or deficiencies in giving birth to children. We do not deny sterile (heterosexual) couples the rights to make use of this form of technology. In fact, the invention and perfection of this form of technology aims at serving their needs. To consider whether homosexuals have the same rights in this aspect is an important issue, for society should attend to their urges to have the same rights as heterosexual people categorically. To allow homosexuals the rights to use reproductive technology is a higher level of respect of their rights. For the respect of the rights of homosexuals to live according to their own sexual inclination is far easier to conceptualize than their rights to organize families and to procreate children. The latter considerations necessarily involve more premises that are in need of justification. It is the purpose of this article to plead for this higher level of rights through a series of rational justifications and argumentation.

II. THE ANALYTIC APPROACH

The approach of this article is straightforward and simple. The point of departure is that of adoption. If adoption is morally and legally acceptable, then by analogy, it will be argued that the borrowing of sperm and eggs, as well as the use of reproductive technology to achieve giving birth to children, should also be morally and legally justifiable. The argument will then go further to justify heterosexual couples and

Julia Tao Lai Po-wah (ed.), Cross-Cultural Perspectives on the (Im)Possibility of Global Bioethics, 265–276.

homosexual couples as having the same rights regarding this aspect. (Although it is rationally argued out, it may not be legally admissible in our social reality, and this has political implications in the sense that legal reform and policy change will have to take place.) To arrive at this conclusion, one has to examine at least the following viewpoints:

   a. Why should there be justice as equal treatment of homosexuals and heterosexual sterile couples, where the use of reproductive technology is concerned? This is a question of basic rights.
   b. Will the allowance of homosexuals to utilize reproductive technology lead to undesirable social consequences? Will this lead to the encouragement of homosexuality? Will children of homosexuals be induced to exhibit sexual inclination or behavior that is different from children of heterosexual couples? Will children of homosexuals become subjects of social derision and discrimination? These are questions of social consequences.
   c. Allowing homosexuals to utilize reproductive technology may be a kind of preferential treatment, and not just equal treatment.[1]

III. ANALOGY BETWEEN REPRODUCTIVE TECHNOLOGY AND ADOPTION

Adoption is a generally accepted social practice. It is not immoral to adopt children, when the natural parents agree to the adoption, and the adoptive parents can provide for the welfare of the children. The main concern of adoption is to provide the child of adoption the best possible welfare. Another important criterion is that the child is given up to adoption not to satisfy the pecuniary interest of the natural parents, or else there would be the moral problem of natural parents selling their children irrespective of the rights or happiness of the latter, which is immoral.[2]

Allowing the adoption of children in a society is comparable to allowing the most generous form of utilizing reproductive technology, for the adopted child has neither a genetic nor a gestational relationship to the adoptive parents. Likewise, borrowing sperm and egg, making use of in vitro fertilization and engaging surrogate mother to give birth to a child will lead to the same lack of genetic and gestational relationship between adoptive parents and child.

The immediate question asked by some ethicists is: If a couple is unable to give birth to a child except by means of borrowing sperm and egg, in vitro fertilization and surrogate motherhood, why shouldn't they simply adopt a child![3] This is questioning the rights of this couple in utilizing these forms of technology. A homosexual couple will have to encounter even more objections. I would like to investigate the nature and validity of this kind of query: To what extent is it related to the equal rights of homosexual couples in utilizing these forms of technology, and to what extent does it involve only prudential considerations? Then there is the consideration of whether the arguments against the use of these forms of technology by homosexual couples can hold. Finally, I shall also add a short projection of the use of these forms of technology by homosexual single persons in light of the fact that in some states in the United States, adoption by single parents is allowed.

IV. THE SOCIAL STATUS OF HOMOSEXUALS

Nowadays, I do not think that one still has to argue about whether homosexuals are normal persons, and whether they have the rights to follow their natural inclinations and to lead their own preferred ways of living. However, following one's natural inclinations and the consequent way of life is one thing; having the same rights like any normal individual is another, when such rights embrace the rights to organize families, which imply marriage and having children.[4] Society respects the rights of every individual to get married, organize a family and have children. This is very important because man is a social animal, and the happiness of any individual is considered reliant upon his or her sociability, of which the family system is the cornerstone. Thus it is necessary to emphasize the rights of individuals to organize their families and have children.

Should homosexual individuals be excluded from enjoying the same rights described above, because they are regarded as abnormal and unhappy individuals, due to their "promiscuity, anonymous encounter and humiliation" (Levin, 1997, p. 237)? To plead such an argument is to mistake the cause for the reason. The abnormality and unhappiness of the homosexual relationship can easily be seen as a result as well as a (prejudiced) judgment of the opposing public. It is not fair at all to label homosexuals as promiscuous, and their sexual relationship as anonymous, vacuous and humiliating, before we have granted them the respect and rights to express and organize their relationships openly. Actually, among the rights claimed by the homosexual activists, the rights to a normal familial form of life are very basic. This claim should lead to the promotion of families leading to human happiness. Thus to exclude homosexuals from enjoying the same rights because of the mere factually labeled "unhappy or abnormal" social existence of (some, not all of) them is not justifiable (Levin, 1997).

Finally, if someone can follow my line of argument as far as granting homosexuals the same rights to organize families, there is still the problem of allowing them to use reproductive technology to have children. For to allow them these two kinds of rights would entail certain social consequences that might perhaps be undesirable.

V. THE MEANING AND SIGNIFICANCE OF THE USE OF REPRODUCTIVE TECHNOLOGY BY HOMOSEXUALS

I can think of one line of argument against allowing homosexuals to utilize reproductive technology to beget children. Sterile heterosexual couples should be allowed to make use of this form of technology to compensate for their sterility, because they are potentially endowed to beget children. They themselves are not responsible for their sterility. Homosexuals are categorically different, for they have to accept the fact that their formation of partnership involves two people of the same sex, determined by Nature to be sterile. Allowing them to use reproductive technology would mean the violation of Nature.

The above argument can be analogically expressed like this: Let us suppose that there were three kinds of people on earth: one kind of people born without legs, another kind with weakness in their legs, and the third kind with normal legs, which could be amputated surgically or lost in accidents. Would we allow only the second and third kinds of people to have artificial limbs, because they have the potential to possess normal legs? Can we simply tell the first kind of people that since they are born without legs, they should be happy to put up with reality? How about the fact that mankind does not possess wings, and still tries to fly by means of technological inventions of all kinds? Actually, technology is meant to compensate for all the inadequacies of mankind, and all should be entitled to make good use of or benefit from it. We may even go so far as to claim that since homosexuals necessitate categorically the aid of reproductive technology, this question is highly significant. Furthermore, *to affirm their rights to make use of reproductive technology, society should also consider the legal consequences of rights and duties of homosexual parenthood,* as indicated by Joan Mahoney.

Heretofore, we have found that homosexuals should be *categorically* granted the same rights to happy familial life as other individuals, which also entails the rights to use reproductive technology. The following considerations are only considerations of expediency and consequences, which means, we should respect their rights unless doing so would infringe upon the rights of others.

Why shouldn't heterosexual and homosexual couples simply adopt children instead of making use of this expensive, wasteful and cumbersome technology? Adoption, in comparison, saves many more resources and solves a lot of social problems. I will not go into the question of whether the experimentation on fertilization and implantation of the fertilized eggs, of which the success rate is not very high, entails the murder of a potential life. We should not consider simply that adoption is a means to solving a social problem, whereby the forsaken child can find a happy home. The willingness and the anticipation of happiness of the couple together with the adopted child should be given priority. At least, a liberal position should not deny the needy the rights to use this form of technology, and it is also conceivable that some people may prefer the use of this technology to adoption, for the use of reproductive technology may have certain conceivable advantages. The couple can arrange the best conditions for the fertilization, gestation and birth of the child, and this total participation is a way to express the couple's commitment and concern.

VI. THE ACTUAL CONSEQUENCES OF THE EXISTENCE OF HOMOSEXUAL FAMILIES

Some people object to granting equal familial rights to homosexuals, because they consider homosexuals and their families to be bad social models for the young.[5] It is also doubtful whether homosexual couples are able to establish permanent relationships, comparable to heterosexual couples. Homosexuals and their children, no matter whether the latter are adopted or born by means of reproductive technology, may become subjects of social discrimination or derision, which is a usual phenomenon. These are certainly undesirable consequences of allowing homosexuals to establish families with children, which merit discussion.

The first question to be answered concerns homosexuals as role models. If it is generally more and more recognized and accepted that homosexuality is not a symptom of any disease, but is, rather, one of the possible naturally or socially developed sexual inclinations (I, personally, will not rule out bisexuality as another inclination, apart from heterosexuality), then why should society be troubled by the fear that homosexuals are bad role models? In fact, in a male dominant society, women are constantly discriminated against by the unreasonable social division of labor and power, which is worsened by the double role of social and domestic labor. This type of male supremacy is definitely a worse role model.[6]

A lot of research has confirmed that children brought up in homosexual families have normal development and are socially and psychologically normal. They even demonstrate a keen sense of sexual equality, and are free from the limitations of the traditional social construction of sex roles.[7] If we accept the above research results as indicating that children in homosexual families can have a normal development into adulthood, and so, homosexual parents are not bad social models corrupting their children, then it would be all the more absurd to assume that these families can corrupt the society and its youths.[8]

It is certain that children brought up in homosexual families are sometimes derided or even discriminated against by some members of the society, in which sexual, racial and other kinds of prejudices exist. Homosexual offspring developed by means of reproductive technology can add a further factor for social derision. Should the colored races refrain from having children or members of the misogynic society refrain from giving birth to girls, for fear that these children would be discriminated against? Actually, adopted children also have to face the same kind of social discrimination, but that does not mean we have to abolish the system of adoption to avoid this kind of socially prejudiced discrimination. The question is whether we should avoid affirming basic rights and act in conformity with social prejudices for fear of offending those prejudiced, or should we rather criticize these prejudices, and defend and uphold the rights of those subject to discrimination.

VII. IS THE USE OF REPRODUCTIVE TECHNOLOGY A FORM OF SOCIAL PREFERENCE?

Reproductive technology is certainly very expensive and rather wasteful. As long as there is no political intervention, this technology can still be commercialized and develop into a profitable business. Whoever has more social resources at his or her disposal, can make use of this technology. If we are talking about the possibility of exchanging money for service, then this transaction is not preferential treatment at all. On the contrary, let us take Hong Kong as an example for consideration. The medical community in Hong Kong is still very conservative. Not only are homosexual couples not allowed to use reproductive technology (as a consequence of not acknowledging their equal rights), but also commercial surrogacy is not allowed. This is a far cry from preferential treatment.

I do not think that homosexual couples should be preferentially treated in the sense that we grant them free or subsidized use of reproductive technology, because this form of technology is expensive, and surrogate motherhood does involve a lot of complicated issues of just compensation and cost of service. However, if we are truly convinced that homosexual and heterosexual couples should be treated equally, then society should allow both parties to make use of this form of technology. This actually will have to lead to legal reform, the consequences of which will be defining the rights and duties of homosexuals in marriage, adoption, and the dissolution of marriage, compatible with the existing rights of heterosexual couples. Only this can qualify as a fundamental acknowledgement of equality for homosexuals. In this age of egoism, tendencies to forsake intimate social relationships become a great threat. This threat comes more and more from the heterosexual world. If we truly want to uphold the ideal of having stable social relationships, which are communicative and productive of individual development, emphasizing family values is of primary importance.

The greatest hypocrisy in the general attitude of our society towards homosexuality consists in the stressing of family values as a remedy for the heterosexual world, which is accused of becoming more and more promiscuous and irresponsible, on the one hand; and labeling homosexuals as promiscuous and a threat to family values, and thereby blocking their rights to establish permanent partnerships and families, not to mention having children, on the other. It is high time that we criticized this schizophrenic mentality.

VIII. CONCLUSION

I would like to bring up three more points for further consideration. The first concerns the screening process of adoption and the utilization of reproduction technology. Presently, the process of adoption by familial couples normally involves a certain kind of screening, in which the couple is supposed to be able to demonstrate good means to support the adopted child in order to ensure the child's good development and happiness. On the contrary, there is no social process to screen couples, married or unmarried, when they want to beget children. Actually, if an unmarried woman wants to beget a child, she can also do so without any social control and sanction, and her act is not considered immoral, if the child is given love and good care. I think the most important reason to explain this discrepancy is that in adoption, the child is there and the process of gestation is overcome, which is the most difficult part involving sentiments, emotions and a great sense of intimacy between gestational mother and child. We have reasons to believe that without this involvement and genetic relationship, abuse is more likely to happen in adoption. Another reason is that screening worthy parents is impossible to carry out, for it would imply screening every couple and even single mothers, married or unmarried. The screening of worthy adoptive parents is far easier to conduct. Based on these two considerations, we can take up the question of whether couples utilizing reproductive technology should be screened. It is tempting for us to assume that since the process is relatively easy, it should be carried out. Yet, if the utilization of this form of technology involves a lot of personal commitment and resources, it is superfluous to do any screening, for only couples, homosexual or heterosexual, possessing a lot of social wealth and a lot of concern for having children will start this process. Does this imply that these rich couples are granted more social liberty than the poor? Complaints about this would look similar to complaints about rich people driving large cars and living in big mansions.

The second point concerns the allowance of single-parent adoption in some states of the United States and its analogy to single-parent utilization of reproductive technology. I do not think the homosexual inclination of the applicant for adoption should become a criterion for exclusion, if we follow the above line of argument. Furthermore, equating the nature of adoption with the use of reproductive technology, we should extend the utilization of the latter also to single persons, when we consider the granting of the same rights alone. When the social consequences are also taken into consideration, I think it is less advisable to allow single-parent adoption and consequently the use of reproductive technology by single persons. However, this is only a prudential consideration, because we can easily conceive that two parents are often better than one. Statistically speaking, single-parent families tend also to have more problems than normal families. However, I can also understand that if there are more children devoid of parental care than the number of willing couples to adopt them, single-parent adoption is one possible alternative solution, provided that the single-parent applicant can demonstrate means to achieving happiness for the adopted child. Thus, all single parents will have to undergo stringent screening, before they can adopt any children. Likewise, single persons applying for the use of reproductive technology will have to undergo the same kind of stringent screening.

As a pessimist of personal commitment and an optimist of human communal sociability, I am inclined to think that there should be an alternative solution to the problem of the increase of homosexual and heterosexual single parents, which is actually taking place because of the breaking down of traditional families, and which will certainly become more pressing if single-parent adoption and utilization of reproductive technology are permitted. This alternative solution is to provide a more socially supportive network to single-parent families, and to encourage them to organize communal supportive groupings or organizations among themselves to solve their own problems.

Finally, we come to the third question. We may be alerted by some critics that so far we have been considering the equal rights of the homosexuals. It seems that the rights or entitlements of the unborn are not seriously considered. Should they be given choice? What are their rights that we have to respect? I certainly agree that we should consider very seriously the entitlements and rights of unborn children. The problem is only how this can be put in concrete terms. Is it possible to figure out beforehand what form of life the unborn would find desirable and expect to be prescribed? How do we proceed to consider the rights and entitlements of the not yet born? I think that the best we can do is to see to it that all babies born will be granted equal rights and opportunities, and then on top of that, if possible, a stable family and good material environment which will contribute to their happiness. It is imperative that society provide equal rights and opportunities to the future generation, and society is to blame if it fails to achieve this. To provide the future generation with stable relationships and a good material environment for its happiness is no longer a duty but only an aspiration of society, for this achievement will have to depend upon a lot of contingent factors. The question here remains: How does respect for the rights of homosexuals to use reproductive technology to procreate their offspring contradict the respect for the rights and entitlements of the future generation? Rather, I would conceive this respect for the rights of the homosexuals as part and parcel of the respect for human rights, including the rights of the future generation, namely, not to be discriminated against because of their personal sexual inclination. I fear that the argument underlying the emphasis of choice and rights of the unborn leads to the question of whether the unborn generation would find being born by means of reproductive technology and brought up in a homosexual family desirable. It is similar to the actual case in which many women find life in the present male dominant world quite unbearable. Some might go so far as to say that if they had had the choice, they would have chosen to be born men. Does this mean that we should consider this fact, and then ponder not giving birth to a daughter, for fear of disrespect of her rights and entitlements? Of course not, for the only right measure is to achieve a world of equal rights for all.

I think so far I have made myself very clear that based on equal rights and justice as treatment of equals as equals, homosexuals should be granted rights of adoption and the utilization of reproductive technology. Constraints, if any, are only due to concrete prudential considerations, and I have also demonstrated that a lot of consequential arguments against the allowance of homosexuals having access to this form of technology are rationally untenable.

General Education Center,
The Hong Kong Polytechnic University,
Kowloon, Hong Kong.

NOTES

[1] See Samar, V. J., 1997, p. 65. Samar represents a conservative position. He considers homosexuals bad role models in his fifth argument, and in the sixth he points out that granting the same rights to homosexuals would be open the door for preferential treatment.

[2] Why is giving birth to children for sale immoral? The most important reason is disregard for the happiness of the children, for the commercial transaction is not meant for and cannot guarantee the happiness of the children. This act is a violation of the natural duty of parents to safeguard and provide their children with the greatest possible happiness, and intentionally giving birth to children for sale is to violate this natural duty. Yet, being forced to give up one's child for adoption out of hardships in life is different, because the original intention is for the good of the child, which is frustrated by unforeseeable difficult circumstances.

[3] A very interesting view is that of Joan Mahoney. On the one hand, she realizes that reproductive technology can be very expensive (the cheapest is for lesbians to borrow semen for insemination), so adoption is the most expedient way to have a child in the family. On the other hand, she points to the fact that in cases of adoption or artificial reproduction, when the right of custody is concerned, this right will go to one party alone. There were cases among homosexual families, in which genetic relation and gestation were the criteria for conferring the right of custody. Mahoney considers it necessary to add the criteria of care and best interests of the child. This shows that she is well aware of the reality, in which homosexual families exist, and the rights and duties among the members require specific legal concern or even reform. See Mahoney, J., 1995, pp. 41-42.

[4] See Levin, M., 1997, pp. 233-241. The basic argument of Levin is that homosexuality is abnormal and unhappy. The cause of "homosexual unhappiness is a taste for promiscuity, anonymous encounter and humiliation" (p.237). In the same anthology, Leiser represents the opposite position. He argues that abnormal sexual inclination, that is, inclination that is not heterosexual, should not be regarded as immoral. For him, there is no morally justifiable ground to deduce immorality from homosexual inclination. See Leiser, B. M., 1997, pp. 242-253.

[5] See Levin, M., 1997, pp. 239-240. Levin insists on the need to protect children from the harm of homosexuality through legal legitimization, which will subtly increase the chances of children becoming homosexual. This is comparable, according to the author, with the protection of children from a religious education.

[6] Maggie French mentions that some women, who have freed themselves from the yoke of the unfair domestic division of labor in their former patriarchal family, discover later a better partnership of participation and sharing in this aspect in lesbian families. See French, M., 1992, p. 96.

[7] See Green and Bozett, 1991; Patterson, 1992, 1994, 1995; Tasker and Golombok, 1995. Especially noteworthy is the research of Charlotte J. Patterson, published in 1996, in which she conducted the "Social Competence and Child Behavior," "Self-concept" and "Sex Role" tests upon the children of homosexual families in the San Francisco Bay Area. The results proved that these children were no different from those of 'normal' families.

[8] Pat Romans concludes that lesbian parents are not only good mothers, but they can bring up their children to be "people of the future," for this alternative form of familial structure can patch up the unfavorable traditional family life, and break through the limitations of the traditional male-female roles. See Romans, P., 1992, pp. 105-106.

REFERENCES

Callahan, J. C. (Ed.) (1995). *Reproduction, Ethics and the Law*. Bloomington: Indiana University Press.

Corvino, J. (Ed.) (1997). *Same Sex - Debating the Ethics, Science, and Culture of Homosexuality*. Oxford: Rowman & Littlefield.

French, M. (1992). Love, sexualities, and marriages. In: K. Plummer (Ed.), *Modern Sexualities -- Fragments of Lesbian and Gay Experience* (pp. 87-97). London: Routledge.

Gonsiorek, J. C. & Weinrich, J. D. (Eds.) (1991). *Homosexuality: Research Implications for Public Policy*. Newbury Park: SAGE.

Green, G. D. & Bozett, F. W. (1991). Lesbian mothers and gay fathers. In: J. C. Gonsiorek & J. D. Weinrich (Eds.), *Homosexuality: Research Implications for Public Policy* (pp. 197-214). Newbury Park: SAGE.

Gruen, L. & Panichas, G. E. (Eds.) (1997). *Sex, Morality and the Law*. New York: Routledge.

LaFollette, H. (Ed.) (1997). *Ethics in Practice – An Anthology*. Oxford: Blackwell.

Laird, J. & Green, R. J. (Eds.) (1996). *Lesbians and Gays in Couples and Families*. San Francisco: Jossey-Bass Publishers.

Leiser, B. M. (1997). Homosexuality, morals and the law of nature. In: H. La Follette (Ed.), *Ethicsin Practice -- an Anthology* (pp. 242-253). Oxford: Blackwell.

Levin, M. (1997). Why homosexuality is abnormal? In: H. La Follette (Ed.), *Ethics in Practice -- an Anthology* (pp. 233-241). Oxford: Blackwell.

Mahoney, J. (1995). Adoption as a feminist alternative to reproductive technology. In: J. C. Callahan (Ed.), *Reproduction, Ethics and the Law* (pp. 35-54). Bloomington: Indiana University Press.

Nardi, P. M. & Schneider, B. E. (Eds.) (1998). *Social Perspectives in Lesbian and Gay Studies – A Reader*. London: Routledge.

Patterson, C. J. (1992). Children of lesbian and gay parents. *Child Development, 63*, 1025-1042.

Patterson, C. J. (1994). Children of the lesbian baby boom: Behavioral adjustment, self-concepts, and sex role identity. In: B. Greene & G. M. Herek (Eds.), *Lesbian and Gay Psychology : Theory, Research, and Clinical Applications* (pp. 156-175). Newbury Park: SAGE.

Patterson, C. J. (1995). Lesbian mothers, gay fathers, and their children. In: A. R. D'Augelli & C. J. Patterson (Eds.), *Lesbian, Gay and Bisexual Identities over the Life-Span* (pp. 262-290). New York: OUP.

Patterson, C. J. (1996). Lesbian mother and their children. In: J. Laird & R. J. Green (Eds.), *Lesbians and Gays in Couples and Families* (pp. 420-437). San Francisco: Jossey-Bass Publishers.

Plummer, K. (Ed.) (1992). *Modern Homosexualities -- Fragments of Lesbian and Gay Experience*. London: Routledge.

Romans, Pat (1992). Daring to pretend? Motherhood and lesbianism. In: K. Plummer (Ed.), *Modern Sexualities -- Fragments of Lesbian and Gay Experience* (pp. 98-107). London: Routledge.

Samar, V. J. (1997). A moral justification for gay and lesbian civil rights legislation. In: L. Gruen & G. E. Panichas (Eds.), *Sex, Morality and the Law* (pp. 64-74). London: Routledge.

Stiers, G. A. (1999). *From This Day Forward – Commitment Marriage, and Family in Lesbian and Gay Relationship*. New York: St. Martin's Press.

Tasker, F. & Golombok, S. (1995). Adults raised as children in lesbian families. *American Journal of Orthopsychiatry, 65*, 203-215.

GEORGE KHUSHF

THE DOMAIN OF PARENTAL DISCRETION IN TREATMENT OF NEONATES: BEYOND THE IMPASSE BETWEEN A SANCTITY-OF-LIFE AND QUALITY-OF-LIFE ETHIC*

Treatment decisions for severely compromised neonates raise important questions about what lives should be saved, at what social and economic cost, and whether there are certain lives that are not worth living. Deliberation on these issues can be framed from the perspective of an individual involved in the care (micro-ethics), a religious or cultural community that would provide guidance for such an individual (inter-ethics), or from a broader social or policy perspective (macro-ethics). The resolutions given at these diverse levels can be in tension with one another, and each by itself is incomplete. Values of parents, culture, religion, and society may compete, and each level (micro-, inter- and macro-) may weight these values differently. In modern societies there are also pressures to discount more traditional religious or family values in favor of social or medical values that influence the current configurations of healthcare. Often the tension between these value systems is not even appreciated by the health care workers who make treatment decisions. In the face of these complex realities of medical decision-making, questions need to be asked about who should make decisions and how they should be made. Should treatment decisions on behalf of neonates be made by parents, family, physician, a cultural or religious group, or society? And what ethical frameworks can be used to guide deliberation?

Peter Singer and Helga Kuhse have suggested that there are two competing ethical systems that can be used to answer these questions, a sanctity-of-life ethic that values all human life equally, and a quality-of-life ethic that places value on some characteristic such as rationality or capacity for pleasure, and uses this to assess quality (Singer, 1983; 1993; Kuhse, 1987; 1991). I argue that their contrast misses the true challenge of health care decision-making. The appropriate contrast is between the cultural and religious systems that give purpose and orient human life, on one hand, and the concrete needs of decision-making in health care, with the demand for fine lines and cost-benefit calculations, on the other. A

Julia Tao Lai Po-wah (ed.), Cross-Cultural Perspectives on the (Im)Possibility of Global Bioethics, 277–298.
© 2002 *Kluwer Academic Publishers. Printed in Great Britain.*

sanctity-of-life ethic seeks to account for the depth, integrity and rich texture of human life, while a quality-of-life ethic works with day to day cost-benefit calculations and addresses instrumentalities for proximate ends.[1]

Singer and Kuhse point to an alternative way of interpreting the contrast when they closely associate the sanctity-of-life ethic with the Christian tradition. Implicit in their association is the assumption that the sanctity-of-life position is tied to an outmoded, traditional, mythological worldview, while their quality-of-life ethic is rigorous and scientific. This contrast is developed by them for purely rhetorical reasons (they think tradition and narrative have no claim in rational discourse), but it still has value. Advocates of the sanctity-of-life do often speak of a 'mystery' to life, with its 'alien dignity' endowed by God. Some like Paul Ramsey (1970) are explicit about the importance of the theological mode of discourse, and they situate their deliberation within the context of a particular communal tradition. I further develop this contrast between forms of discourse, and attempt to link the sanctity vs. quality dispute to the broader question of who should decide, and how that decision should be developed. The real challenge is to integrate these considerations, so that broader cultural and religious values can concretely direct health care decision-making. The appropriate integration involves a confluence of micro-, inter- and macro-ethical perspectives.

In this essay I do not provide detailed action guides for determining when treatment for neonates is indicated. Rather, I attempt to bring into view the kinds of considerations that are relevant, so that a reductionistic response can be avoided. Singer and Kuhse advocate a utilitarian system that opts out of genuinely ethical deliberation, and loses the capacity to work through the challenges posed by new medical technologies. They set two legitimate strands of reflection in opposition to one another. I challenge their system of practical reasoning by reinstating the problem of decision-making in its full scope and complexity.

My argument proceeds as follows. First, I consider a common way of formulating the alternatives in Western bioethics; namely, the contrast between a sanctity-of-life and quality-of-life ethic. I suggest that both of these approaches are inappropriate, because neither fully appreciates the full depth of the problematic. After this discussion about ethical theories, I consider concrete attempts to formulate a policy regarding treatment of neonates in the United States. The alternatives in the dispute over the so-called 'Baby Doe Regulations' represent the two insufficient approaches found in the earlier theoretical contrast. A third way is needed. I then consider a pure proceduralist solution, and suggest that such an approach is insufficient in the case of children and other dependent third parties. In the fourth section, I consider those who think that medicine can unilaterally determine what treatments are 'futile', and suggest that that too is insufficient. Each of these diverse approaches is incomplete by itself, because each addresses only one aspect of the broader problem.

Since the decision about treatment of a severely compromised neonate involves the individual child, family, medicine, religious and cultural community, society, and state, any appropriate response must account for the role and responsibility of each of these domains. In the concluding sections of this paper, I consider how such a response can be formulated. I suggest that certain minimal standards of care, referred to as "ordinary care," must be provided. However, considerable latitude for parental decision-making should be sustained. The recommended balance between social policy, medical norms, and parental discretion provides an indication of how health care decisions in modern societies should be addressed.

I. THE SANCTITY VS. QUALITY-OF-LIFE: FRAMING THE PROBLEM

Helga Kuhse, one of the most prominent representatives of the quality-of-life ethic, opens an essay on quality-of-life as a decision-making criterion in perinatology with the following characterization of the options for bioethical reflection:

> There is a school of thought – I want to call it the 'sanctity-of-life view' – which rejects quality-of-life considerations as a criterion for medical decision-making. According to the sanctity-of-life view, all human lives, regardless of their quality or kind, are equally valuable and inviolable. Consistently applied, the sanctity-of-life view would entail that life and death decisions, in the practice of medicine, must not be based on the quality of life in question.
>
> The sanctity-of-life view locates the value of life in human life qua human life, that is, in the continued existence of a human organism. The other approach – the 'quality-of-life' view – locates the value of human life in some valuable characteristic or characteristics, such as

self-consciousness, rationality, the capacity to relate to others, the
ability to experience pleasurable states of consciousness, and so on.

I share ... the view that the 'sanctity-of-life' view is an implausible
ethical doctrine. It has a philosophically unsound basis and,
consistently applied, would require health-care professionals to
prolong the lives of all patients with equal vigor – even if their doing
so would not be in, or would be contrary to, the patient's best interests
(1995, pp. 104-105).

Note how the options are presented in this overview. Either (1)
everything must be done to sustain life, because the value is in the bare
fact of existence, or (2) one locates the value of human life in some
quality such as consciousness or rationality, and thus some lives are of
greater worth than others; further, some lives are not worth living, and
thus it is in their 'best interest' to die (a corollary that Kuhse and others
develop in some detail in their writings). However, are these the only two
alternatives? Is it possible to argue that there is a mystery and
inviolability to human life, such that one can never actively kill, while
simultaneously allowing for limits on treatment and incorporating
consideration of the quality-of-life?

Before addressing this question, I would like to note that Kuhse and
other advocates of the quality-of-life position do not properly indicate
what is at issue for the more prominent representatives of the sanctity-of-
life position. The issue is not simply a question of where one sees the
value of human life residing, whether in the bare fact of the biological
organism or some other quality such as rationality. To the contrary, for
Paul Ramsey (1970), Edward Shils (1968), Daniel Callahan (1970), Leon
Kass (1991), and other prominent representatives of the sanctity-of-life
position, one does not simply have a contrast between vitalism and
personalism.[2] There is a more foundational dispute over whether one can
definitively answer where exactly the value of human life resides. For
advocates of the sanctity-of-life position, it is a question of mystery vs. a
scientific and ethical reductionism, and both sides of the contrast (as
formulated by the quality-of-life advocate; namely, the vitalist and
personalist) would be viewed as a form of the reductionism that is
rejected. The proper contrast, from the perspective of scholars like
Ramsey, also entails a contrast between different disciplinary modes of
reflection.

The ethical issues in peri- and neonatology take place at the
intersection of the biomedical sciences and broader philosophical and

religious reflection on the nature and purpose of human existence; i.e., at the juncture of the sciences and humanities. Philosophical and religious reflection can be framed in a form of discourse that is well suited for the appreciation of the mystery and uniqueness of human life, but often not so well suited for the day to day demands faced by the modern person. To just provide one example, Paul Ricoeur, a French philosopher in the phenomenological tradition, has argued that there are certain aspects of experienced temporality that cannot be speculatively resolved (1984, 1985, 1988). He argues that it is through narrative, rather than in speculative or scientific discourse, that we can come to some understanding with the mystery of humanity; and it is through poetry, metaphor, and narrative that we struggle with the boundaries of emergent and evanescent life. Narrative is thus essential and irreducible.

In this case, when one speaks of the sanctity or mystery of human life, one does not at all refer to some simplistic vitalism or make some affirmation about the illicit character of withholding or withdrawing treatment. Instead, positively, one is making an affirmation about the need to appreciate the importance of narratological modes of discourse and avoid a speculative or scientific reductionism, which assumes that the depth and complexity of human life can be sufficiently grasped in terms of the categories of an instrumental, utilitarian rationality or the biomedical sciences alone. Life, especially human life, is much richer, and it evades the types of quality-of-life criteria that are often advanced as a sufficient basis for discrimination. Further, it is this mystery, signifying a remainder or plenitude that resists all totalizing and categorizing activities, that allows for a value and worth that is equally predicated of all individuals, irrespective of mental or physical capacity. This does not mean that quality-of-life cannot be accounted for in treatment decisions; but it does mean that one cannot move directly from an assessment of quality-of-life to a determination of the value of the life in question. That is the fallacy of those that advocate the quality-of-life ethic.

Summarizing, among the profounder representatives of the sanctity-of-life position there is an appreciation of the deeper philosophical and theological import of the issues associated with personhood and involved in peri- and neonatal decision-making. Further, there is an appreciation of the mystery and depth of human existence, and thus of the need to draw on the full richness of various modes of discourse, especially the poetic, metaphorical, and narratological, when attempting to account for and articulate the meaning and purpose of human life. This, however, is obviously just one side of the problematic, and it is here, in recognizing this one-sidedness, that representatives of the quality-of-life position have some important things to say.

In peri- and neonatology, the point of departure for reflection is the demand and responsibility for concrete action. As a result of the developments in modern medicine, we can now directly alter the conditions of emergent life. We can make probabilistic judgments regarding success, quality-of-life, etc., and then are forced to weigh the benefits and costs of alternative treatment options. In this context, where physician and family must categorize and decide, the language of poetry and narrative does not seem to be of much assistance. That is the problem and paradox of decision-making in that context: (1) we must decide, and (2) in that decision we draw on fundamental insights regarding the nature and purpose of humanity, and (3) the deepest insights are often articulated in poetic and narrative modes of discourse that are irreducible, but (4) such modes of discourse are poorly adapted for providing the types of concrete guidance needed in the current medical context. Thus, for those who do draw on the rich philosophical and religious traditions of reflection, the result is often a paralysis, arising from the incommensurability between the needs of the moment and the resources for addressing them. In the end, *de facto*, the sanctity-of-life advocate often decides not to decide, not to categorize or calculate, and thus, to do everything, because all life is of equal value. What is sometimes characterized as the sanctity-of-life position is actually the result of the failed attempt at appropriately responding to the demands of the current biomedical predicament. One affirms the mystery and depth of human life, and at that expense, flees the realities of the concrete context, and the need to categorize, calculate and decide.

Kuhse and other advocates of the quality-of-life ethic rightly recognize this flight and failure. However, they do so at the expense of an alternative reductionism. After providing a simplistic characterization of the sanctity-of-life ethic and arguing that one cannot make concrete decisions regarding treatment limitations from such a perspective, they go on to develop everything in terms of biomedical and reductionistically utilitarian categories. In doing so, they have not been able to avoid the deeper metaphysical and philosophical issues surrounding the nature and purpose of humanity, or bracket them so they are individually relative. Instead, they bring their own metaphysic, one that is, I would argue, very shallow, missing the richness and depth that is better accounted for in the narrative and poetic modes of literary, philosophical, and religious modes of reflection. Further, in this flight to a purely instrumental rationality, there is a concomitant flight from ethics itself, although the nature of ethics is reconstructed in the image of that rationality. People become pleasure machines, to be turned off when one does not have the requisite capacity to accrue the appropriate hedon units. The result is hubris, which assumes that we can concretely assess which lives are worth living and allows for the direct termination of those that do not satisfy our criteria. Any distinction between killing and letting die, terminating people and terminating treatment is completely obliterated, and all decisions are based on quality-of-life criteria that indicate who should live and who should die. Admittedly, the concomitant algorithms are quite useful; they do enable one to give clear answers to concrete questions. However, minimally, one would like to see a little dose of skepticism regarding the capacity to determine and specify, and there is good reason to question the sufficiency of the solutions provided. Clear but wrong guidance is not preferable to ambiguity and uncertainty.

Instead of solutions, one is thus thrown back on the fundamental problem. How can one simultaneously account for (1) the mystery and depth of human life, and the modes of literary, philosophical, and religious reflection that enable one to come to terms with this, and (2) the concrete needs associated with medical decision-making in peri- and neonatology; needs that simultaneously call for the broader philosophical insights as well as the categorization and calculation for which those insights, given their form, are too often ill suited. The two standard positions, seen in the sanctity-of-life and quality-of-life ethics, each account for but one side, and thus are in different ways reductionistic.

II. A CONCRETE FORMULATION OF THE PROBLEMATIC: THE BABY DOE CONTROVERSY

The alternative ethical positions can be operationalized in neonatology by considering the two sides in what has been termed the 'Baby Doe controversy'. Baby Doe was a Downs Syndrome infant born in Indiana with an esophageal atresia. Although there was some dispute about the likely success of surgical intervention, most believe that the condition was not difficult to treat. However, in consultation with their physician, the parents decided to have treatment withheld and the baby died five days later. It is very likely that the reason for the treatment withholding is that the baby was diagnosed with Downs Syndrome, and thus would not be normal.

In response to this case, the Administration of President Reagan and the United States Congress, after pursuing various options unsuccessfully, amended the statutes regarding child abuse, so that the failure to provide treatment in response to a life-threatening situation would be construed as a form of child abuse. With 'Baby Doe Regulation', treatment decisions in neonatology became more directly regulated than other areas of medicine in the United States, and considerable discretion was taken away from parents and physicians in the case of infants needing treatment.[3]

At the beginning, many physicians interpreted the regulations to mean that virtually all treatment possible was now required and this meant aggressive, high technology intervention. *De facto*, the Baby Doe regulations were regarded as an operationalization of a sanctity-of-life position that excluded the legitimacy of decisions regarding the withholding or withdrawing of treatment, and did not allow for quality-of-life judgments (Kopelman, Irons, & Kopelman, 1988). This seemed unsatisfactory, often requiring what seemed to be cruel, aggressive management of severely compromised neonates, and it took important decision away from parents. However, the alternative seemed equally problematic. Although those favoring a quality-of-life ethic were very comfortable with the decisions that led to the death of Baby Doe (Singer, 1983), many people rightly felt that it was an inappropriate infanticide, involving an assessment that Downs Syndrome babies do not have lives worth living, and playing on a technicality regarding the legitimacy of withholding treatment to realize the illicit end.

Practically, the question thus becomes: is there a way to prevent the particular types of quality-of-life judgments found in the Baby Doe case (i.e., judgments regarding what lives have worth), while simultaneously avoiding the elimination of all discretion regarding treatment decisions that one finds in the early interpretations of the Baby Doe regulation? The broader challenge of mediating a sanctity-of-life and quality-of-life ethic can thus be concretely formulated as an escape from the Baby Doe dilemma. Is there a discretionary space for parental decision-making? Can one operationalize a third way?

III. THE INSUFFICIENCY OF A PURELY PROCEDURAL SOLUTION

In the case of adults we have in liberal societies a readily available means of escaping the dilemma. It is recognized that we live in a pluralistic society, where there are diverse values with very different implications for end-of-life decision-making. For adults, we thus turn to a *procedural solution*, which makes unnecessary the state legislation of one particular substantive view. In informed consent doctrine, and, by extension, in advance directives, we allow each person to determine what is in his or her own *best interest* (Engelhardt, 1996).

However, in the case of an infant, this option is, at best, available in a more restricted way. To the degree that we allow parents and their values priority in the decision-making process, the procedural solution can play a role. But we still must resolve where the limits on such discretion lie, and this will involve a determination of "best interest" that is independent of the parents' values.[4] For example, in the case of Baby Doe, was it appropriate for the parents to withhold treatment because the child had Downs Syndrome? When may a physician override a parents' wish, and when may the state, in the name of the child's interests, place limits on what both physicians and parents may do with the child? As will be clear shortly, I seek to provide parents with considerable discretion in life-and-death decision-making, but not without limit. For example, I do not think infanticide should be allowed. The Baby Doe case, as I understand it, crossed the limit of parental discretion. Thus, independent from the procedural solution and defining its appropriate domain, there must be some substantive account of how the outer bounds are delineated.

Recognizing this important difference between treatment decisions for infants and adults, we can acknowledge a certain wisdom in placing the Baby Doe restrictions under the rubric of child abuse statutes. Although I would want to question the substantive content given to such statutes, arguing that much greater discretion should be available to parents, I do think there are certain limits. And parents have a positive obligation, which does involve sustaining certain basic conditions required for life. In the case of Baby Doe, it was political expediency and the failure of attempts to regulate neonatology by way of civil rights statues (discriminations against the handicapped) that led to the use of the child abuse statutes (Barnett, 1989). However, it was a fortuitous development, since the child abuse statutes allow in principle (although not in fact) considerable discretion on the part of the parents, setting a lower bound on allowable behavior. The lower bound defines what the state will regard as "obligatory" or "ordinary" care, while sustaining for parents the capacity to define what is appropriate beyond the basic minimum. One thus finds the conceptual framework for further development of the Baby Doe regulation. If, on the other hand, the civil rights statutes had been used, then the development of a domain of parental discretion would have been much more difficult.

IV. BEYOND MEDICAL DECISION-MAKING: THE INSUFFICIENCY OF FUTILITY ASSESSMENT

When the Baby Doe regulations were formulated in 1984, the United States was just coming to closure on what may be regarded as the first stage of a broader debate surrounding the limitations on treatment. In the 1970s, popularized by cases such as that of Karen Anne Quinlan, there was an imperative in medicine to do any and everything that could possibly be done. It was thought by many that the withholding or withdrawing of treatment (on adults or infants) was the same as killing, and thus illicit (Amundsen, 1978). Gradually, a consensus emerged that this was a false characterization; that it was indeed appropriate to respect wishes to limit medical services, and this was a part of a broader right to live one's life according to one's own values. Allowing limits on treatment did not imply that one was taking or could actively take a person's life (active euthanasia or assisted suicide). Thus the first stage of the debate was framed by an intrusive medical establishment that wanted to do everything, and by patients and families that wanted to resist this medical paternalism (Childress, 1982 provides a prominent response on this issue).

In the second stage of the debate, beginning in the mid 1980s, there was a reversal. Now physicians sometimes felt that at certain times treatment was no longer appropriate, but there were patients who demanded that any and everything be done. Instead of 'refusal directives' that limit medicine, patients or their surrogates advanced 'request directives' that some physicians thought were unreasonable (this stage is reviewed in Khushf, 2002). Recently, in the U.S., there were some prominent cases in neonatology where this was the case. Thus, for example, in Virginia, there was an anencephalic child, whose mother demanded aggressive treatment, including extensive NICU care.

The rubric of 'futility' or care that is 'not medically indicated' has been developed as a response to such request directives (Engelhardt and Khushf, 1995). Through this concept, physicians now attempt to place limits on what patients can demand of them, and they do so in the name of the integrity of medicine.

When the Baby Doe Regulations were written, they incorporated the language of futility and medical indication. The withholding of medically indicated treatment constituted child abuse, and:

> ... [T]he term "withholding of medically indicated treatment" means the failure to respond to the infant's life-threatening conditions by providing treatment (including appropriate nutrition, hydration, and medication) which, in the treating physician's or physicians' reasonable medical judgement, will be more likely to be effective in ameliorating or correcting all such conditions, except that the term does not include the failure to provide treatment (other than appropriate nutrition, hydration, and medication) to an infant when, in the treating physician's or physicians' reasonable medical judgement, (A) the infant is chronically and irreversibly comatose; (B) the provision of such treatment would (i) merely prolong dying, (ii) not be effective in ameliorating or correcting all of the infant's life-threatening conditions, or (iii) otherwise would be *futile* in terms of the survival of the infant; or (C) the provision of such treatment would be *virtually futile* in terms of the survival of the infant and the treatment itself under such circumstances would be inhumane (cited in Kopelman, Irons, & Kopelment, 1988, p. 67; my emphasis).

Note that at that time, futility was not yet a technical term. However, as a result of the subsequent development of the debate regarding treatment decisions, the exceptions built into those regulations are now often interpreted in the light of current literature on futility.

If the Baby Doe regulations themselves can be closely aligned with a sanctity-of-life ethic that does not allow for quality-of-life assessments, the use of futility assessment can be regarded as the reintroduction of just such quality-of-life judgments. Futility is simply the inverse of utility; it signifies those treatments that yield no or only marginal benefit. The key question is then how one determines what counts as a benefit, and how marginal a benefit must be to count as 'virtually futile'.

In the literature on futility, one can identify several distinct responses to that question (these are surveyed in Engelhardt and Khushf, 1995):

(1) Probablistic – that which has a small likelihood of succeeding (ranging from less than 20% to 0%);

(2) Quantitative – when the benefit obtained will have a short duration, or when the life itself can not be extended beyond a limited time period; e.g., not to live longer than one week as a result of a given treatment intervention;

(3) Qualitative – when the quality-of-life obtained falls below a certain threshold, for example when an individual is severely retarded or handicapped, or in a vegetative state;

(4) Physiologic (pure) – when a given intervention cannot realize the end for which the intervention is taking place;

(5) Nonmedical instrumentality – when an instrumentality (perhaps associated with alternative medicine) is unproven or irrational by the standards of modern allopathic medicine, and thus requires that one work within a framework that is not sanctioned by the profession;

(6) Wasteful – care that has a cost (financial and social) that far outweighs the benefit.

It is argued that futile treatments can be resisted in the name of the integrity of medicine – an integrity that can be violated in one of two ways. First, there can be a violation of the means or instrumentality of medicine, by requiring a nonscientific or unproven means to realize an end that is generally regarded as legitimate. Second, there can be a violation of the end of medicine, by requiring the use of a proven means for the realization of an inappropriate end. An example of the former involves the use of laetrile for cancer or CPR when it is clear that it will not work in resuscitating the patient (so called "physiological futility"; see Halevy, Neal, & Brody, 1996 for an assessment of these). An example of the latter involves the use of drugs for assisting in suicide, or, for many, the use of an ICU to sustain a person in a persistent vegetative state.

In a recent publication on futility, Nancy Jecker and Roberta Pagon (1995) sought to develop futility concretely as a basis for withholding and withdrawing treatment in peri- and neonatology, primarily through the use of probablistic futility (called quantitative futility) and qualitative futility. These roughly address the means and ends of medicine, since the former considers those interventions that would not be effective in bringing about their end, while the latter focuses on treatment that violates the ends of medicine by sustaining life that has a quality that "falls clearly below a minimally decent level," as in the case of one in a vegetative state.

Norman Fost (1995), in a rejoinder to Jecker and Pagon, points to the difficulty in using such assessments as a basis for providing physicians with unilateral decision-making authority. He argues, and I think rightly so, that such assessments cannot be made on the basis of medical science alone; to the contrary, they involve variable value assessments and cost-benefit calculations, which are not independent from deeper commitments regarding the nature and purpose of human life. For example, Jecker and Pagon argue that if a treatment has a less than 1 in 100 chance of succeeding, then it is quantitatively futile. But where do they get that number? Why not 1 in 200? In the case of some types of screening (PKU screening) a chance of 1 in 10,000 is considered worthwhile, while in the case of other, extremely expensive treatments, a chance of 1 in 50 may not be high enough. There are thus cost-benefit calculations involved in any choice of a probability, and one should be explicit about this.

Fost is even more critical of qualitative futility, since it involves judgments about which medicine cannot claim any particular privilege. What scientific basis would enable a physician to say that a person in a vegetative state should not live, and it is therefore inhumane to treat such a person? Fost thus concludes by quoting Troug, et al. that "the rapid advance of the language of futility into the jargon of bioethics should be followed by an equally rapid retreat" (cited in Fost, 1995, p. 80). Following Kass, he also notes that "technology cannot do moral work." Similarly, the ethical issues associated with treatment of premature infants cannot be resolved simply by appealing to medicine.

There is also a broader problem associated with the use of futility assessment in neonatology (at least in the United States). Futility is largely developed as a basis for resisting patients' request directives; the current literature on the topic thus implies that a patient wants something done that the physician thinks is inappropriate. However, in the case of neonatology, it is not just the parent that wants something done. The state, in the name of protecting children against abuse, requires that treatment be provided, and 'futility' or 'virtual futility' is taken as the exception, delineating when the state's general demand may be resisted. The implication is that discretion is taken away from parents completely; either the state requires treatment or the physician can resist the treatment in the name of medical futility. The only exceptions to treatment are thus those where a physician can unilaterally decide.

Recognizing that 'medical indication' and 'futility' involve complex assessments that include variable value commitments, it is not at all clear why physicians should have the final word in making such determinations. These issues are better addressed procedurally by a process of negotiating values differences, rather than by a substantive, unilateral resolution by physicians (Halevy & Brody, 1996).

V. CONDITIONS FOR ESCAPING THE BABY DOE DILLEMMA: THE ORDINARY/EXTRAORDINARY CARE DISTINCTION

The general problems associated with limiting medical interventions are not new, although they are raised with a special force by modern science and technology. Already in the sixteenth century, in the wake of the birth of modern, empirical medicine, the Roman Catholic tradition addressed the basic problem by way of its distinction between ordinary and extraordinary care, and a related distinction between proportionate and disproportionate care (Wildes, 1995 provides a nice review of the meaning and usefulness of this distinction).

The question addressed by these distinctions was raised as follows. If you are offered a medical intervention, are you obligated to accept it? When, if ever, can treatment be declined? The answer was: you are only obligated to accept ordinary care; extraordinary care can be rejected. The very use of the distinction thus acknowledges that one is not obligated to sustain life under any and all circumstances. Some types of treatment can be declined. The distinction between proportionate and disproportionate care was then used to address the types of treatment that were appropriate, given the needs and resources of the individual in question. These deliberations were thus context dependent and required a prudential reasoning that was necessarily situational.[5]

Presupposed in this discussion was a positive obligation to preserve one's life. However, unlike the prohibition against taking one's life, which is absolute, the obligation to preserve life is limited in several ways, constrained by one's ability and other obligations. The distinction between ordinary and extraordinary care is used to delineate the limits on one's positive obligation.

In the Catholic tradition it was assumed that there is not an absolute norm that determines ordinary care for all in the same way. Minimally, this obligation was dependent upon three things that varied: (1) the resources of the individual and society; those who were of more limited means would not be required to expend as much as those of greater means; (2) the psychological and spiritual abilities of the individual; for example, those who could endure less pain or had greater fear were not required to undergo painful treatments that might be obligatory for others; and (3) the other obligations of the individual; for example, one may choose to decline expensive treatment, in order to preserve an inheritance for one's family. One may also advance more important spiritual values at the expense of health, which is a relative end, not the *summum bonum*. In the case of each individual, the line between ordinary and extraordinary care thus depends on a careful consideration of all the factors pertaining to that particular case.

Historically, the ordinary/extraordinary care distinction has been used almost exclusively to address micro-ethical issues associated with a particular individual's obligation to get medical treatment. However, there have been some recent attempts to use this distinction in a broader way to address macro-ethical problems in health policy (Wildes, 1995, pp. 114-116). When used in this way, the basic benefits package consists of 'ordinary care', and it should be provided to all people. The care that is not part of the basic package is extraordinary. Such care may constitute a second tier of benefits for those who have the resources to obtain them on their own, but a state is not obligated to provide those treatments.

There is some value to extending the traditional discussion in this way. Some of the same considerations that play a role in individual decision-making regarding treatment also play a role in social decision-making about health policy. As in the case of an individual, there will be no clear line for all societies, delineating the benefits package that constitutes ordinary care. The resultant line will depend on diverse factors, such as the social resources available and other social obligations, such as those related to education or the environment. Under some socio-economic conditions (e.g., in great poverty), no treatment may be available for a child such as Baby Doe. What is ordinary and obligatory in one context may be extraordinary and nonobligatory in another.[6]

An important difference exists between micro-ethical and macro-ethical uses of this distinction. In the former case, 'ordinary treatment' is what an individual is obligated to obtain. In the latter, it is what a state is obligated to provide. But an individual is not obligated to obtain whatever the state is obligated to provide. Thus, individuals may find it fully appropriate to reject treatments that are a part of the state's benefits package, regarding such 'ordinary treatment' (from the state's perspective) as 'extraordinary'.

In modern, western states, it is generally recognized that individuals should have considerable discretion in making treatment decisions, because of the need to assess the merit of treatment in the light of patient values and context. Patients are the ones in the best position to make such assessments for themselves. Thus, from the perspective of the state, there is a minimal obligation of the patient to obtain treatment. It is assumed that patients will act in their own interest, and obtain the individually obligatory care from the benefits package that the state makes available.

In the case of dependent third parties, however, the case is more difficult. Parents are supposed to act on behalf of the interests of children, who cannot speak for themselves. Yet, as we saw in the case of Baby Doe, parents sometimes act against those interests, sometimes for reasons that may be inappropriate, reasons such as the disvaluing of the life of a Downs Syndrome child. In order to guard against such abuse of parental authority, the state has an interest in establishing a higher threshold of ordinary care. But this does not mean that parents should be required to obtain all care that is offered as a part of a state's benefits package. We must thus find a way to open greater space for parental discretion, and this means we must find an alternative to the impasse in the Baby Doe dilemma.

VI. DEVELOPING AN APPROPRIATE RESPONSE

Bioethicists have sought simple and unambiguous answers. However, it is clear that treatment decisions for severely compromised neonates involve multiple levels, and cannot be resolved by a simple ethical principle. While the sanctity-of-life should be respected, it alone cannot answer all questions about life-sustaining treatment. Similarly, quality-of-life considerations are important, but they must be situated in a broader context of meaning. Pure procedural solutions will not work, because there is an additional consideration about the limit of parental discretion, and this involves a substantive resolution. We also cannot simply turn to medical judgments (e.g., regarding futility), because there are important values issues that go beyond the competence and prerogative of medicine. While each of these approaches involves relevant considerations, the problem is too complex to be addressed by any of them alone.

I cannot here consider in detail the elements of an adequate response. That must be the next step. However, I would like to close by briefly pointing toward what I believe is required.

First, decisions about treatment for severely compromised neonates involve fundamental assumptions about the meaning and purpose of human life, and thus must be situated within the context of those religious, philosophical, and cultural frameworks that provide our basic understanding about life. When parents are involved in making decisions on behalf of their children, they do not just want the abstract policy resolutions that have been developed by most Western bioethicists. Life-sustaining decisions are particularist, and the rich interpretive context that informs such decisions should be appreciated.

While this claim may seem relatively obvious, it is often unappreciated in those concrete circumstances in which decisions must be made. Although the field of bioethics in the West was initially informed by theological reflection, it moved in a philosophical and policy direction, seeking those resolutions that are universally accessible and independent of religious or otherwise particularist commitments. While there is some legitimacy to this movement when considering the justification of state policy, it is inappropriate for providing concrete guidance to the parents who are making the decisions. For them, the decisions should be formed by the richer religious and cultural commitments that inform their lives. Unfortunately, bioethicists have not provided much direction for those concrete decisions, because the particularist reasoning associated with these decisions has been marginalized in the field. Instead, a minimalist, analytically based reasoning is put forth by many as an answer, not just to the general policy questions, but also to the needs associated with concrete decision-making. This insufficient response is especially apparent among those who advocate a quality-of-life ethic.

The challenge to different religious and cultural communities is thus to think through bioethical issues in the context of the full commitments of that community. A middle space is needed between the micro-ethic of individual-individual relations, on one hand, and the macro-ethic of social obligations, on the other. Communities at the middle, inter-ethical level should resist the temptation to focus on the macro-policy level. They should formulate the norms that can guide communal members faced with the need to make decisions in the context of modern health care systems (Khushf, 2001). As noted at the beginning of this presentation, treatment decisions take place at the intersection of the practical domain of medicine, on one hand, and the rich interpretive frameworks of purpose and meaning, on the other. Without concrete guidance from communities, those faced with these decisions tend to either emphasize traditional answers and fail to be responsive to the demands of the situation, or be guided by the demands, and fall back on a technologically informed instrumental rationality such as that found in quality-of-life ethics, thus losing the depth and richness associated with their religious or cultural community.

Preservation of a discretionary space for parents in making treatment decisions for their children involves a recognition of the particularist character of those decisions, and thus of the need to avoid a medical or social reductionism. Parental discretion is a direct correlate of individual and religious liberty.

It should also be recognized, however, that such liberty is not absolute. The state has a legitimate interest in preventing certain kinds of harms. Thus, for example, a parent should not be able to withhold care simply because a child is female and a male was desired. In reference to the Baby Doe case, parents should not be allowed to decline a clearly indicated surgical intervention simply because they do not want a child with Downs Syndrome.

The difficult problem concerns the determination of the concrete limits and obligations that the state enforces with respect to parents. Western and 'global' bioethicists have focused upon this problem. Bioethics has been developed in terms of biopolitics.

While I cannot address the various resolutions here, it is worthwhile to note there are two extremes. On one side, physicians and the state unilaterally decide, and parental discretion is eliminated. This is more or less the situation in the United States.[7] On the other side, there are no limits placed on parental discretion. Both of these are insufficient as a response. What is needed is a proper balance, which prevents inappropriate kinds of quality-of-life judgments while preserving the liberty to configure life in the context of robust communities.

In the end, the lines drawn will partly depend on pragmatic considerations, and they will reflect the outcome of contingent constellations of social values and influence. This is not wrong, as long as the *de facto* resolution is responsive to the complexity of the issue and avoids certain kinds of morally illicit actions.

Medicine and the medical professional will likely play a role in determining the kinds of obligations parents have. However, it should be recognized that medical decision-making is not value free. Today we are tempted to find 'scientific' resolutions to all problems, and thus construct ethical disputes as if they were scientific controversies. This leads to an inappropriate medicalization of life. While physician assessment is relevant, it should be situated in a broader context. When medicine is given a decisive role, we should at least recognize that certain values are being privileged at the expense of others.

Finally, it should be recognized that there will be complex relations between the particularist bioethical resolutions and the biopolitical ones. Often, they will be linked by the traditions of a majority or influential minority. It will be important to consider how these traditional solutions are revised and developed, so they are responsive to the new challenges posed by modern science and technology.

In sum: let each domain – individual, communal and social – have its place. The kind of prudential balancing and deliberation I advance may seem obvious, but the desire for clear and simple solutions can tempt us toward a reductionistic response. In the face of the multiple reductionisms that inhabit current bioethical discourse, I seek to bring into view the complexity of the problem. Only when this is done can we move toward an authentic solution.

Department of Philosophy and Center for Bioethics
University of South Carolina
Columbia, South Carolina, USA

NOTES

* Thanks to Renzong Qui for helpful comments on an earlier draft of this essay.

1 There are two prominent ways of interpreting the contrast between the sanctity-of-life and quality-of-life ethic. The contrast can be seen as a variant of the one between a deontological and utilitarian ethic (Reich, 1978). Deontic constraints on action (e.g., regarding the taking of life, treating a person as means, and violating autonomy) are associated with the sanctity-of-life ethic, while the maximization of some hedon unit (e.g., satisfaction of preferences) or QUALYs are associated with the quality-of-life ethic. Grisez and Boyle (1979) provide a strong philosophical formulation of the sanctity-of-life position, qua deontological ethic.

2 Joseph Fletcher was the first to develop the contrast in this way, and Singer and Kuhse follow him directly. Fletcher formulated his position as a response to Bonnell (1951), who drew on the sanctity-of-life as a Christian principle that opposes euthanasia. Fletcher (1951) associated Bonnell's position with a "vitalism," and argued that it was not Christian, but rather closer to an Eastern philosophy. Fletcher opposed the affirmation of "mere life" with his principle of "personality." His 1951 essay was later incorporated into his influential

Morals and Medicine (1954), and from there made it into the mainstream of bioethical reflection.

³ As documented in Kopelman, Irons, & Kopelman (1988), many physicians interpreted Baby Doe regulations as a major intrusion into clinical decision-making, and they altered their care accordingly, sometimes providing treatment that they felt would not be in the child's best interest. This interpretation of the regulations is disputed; for example, Barnett (1989) argues that the regulations actually have little impact on the care a physician can provide. For the purposes of our discussion, it is sufficient to focus on the intent of the Reagan administration in formulating the regulations and on the interpretation of them by most physicians.

⁴ The issues are well surveyed in Kopelman and Moskop (1989). Several essays in this volume also raise additional questions about how other familial values (e.g., cost, burden on family, etc.) are to be weighted together with issues of best interest. These are important concerns, but they do not alter the basic argument that I make in this section, namely, that at some stage, there will be a limit placed on parental discretion, and this will require a move beyond parentally relative determinations of best interest.

⁵ Here I disagree with Childress (1982, p. 166) who argues that "[t]he language of ordinary and extraordinary means should be discarded because of its imprecision and its tendency to support strong paternalism." Childress does not sufficiently address the role this distinction can play in making decisions for children, and his analysis does not acknowledge the role of prudence, proportionality, and the contextual character of deliberation found in the original distinction (the reason for its "imprecision"). His criticism of Ramsey (pp. 164-165) does not sufficiently consider the problem of integrating deontic constraints with the "ratio of benefits and burdens" he advocates for persons. As a deontologist, Childress associates deontic constraints with respect for persons. Ramsey introduces additional concerns that have a bearing on how people can relate to themselves. Similar problems are found in the criticisms made by Beauchamp and Childress (1990) against the ordinary/extraordinary care distinction.

⁶ When I presented a draft of this essay in Hong Kong, Professor Renzong Qiu, who was the commentator, rightly observed that for many countries the Baby Doe controversy does not even exist, since there are not enough resources for the most basic care. In those countries, the failure to provide care for an infant with immediate need of medical care would not imply anything illicit on the part of parents or those who are stewards of the resources for care. He asked whether the whole problem I was addressing (along with its solution) was of interest to a person in a context of poverty. How would the altered context influence my analysis?

In response, I would suggest that even minimal care can be extraordinary in some cases, and that parents still have positive obligations in cases of extreme poverty. Further, I used the Baby Doe case to illustrate the themes involved in making decisions about care, and would fully accept the claim that what is illicit in one case (e.g., in the United States regarding care for Downs Syndrome children) may be fully licit in another case (e.g., where there are no resources for the basic operation needed). Further, the core issues and problems of balancing between traditional systems of thought and the demands of modern health care are present within Eastern as well as Western cultural systems; consider, for example, the problem of futile care addressed from a Confucian perspective (Hui, 1999). Hui's analysis involves many of the same concerns I address in my essay, and I think he would have similar reservations about the quality of life ethic that I criticize. While I believe the problems and central categories of my analysis are cross-culturally relevant, I also think that the emphasis of ethical analysis completely changes when one moves to contexts of extreme poverty. Those who do not even have their basic needs met are not going to be worried about whether it is appropriate to decline a surgical intervention that is medically indicated. To this extent, I accept Renzong Qiu's criticisms about the limited relevance of the arguments.

[7] This is at least how many physicians perceive the situation in the United States (Kopelman, Irons, & Kopelman, 1988). Barnett (1989) gives good reasons to believe that this perception is inappropriate. The earlier context of decision-making for neonates is seen in Duff and Campbell (1973).

REFERENCES

Amundsen, D. (1978). The physician's obligation to prolong life: a medical duty without classical roots. *Hastings Center Report, (August)*, 23-30.

Barnett, T. (1989). Baby Doe: Nothing to fear but fear itself. *Journal of Perinatology, 10(3)*, 307-311.

Beauchamp, T & Childress, J. (1990). *Principles of Biomedical Ethics, 3rd Ed.* New York: Oxford University Press.

Bonnell, J. (1951). Sanctity of life. *Theology Today, 9 (May-June)*, 194-201.

Brody, B. (1988). *Life and Death Decision Making.* Oxford: Oxford University Press.

Callahan, D. (1970). *Abortion: Law, Choice, and Morality.* New York: The Macmillan Co.

Childress, J. F. (1982). *Who Should Decide? Paternalism and Health Care.* Oxford: Oxford University Press.

Duff, R. & Campbell, A. G. M. (1973). Moral and ethical dilemmas in the special-care nursery. *New England Journal of Medicine, 289(25)*, 890-894.

Engelhardt, H. T., Jr. (1996). *Foundations of Bioethics, 2nd Ed.* Oxford: Oxford University Press.

Engelhardt, H. T., Jr. & Khushf, G. (1995). Futile care for the critically ill patient. *Current Opinion in Critical Care, 1*, 329-333.

Fletcher, J. (1951). Our right to die. *Theology Today, 8 (May-July)*, 202-212.

Fletcher, J. (1954). *Morals and Medicine.* Princeton: Princeton University Press.

Fost, N. C. (1996). Futile treatment II: Decision-making in the context of probability and uncertainty. In: A. Goldworth, et al. (Eds.), *Ethics and Perinatology* (pp. 70-81). Oxford: Oxford University Press.

Grisez, G. & Boyle, J. (1979). *Life and Death with Liberty and Justice: A Contribution to the Euthanasia Debate.* South Bend: University of Notre Dame Press.

Halevy, A. & Brody, B. (1996). A multi-institutional collaborative policy on medical futility. *Journal of the American Medical Association, 276(7)*, 571-574.

Halevy, A., Neal, R., & Brody, B. (1996). The low frequency of futility in an adult intensive care unit setting. *Archives of Internal Medicine, 156 (Jan 8)*, 100-104.

Hui, E. (1999). A Confucian ethic of medical futility. In: R. Fan (Ed.), *Confucian Bioethics* (pp. 127-163). Dordrecht: Kluwer Academic Publishers.

Jecker, N. S. & Pagon, R. A. (1995). Futile treatment I: Decision-making in the context of probability and uncertainty. In: A. Goldworth, et al. (Eds.), *Ethics and Perinatology* (pp. 48-69). Oxford: Oxford University Press.

Kass, L. (1991). Death with dignity and the sanctity of life. In: B. Kogan (Ed.), *A Time to be Born and a Time to Die: The Ethics of Choice* (pp. 117-145). Hawthorne: Aldine De Gruyter.

298 GEORGE KHUSHF

Khushf, G. (2002). Beyond the question of limits. In: H. T. Engelhardt, Jr. & M. Cherry (Eds.), *Moral Issues in Limiting Access to Medical Treatment*. Washington, D.C.: Georgetown University Press.

Kopelman, L., Irons, T., & Kopelman, A. (1988). Neonatologists judge the "Baby Doe" regulations. *New England Journal of Medicine, 318 (11)*, 677-683

Kopelman, L. & Moskop, J. (1989). *Children and Health Care: Moral and Social Issues*. Dordrecht: Kluwer Academic Publishers.

Kuhse, H. (1987). *The Sanctity-of-Life Doctrine in Medicine: A Critique*. New York: Oxford University Press.

Kuhse, H. (1991). Severely disabled infants: sanctity of life or quality of life? *Baillieres Clinical Obstetrics and Gynecology, 5(3)*, 743-759.

Kuhse, H. (1995). Quality of life as a decision-making criterion II. In: A. Goldworth, et al. (Eds.), *Ethics and Perinatology* (pp. 105-119). Oxford: Oxford University Press.

Ramsey, P. (1970). *The Patient as Person*. New Haven: Yale University Press.

Ricoeur, P. (1984, 1985, 1988). *Time and Narrative, vols. 1-3*, K. McLaughlin & D. Pellauer (Trans.). Chicago: University of Chicago Press.

Reich, W. (1978). Quality of life. In: W. Reich (Ed.), *Encyclopedia of Bioethics, vol 2.* (pp. 829-840). New York: Free Press.

Schneiderman, L. J. & Jecker, N. S. (1995). *Wrong Medicine: Doctors, Patients, and Futile Treatment*. Baltimore: The Johns Hopkins University Press.

Shils, E. (Ed.) (1968). *Life or Death: Ethics and Options*. Portland: Reed College Press.

Singer, P. (1983). Sanctity of life or quality of life? *Pediatrics, 72 (1)*, 128-129.

Singer, P. (1993). *Practical Ethics, 2nd Edition*. Cambridge: Cambridge University Press.

Wildes, K. (1995). Conserving life and conserving means: Lead us not into temptation. In: K. Wildes (Ed.), *Critical Choices and Critical Care* (pp. 105-118). Dordrecht: Kluwer Academic Publishers.

DERRICK K. S. AU

ETHICS AND NARRATIVE IN
EVIDENCE-BASED MEDICINE

I. INTRODUCTION

The health care landscape in the 21st century will be busily shaped by new technology and novel interventions, such as functional neuroimaging, gene therapy, and stereotactic surgery. It will be just as importantly shaped by new paradigms and concepts. A new paradigm of scientific medicine born in the last decade of this century holds promise for medical professionals and health care managers alike. Evidence-based medicine (EBM), as it has come to be called, is fast becoming a unifying concept in health care. Evidence-based practice, evidence-based guidelines and protocols, evidence-based health care management, and evidence-based purchasing have all been strongly promoted. It is said to be "a new engine of health care" (Goodman, 1998, p. 117). In many respects, it is not just a concept or a methodology, but a movement.

EBM is being rapidly institutionalized, with the Cochrane Collaboration and a number of international centers taking the lead. The Cochrane Collaboration is hailed as "an enterprise that rivals the Human Genome Project in its potential implications for modern medicine" (Naylor, 1995). The Cochrane Collaboration is named after Archie Cochrane, who as early as 1972 posed a challenge to the medical community to establish a repository of clinical trials results for use by medical practitioners. It is hoped that evidence will guide clinical practice in a systematic and predictable manner. The so-called 'Cochrane's challenge' is also interpreted as an ethical challenge: "Bad consequences follow from uninformed decisions." If uninformed decisions necessarily lead to bad outcomes, then good (ethical) practice can only come out of evidence-based decisions (Goodman, 1998, p. 130).

The relationship between objective scientific evidence and good clinical practice is less than straightforward. Is scientific medicine the best medicine? What constitutes good clinical practice? Is there a golden standard that can be applied to all kinds of therapeutic interventions? Is evidenced based medicine compatible with lay views of healing? How

Julia Tao Lai Po-wah (ed.), Cross-Cultural Perspectives on the (Im)Possibility of Global Bioethics, 299–315.
© 2002 *Kluwer Academic Publishers. Printed in Great Britain.*

will EBM re-shape the doctor-patient relationship? And what new ethical issues may arise as medicine is being redefined?

EBM is not just a scientific methodology. As a movement it has its underlying philosophical presuppositions. Evidence (as data and facts) may be held to be objective and neutral; evidence-based medicine as a movement may not, however, be value-free. The implementation of EBM is a deliberate and methodical process; the underlying philosophy and values may however be unconsciously assumed. The ethical implications of its application need to be considered.

II. THE ORIGINS AND CENTRAL IDEAS OF EBM

That EBM is hailed as a new movement in the 1990's must, at first glance, seem enigmatic. Modern medicine has always emphasized scientific methods and objective evidence. Its roots in anatomy, physiology, pathology and biochemistry are founded upon meticulous and objective observations and experiments. As early as the 19th century, special clinics were set up in France and England to study specific diseases patterns, leading to advances in diagnosis and clinical knowledge (Booth, 1993, p. 207). The root of EBM can be traced to the 1830's, when the French physician Pierre Louise proposed to use a "numerical method" to determine clinical efficacy (Jonsen, 1990, p. 113). In what sense is EBM a new movement?

The core ideas of EBM are simple: Clinical decisions should be based on the best available scientific evidence; evidence should be critically appraised (e.g., by systematic reviews); biostatistical ways of thinking should be used; performance of clinical practice should be evaluated (Hope, 1995). These ideas together spell out a clear and objective methodology, with which the advocates of EBM embark on a movement to liberate medical practices from subjective biases and personal opinions. The EBM methodology is vigorously standardized and formatted, for ready application to clinical practice, management, and resources allocation.

What makes EBM new and powerful may be the following:
(1) A renewed emphasis on quantitative and statistical analysis: The EBM engine is computational – powered by sophisticated statistical methods such as 'meta-analysis'. Advances in computational science and information systems render possible the efficient pooling and

processing of tremendous amounts of data and information. Experts' opinions vary; mathematics cannot lie.

(2) Making the ranking of clinical evidence comprehensible: Clinical evidence is ranked according to how objectively reliable it is and how likely it is to be free from the contamination of bias (Langhorne and Dennis, 1998, p. 9).[1] Based on such rankings and variations of rankings, research findings and recommendations of best practice can be graded (Grade A to Grade C) (Sackett, 1986). Thus, extremely sophisticated statistical analyses are converted into simple ranking categories easily comprehensive to professionals and managers. The simplified scale of objectivity is reassuringly rational.

(3) Moral impetus: The EBM movement upholds a strong belief in stamping out 'variations in practice'. Evidence, being objective, is assumed to be universally valid. It is therefore considered unethical for patients with the same disease condition to be treated differently. The drive to implement uniform practice has a clear moral impetus.

III. THE UNDERLYING PHILOSOPHY

The basic tenet of EBM is simple and sound – good quality information helps clinical decision making and is the basis for proper informed consent. There is however a tendency to deify evidence. The ideal evidence, being totally objective and bias-free, helps to elevate the credibility of medicine to the level of the physical sciences. Physicist John Barrow summarizes the presuppositions of science: (Wolpert, 1993, p. 107):

- there is an external (objective) world separable from our perception;
- the world is rational (logical);
- the world can be analyzed locally – that is, one can examine a process without having to take into account all the events occurring elsewhere;
- there are regularities in nature;
- the world can be described by mathematics;
- these presuppositions are universal.

EBM eagerly even if not deliberately embraces these presuppositions. Diseases and the efficacy of interventions are reduced to specific questions that can be tested by controlled trials, which assumes that variables in diseases can be studied one at a time (i.e., locally analyzed).

EBM, more than ever before in clinical medicine, relies on mathematics to build its case. As disease is viewed as an objective entity and as mathematical method is universal, the conclusions from mathematical analysis of clinical treatments are assumed to be universally valid.

Biomedicine is the youngest of the sciences. Kleinman (1993, pp. 17-18) noted that biomedicine seems to hold a peculiarly "anxious strictness" against non-mechanistic and non-materialistic narratives of illness. Even modern physics does not reject alternative views of the world in the strict manner that medicine holds onto its mechanistic assumptions. EBM in its prevailing form appears to further systematically disseminate this 'anxious strictness'.

Kleinman (1993, Ibid.) further suggests that the objectification of the patient's illness, at its extreme, is reductionistic, positivistic, and ultimately dehumanizing. Little (1998) takes this criticism of objectification further and traces the culprit to the French rationalist philosopher René Descartes (1596-1650). He argues that the Cartesian view has forever dichotomized mind and matter into two unconnected worlds. The human body belongs to the objective, material world; diseases are studied and taught in medicine as if it were a separate entity from human experience.

No doubt the mechanistic Cartesian view has had substantial and wide-reaching influence to this date. In medicine, pathogeneses are described in terms of structural flaws and causal trains (genetic aberrations is a notable example), and efficacy in treatment must be rationalized by chains of mechanism. But it was hardly unique of Descartes to comprehend the world in objective and mechanistic terms. It was also Newton and the whole of classical physics, and it was also the presupposition of modern technology, which is so powerful in shaping the daily life of modern man. Descartes never proposed to relegate human experience (including suffering an illness) to the realms of Matter. And it has been said that he postulated the pineal gland as a site of mind-body interface, evidently in an attempt to unify the subjective and the objective (Kenny, 1994, p. 113). The problem with biomedicine is not just the dichotomy of mind and body, it is the insistence that disease is an objective entity, and that the experience of illness can be reduced to materialistic descriptions. Descartes is as interested in reflections on the Mind as studies of the objective world of Matter; biomedicine largely has done away with reflexive approaches, and EBM epitomizes the deification of the objective. To blame Descartes for the misdemeanors of modern medicine may in itself be a reductionistic approach to trace a single causal mechanism.

Evidence-based practice demands the universal application of objective evidence to clinical decisions in order to benefit the greatest number of patients (i.e., biostatistical thinking for the total patient population). To fully achieve such a benefit at the population level, variations in clinical practice must be abolished. By enforcing universal control of clinical practice through EBM, the greatest measurable health gain should be attainable. Bentham, the father of utilitarianism, is said to have conceived an idea of "a central control tower that would control all activities" in social institutions, in order to ensure the best outcome on a macro-scale (Jonsen, 1990, p. 110). The utopia of EBM bears discomforting resemblance to such a utopia.

IV. EBM CRITICIZED AND DEFENDED

EBM has been criticized and debated, and the criticisms are on two rather opposite counts. Firstly, EBM is criticized as being 'too democratic', in the sense that all pieces of evidence are treated equally in systematic reviews, without regard for the relative merits of different studies. Qualitatively dubious studies are counted the same way as carefully conducted ones. In one EBM systematic review of 127 trials, only 15% had reported clearly how randomization was generated. This casts doubt on the validity of conclusions from EBM reviews (Loannidis and Lau, 1998). Concern about methodological quality is technical; what is more worrying is that studies sponsored by pharmaceutical companies may carry subtle bias, e.g., the companies may be forewarned of the design of outcome scales, allowing manipulations to achieve good outcome. Subtle bias may also be present in the premises and conduct of clinical studies, despite randomization. Some clinical studies are of questionable ethical standards. A hypertension treatment trial widely quoted in systematic reviews did not require patient consent in conducting the study (Kerridge et al., 1998). In the democratic kingdom of statistics, all data are treated as numbers of equal worth.

On the other hand, EBM has been criticized as being tyrannical. Despite earnest efforts in teaching EBM, it is doubtful whether the average medical journal readers are able to critically read or even comprehend fully the sophisticated statistical analysis in large meta-analysis (Jonsen, 1990, p. 114). Systematic reviews are entrusted to academic research centers. The conclusions of the systematic reviews are trusted to be valid. With older-day qualitative reviews the individual clinician can, with some practice, intelligently debate with the reviews – their quality, relevance, and the validity of their conclusions. Now physician readers are powerless in face of the powerful construct of statistical analysis. They become passive recipients of EBM recommendations. EBM has freed physicians from the old tyranny of the 'experts', whose opinions now rank low on the EBM ladder of evidence, but it may have unintentionally replaced it by a new tyranny of statistics.

EBM is not meant to be tyrannical. Marinker (1994, p. 5) noted that "evidence does not always point in the same direction, nor are the basic building blocks of clinical discourse quite as factual as planners might wish." It is not good practice, nor is it even ethical, to manage individual patients strictly according to protocol, without regard to the patient's full clinical and social picture. Sackett (1996, p. 71), one of EBM's staunchest advocates, has cautioned that "without clinical expertise, practice risks becoming tyrannized by evidence, for even excellent external evidence may be inapplicable to or inappropriate for an individual patient." Guidelines should not be confused with decision rules (Goodman, 1998, p. 130). Heterogeneity in medicine is a clinical reality; the real task is to take a close look for explanations of variation, not to get rid of it (Loannidis and Lau, 1998).

'Tyranny' and 'democracy' aside, EBM has been criticized for encouraging a systemic bias in the so-called 'evidence-based purchasing' of health care (Hope, 1995). Managers of health services tend to view EBM uncritically because it seems conveniently 'objective'. A comprehensive text written for health care managers calls for the development of an "evidence-based organization," the culture of which is "an obsession with finding, appraising and using research-based knowledge in decision making"; the head of the organization should be an "evidence-based chief executive." A section of the book is simply titled "evidence-based everything," in which the author stresses the need "to ensure that a scientific approach is taken to all aspects of the work of a health service provider because cost pressures may be generated by multiple small changes" (Muir Gray, 1997, pp. 156-8). It is interesting that the apparently objective and 'value-free' scientific approach of EBM is linked to fiscal pressure to restrain spending. Managers have seized the powerful tool of EBM to justify cuts in specific health services simply because they are 'not proven by evidence'. Allocation of resources to minority patient groups may also be sacrificed because they do not fit into the 'best-buy' shopping list.

Health care policy obsessively guided by 'evidence' has an intrinsic systematic bias. Acute interventions such as drug treatments are easy to evaluate by controlled trials; not so for a chronic illness with a fluctuating course, such as multiple sclerosis. Some interventions take the form of close personal contacts, such as rehabilitation therapy and hospice care; the focus of 'objective efficacy' may be misplaced. In clinical research, potentially profitable new interventions are enthusiastically sponsored, whereas clinical problems affecting small minority groups are under-researched. Many clinical trials exclude patients of very old age and those with multiple co-morbid conditions. Research into unprofitable diseases prevalent in the economically disadvantaged countries is generally sparse. Systematic reviews are basically 'cross-sectional' scannings of the currently available research findings. It is subject to the same 'cohort bias' that EBM tries so hard to dispel.

As is the case in many social movements, the leaders of the movement are less dogmatic than fervent followers. Sackett (1997, p. 5), for instance, is adamant that EBM should not be 'hijacked' by purchasers and managers; but such calls for preserving the non-dogmatic meaning of EBM are relatively powerless compared to the rapid and crude implementations of 'evidence-based purchasing'. The Australian health minister Wooldridge has been quoted as stating that that "[we will] pay only for those operations, drugs and treatments that according to available evidence are proved to work" (Kerridge et. al., 1998). He was merely being candid with his policy. Many other policy makers would be happy to borrow the ideas of EBM without explicitly acknowledging it.

One final criticism of EBM has to do with its claim of universality. The safety and efficacy of clinical interventions vary among different countries and ethnic groups. This may be due to different biological constitutions, or socio-cultural factors. Woo and Chan (1998) have cautioned against direct extrapolation of international findings to clinical practice in Hong Kong. For instance, Chinese diabetic patients appear to respond differently (than Caucasians) to a new class of anti-hypertension drugs called ACEI.[2] In community care, local studies found that elderly residing in nursing homes were satisfied with the residential care, in contrast to international impressions that care at home is preferred. The first example illustrates biological differences, the second socio-cultural.

Evidence in medicine is not as universal as the EBM advocates would have us believe, which is why local research is essential. But even if the importance of local research is recognized, it is unlikely that many clinical trials can be repeated locally. The scale and the number of clinical trials are simply too big. One simple clinical question can take years for a large multi-center trial to answer. Available local evidence will never catch up with clinical questions.

V. THE NARRATIVE OF EBM

Many of the above concerns about the misuse of EBM can be addressed by a moderate and less rigid approach in practice. Inappropriate use of EBM can be identified and refuted. Yet, a more fundamental issue with EBM remains. Who defines the clinical questions? Before a literature search gets started, a clinical question must be first defined. The definition of a clinical question can be subtly shaped by the conductor of the review. In a thoughtful example quoted by Hope (1995), a systematic review provides strong evidence that prophylactic antibiotics reduce the frequency of infection following Cesarean section, thereby recommending that all women undergoing this mode of delivery should be given prophylactic antibiotics. However, a patient representative looking through the data found that infection rates in different studies and different communities are widely different. The apparently value-free rational recommendation of prophylactic antibiotics has concealed the equally important problem of aseptic practice.

Or take the case of stroke as an example. Stroke, or cerebrovascular accident, is a prevalent disease with grave consequences. Specially organized stroke units can improve the outcome for stroke patients, reducing mortality and disability. A succinct and fair overview of the subject is provided by a book entitled *Stroke Units: An Evidence-based Approach* (Langhorne and Dennis, 1998). The book faithfully applies the full EBM methodology, with a historical review of stroke unit development, clinical questions answered by systematic reviews of evidence, and evidenced based conclusions. The focus of interest is typical of EBM. In a review of the history of stroke units, landmarks consist exclusively of advances in EBM – the first randomized control trials published in 1962; the first systematic review in 1993 to prove that stroke unit does prevent deaths; the first systematic review in 1997 showing a reduction in dependency, and so on. In the review, 'stroke unit' is treated as if it were a single physical entity (like a tonsillectomy operation or an angioplasty procedure). Scanty description is made of individual stroke units, and no historical review is provided on the 'life-story' of this model of care.

The narrative of EBM is austere and sometimes bland. Stroke unit is considered to be a 'black box' for the purpose of statistical analysis (Langhorne and Dennis, Ibid, p. 49). Is the content of an effective stroke unit really a black box? Or is it a black box because randomized controlled trials and systematic reviews are not quite the appropriate instrument to look inside the box? A telescope 'sees' the galaxies but not the microcosm of the human cell. Stroke is a multi-faceted illness, and not all facets are objective. Clinicians – therapists, nurses, doctors, and social workers – can develop insight on how best to assist the patient in navigating through the recovery process. Some systems of care work better than others. For instance, a stroke unit with good teamwork has better outcomes than one with inter-disciplinary rivalry. Not all empirical knowledge in clinical practice is quantifiable, nor are they necessarily 'objective'. By restricting knowledge to objective evidence, EBM becomes a self-fulfilling methodology – clinical questions are defined in such a way to ensure that evaluation by objective and quantitative methodology will work; clinical questions that cannot be so defined are, by default, non-evidence-based. That empirical knowledge has a wider basis than quantifiable evidence is not fully considered.

A wide chasm exists between the scientific and methodical language of EBM and the emotive language of the patient. The patient tends to ask personal questions: Why did the stroke happen to me (even though I do not smoke, etc.)? Will I be able to go back to work? Should I try acupuncture or herbal medicine? A substantial responsibility of the medical profession is to address such personal concerns. The language of EBM cannot be readily translated into a language of healing. In the case of stroke care, meta-analysis of 'stroke units' is feasible precisely because they come in quantifiable 'units'. EBM is valuable in answering the pre-defined questions, but may be too general and too 'macroscopic' to be of help in dealing with the personal illness experience of the patient.

Are we expecting too much from EBM? Patients may of course have experiential anxiety and legitimate personal questions, but these questions may not fall into the realm of clinical science. One may also argue that the art of medicine should after all supplement the science.

Bringing in the 'art of medicine' however does not solve all the problems for EBM. EBM strives to reduce variation in clinical practice. That variation may actually constitute much of the so-called 'art of medicine' – physicians make adjustments according to the individual patient's total condition. The core questions must be these: Has EBM adequately defined the scope of clinical medicine? Is it defining clinical medicine so narrowly that valuable clinical knowledge (such as that based on detailed studies of individual cases) becomes dismissed as unscientific or non-evidence-based? What defines validity in clinical medicine? Who defines it?

The patient will not celebrate the same EBM landmarks as EBM advocates, and is in a relatively powerless position to dispute any 'evidence-based' clinical decisions. Finding the language of EBM too cold and impersonal, unsatisfied patients would turn elsewhere for answers to their questions. Not enough of the art of medicine is left to attract them. Many patients suffering from chronic illnesses will seek alternative treatments such as acupuncture, herbal medicine, and chiropractic. Alternative medicine attracts patients because of its alternative paradigms. EBM is much more distant from the patient's narrative.

VI. EBM AND ALTERNATIVE MEDICINE

Despite the tremendous power of modern day medical technology, alternative medicine is thriving. A review of 25 surveys conducted on the utilization of complementary and alternative medicine in the U.S. shows that 43% of mainstream physicians refer their patients for acupuncture, and 40% for chiropractic treatments (Astin et al., 1998). In Hong Kong, even though mainstream physicians are not yet allowed to refer patients to alternative medicine practitioners, spontaneous utilization rates of Traditional Chinese Medicine (TCM) range from 36% to 55% in a three-year period before various surveys (*Working Party on Chinese Medicine Interim Report*, 1991,p. 4).

Mainstream physicians' responses to the popularity of alternative medicine range from healthy skepticism to outright rejection. The rise of the EBM movement provides a convincing yardstick for evaluating alternative medicine. In a balanced treatment of the subject by an editorial of a major medical journal: "As physicians whose job description requires us to help people, we cannot reject 'out-of-hand' any proposed therapies just because they did not originate in modern mainstream medicine ... Promising unconventional therapies must be subjected to the same level of scientific scrutiny that we now require for drug therapies introduced by 'mainstream' physicians ... (they) should be subjected to appropriately designed clinical trials" (Dalen, 1998).

No doubt, some alternative treatments can be evaluated by randomized controlled trials (RCT); yet many more seem resistant to assessment by the objective methodology of EBM. Kaptchuk and Eisenberg (1998) note that alternative medicine "generally lacks the biomedical notion of clinical experiments under artificially and deliberately controlled conditions," exemplified by RCT. This may mean either (1) that alternative medicine has a lot of catching up to do, or (2) that the RCT methodology has some fundamental conflict with the paradigms of alternative medicine. For instance, TCM adopts a holistic diagnosis approach and individualized tonic regimen of herbal prescriptions (often more than a dozen herbs in one prescription). Evaluation of efficacy of a single prescription for a large number of patients contradicts the basic tenets of TCM. Many alternative treatments are hands-on (e.g., acupuncture, chiropractic), and the placebo effect is difficult to exclude.

Is it ethical to prescribe treatments that have not been 'proven' by the EBM methodology? On the other hand, is it ethical to insist on using an objective and reductionist methododogy, exemplified by controlled trials, as the only yardstick? McNeill (1998, p. 104) argues that "to treat objectivity as the only valid view is a form of fundamentalism, a belief, which is as pernicious as any other fundamentalist belief."

The difference in paradigm between EBM and alternative medicine is best illustrated by their strikingly disparate views towards the 'placebo effect'. In EBM, objective assessment of clinical efficacy demands the control and exclusion of the placebo effect. Kleinman (1993, p. 19) considers it peculiar "that practitioners of Western medicine are trained in a radically sceptical method that ought (sought?) to diminish the placebo response in their care." Alternative medicine, on the other hand, makes no distinction between a 'real' healing effect and the 'placebo effect' – whatever helps the patient is 'real'. This is not just crude empiricism. Holism presupposes the removal of distinction between the 'objective' and 'subjective'. As Kaptchuk and Eisenberg (1998) observed, in the world of alternative medicine, "truth is experiential ... Nature is not separate from human consciousness," nothing is 'only in the heads' of the patient. Whereas EBM meticulously excludes and rejects the placebo effect from clinical practice, alternative medicine embraces it unapologetically. The alternative medicine approach is thus experientially closer to the patient than EBM.

A valid criticism of alternative medicine is this: If objective evaluation methods are rejected, it will be difficult to distinguish good practice from quackery. From the EBM perspective, quackery may be defined as the deliberate use of a placebo to exploit the patient's trust. To practice quackery is unethical. This may be viewed as a 'strong' position of EBM, i.e., 'to practice non-evidence-based medicine is unethical'. In contrast, from the point of view of alternative medicine, to abandon the use of beneficial treatment (whether called a 'placebo' or not) may be unethical. From a therapeutic point of view, one can err on either side. Given its holistic presupposition, it is understandable that alternative medicine tends to refuse to dissect a clinical problem into 'controlled' parts for study.

Taking a rather more 'meta-' stance, one may ask of EBM: If science is about making falsifiable hypotheses, is the EBM assertion that 'all medical interventions can be evaluated by EBM' falsifiable? In other words, what does it take for EBM to admit that its methodology is not applicable to the evaluation of a particular intervention? If the response is that 'all interventions with real efficacy should in principle be objectively measurable and quantifiable; whatever therapeutic effect that cannot be measured cannot be a real effect', then the EBM paradigm is not falsifiable. What one cannot see (or measure) does not objectively exist. Clearly there is a problem with this positivistic position. Is it then ethical to demand that all medical interventions and healing methods be evaluated in the EBM format, when the basic paradigm of EBM itself is built on belief?

VII. EVIDENCE VS. PATIENT'S NARRATIVE – WHAT NEXT?

Within mainstream medicine, every sign points to a dominating reign of the EBM paradigm for decades to come. The legitimacy of EBM, built upon an ethical impetus (Cochrane's challenge) and sophisticated statistical methods, is too authoritative and profound to be challenged by the average physician. In the foreseeable future, EBM fundamentalists will continue to strive to suppress alternative paradigms and reject 'inferior' evidence. They are uncompromising to the patient's subjective narrative, and look the other way from personal context and meaning in managing a 'case'. A 'case' is viewed as one numerical case of regularity, rather than a particular position in the patient's life (Gadamer, 1993 p. 95).

The language of EBM may be remote from the experience of the patient, but it may still ultimately win. Advances in biological sciences and medical technology will help to mould the perspective of the lay person. Embryology and neonatology defines early-life personhood; cognitive science defines personhood in old age; neurochemistry defines our emotions and psychology; the Human Genome Project defines our constitution. The objective narrative of EBM and related sciences seem destined to win.

Even if the lay people do eventually submit to the powerfully objective narrative of EBM and the biological sciences, illness will nonetheless remain a highly personal and emotional experience. Patients can be rationally persuaded to accept the language of evidence-based medicine, yet emotionally they remain unhealed. The experiential narrative and the intense emotion intrinsic to illness (and also intrinsic to caring for the ill) may be more tenacious than the rational physicians imagine. Patients seek health care not just to receive rational scientific treatments; they are equally seeking professional support for their emotive illness experience, asking for meaningful explanations, and looking for personal advice. As the patients look up at the holy mountain of scientific evidence with awe and perplexity, they may also be estranged from it. When they perceive a mismatch between the illness experience and the science of medicine, they are prone to look elsewhere for answers. What is wrong with me? A meaningful answer has to be sought – it matters little whether that answer is scientifically based or not. Here then is the paradox of EBM in the new Millenium – the fundamentalists of EBM may win the holy war, but the patient may be lost.

This rather doomed outlook is built upon one premise: that dogmas of EBM prevail. There are some reassuring signs that this may not be the case. The future of medicine may not be monopolized by dogmatic insistence on the objective. One interesting proposal calls for integration of the patient's narrative into medicine. This new initiative, called 'Narrative based Medicine', borrows from the field of narrative based ethics. Jones (1999) adapts narrative theory to medical patient care. He quotes Brody (1994) to advocate a new narrative ethics, in which "the doctor must work as co-author with the patient to construct a joint narrative of illness and medical care." Multiple narratives are also invited from family, nurse, friend, and social worker. McNeill (1998, p. 111), following the line of thinking of Derrida and Levinas, proposes a medical ethic "that allows the other person to be present in the moment with his/her experience, emotional expression and story, without interrupting and without defending myself against the possibility of being affected by that person and without attempting to comprehend his/her story in a way that makes it known as an acquisition of mine." The statement is somewhat convoluted, but the idea is interesting: namely that physicians should not hurry to see 'the case' as a mere source of 'data' to be acquired and categorized. Evidence-based intervention must await a full and open narrative from the patient. The patient's narrative will help to place the physicians' objective knowledge into a personally meaningful context. Rather than obsession with 'reducing variations in practice', heterogeneity of illness stories is respected and assimilated into clinical decisions.

One step further in narrative based medicine is to apply the life-history approach to generate empirical knowledge for medical ethics study (Tovey, 1998). Disciplined listening to personal narratives of many patients without imposing a pre-identified framework may lead to useful 'proposition development', i.e., identification of patterns of influence, how life process and illness are shaped by social and cultural factors. An appropriate ethical solution will depend on understanding the meaning of the situation and knowledge of the personal, cultural and social context (Tovey, ibid., p. 177).

The core ideas of narrative based medicine may seem remote from busy day-to-day clinical practice, where every minute counts (for the patient, and also in terms of dollars). Yet, it concords surprisingly well with a core concept in clinical rehabilitation: context and meaning are important. Gadamer (1993, p. 97) argues that the purpose of medicine is to assist the patient in becoming 'well' again, signified by a "return to accustomed form of life." In the language of rehabilitation, recovery is about enhancing 'quality of life' (well again) and community re-integration (return to accustomed form of life).

The main barrier to integrating narrative based medicine and evidence-based medicine may still be a gap in language. The language of evidence-based medicine is uncompromisingly constructed: randomization and control, probability and statistics, objectivity and removal of bias. The language of narrative based medicine is personal, relational, context dependent, focused on personal meaning, and occasionally amorphous. It is easier to incorporate personal narratives into the malleable paradigms of alternative medicine; it is a much more daunting task to square hard evidence-based clinical decisions with the patient's experiential world. To address personal meaning and context fully, EBM would need to develop a more fluid and open language for communication; it needs to accept heterogeneity, and to look beyond the comfort zone of objectivity and universality.

It is premature to speculate what will become of EBM in the end – will it be powerful enough to 'win over' the patient completely, obviating the patient's experiential and narrative needs? Or will it relax its present rigidly formatted language, to accommodate narrative from the patient, or even from alternative medicine? It is hoped that, as EBM evolves in the post-modern era, the original Cochrane's challenge will be amended: "Bad consequences follow from uninformed decisions; *meaning*less outcomes follow from unthinking evidence-driven decisions."

Department of Rehabilitation
Kowloon Hospital, Hong Kong; Email: ksau@ha.org.hk

NOTES

[1]   One popular ranking of levels of evidence is constructed by the U.S. Department and Human Services Agency for Health Care Policy and Research (Langhorne and Dennis, 1998):

   Ia   Meta-analysis of randomized controlled trials;

   Ib   At least 1 randomized controlled trial;

   IIa   At least 1 well designed controlled study without randomization;

   IIb   At least one other type of well designed experimental study;

   III   Well designed non-experimental descriptive studies, such as comparative studies, correlation studies, and case studies;

   IV   Expert committee reports or opinions and/or experiences of respected authorities.

[2]   ACEI is an abbreviation for angiotensin-converting enzyme inhibitor.

REFERENCES

Astin J.A., Marie A., Pelletier K.R., Hansen E., Haskell W.L. (1998). A review of the incorporation of complementary and alternative medicine by mainstream physicians. *Archives of Internal Medicine, 158*, 2303-2310.

Booth, C. C. (1993). Clinical research. In: W. F. Bynum & R. Porter (Eds.), *Companion Encyclopedia of the History of Medicine, vol. 1* (pp. 205-232). London: Routledge.

Brody, H. (1994). "My story is broken; can you help me fix it?" Medical ethics and the joint construction of narrative. *Literature and Medicine, 13.* 79-92.

Dalen, J. E. (1998). "Conventional" and "unconventional" medicine – can they be integrated? *Archives of Internal Medicine, 158*, 2179-2181.

Gadamer, H. (1993). *The Enigma of Health.* Cambridge: Polity Press.

Goodman, K. W. (Ed.) (1998). *Ethics, Computing, and Medicine – Informatics and the Transformation of Health.* Cambridge: Cambridge University Press.

Hope, T. (1995). Evidence-based medicine and ethics. *Journal of Medical Ethics, 21*, 259-290.

Jones, A. H. (1999). Narrative in medical ethics. *British Medical Journal, 318*, 253-256.

Jonsen, A. R. (1990). *The New Medicine and the Old Ethics.* Cambridge: Harvard University Press.

Kaptchuk, T. J. & Eisenberg, D. M. (1998). The persuasive appeal of alternative medicine. *Annals of Internal Medicine, 129*, 1061-1065.

Kenny, A. (1994). Descartes to Kant. In: A. Kenny (Ed.), *The Oxford Illustrated History of Western Philosophy* (pp. 107-192). Oxford: Oxford University Press.

Kerridge I., Lowe M., Henry D. (1998). Ethics and evidence-based medicine. *British Medical Journal, 316*, 1151-1153.

Kleinman, A. (1993). What is specific to western medicine? In: W. F. Bynum & R. Porter (Eds.), *Companion Encyclopedia of the History of Medicine, vol.1* (pp. 15-23). Oxford: Oxford University Press.

Langhorne, P. & Dennis, M. (Eds.) (1998). *Stroke Units: An Evidence-based Approach.* London: BMJ Publishing.

Little, M. (1998). Cartesian thinking in health and medicine. In: P. Baume (Ed.), *The Tasks of Medicine – An Ideology of Care* (pp. 75-95). Sydney: MacClennan & Petty.

Loannidis, J. P. A. & Lau, J. (1998). Can quality of clinical trials and meta-analyses be quantified? *Lancet, 352*, 590-591.

Marinker, M. (1994). Evidence, paradox, and consensus. In: M. Marinker (Ed.), *Controversies in Health Care Policies – Challenges to Practice* (pp. 1-25.). London: BMJ Publishing.

McNeill, P. M. (1998). Reason and Emotion in Medical Ethics: A Missing Eement. In: P. Baume (Ed.), *The Tasks of Medicine – An Ideology of Care* (pp. 96-115). London: BMJ Publishing.

Muir Gray, J. A. (1997). *Evidence-Based Healthcare – How to Make Health Policy and Management Decisions.* London: Churchill Livingstone.

Naylor, C. D. (1995). Grey zones of clinical practice: Some limits to evidence-based medicine. *Lancet, 345*, 840-842.

Sackett D. L. (1986). 'Rules of evidence and clinical recommendations on the use of antithrombotic agents. *Chest, 89*, 2S-3S.

Sackett D.L., Rosenberg W.M.C., Muir Gray J.A., Haynes R.B., Richardson S.W. (1996). Evidence-based medicine: What it is and what it isn't. *British Medical Journal, 312*, 71-72.

Sackett, D. L. et al (1997). *Evidence-Based Medicine – How to Practice and Teach EBM.* London: Churchill Livingstone.

Tovey, P. (1998). Narrative and Knowledge Development in Medical Ethics. *Journal of Medical Ethics, 24,* 176-181.

Wolpert, L. (1993). *The Unnatural Nature of Science*. London: Faber & Faber.

Woo, J. & Chan, J. (1998). Evidence-based medicine: Ethical considerations. *Hong Kong Medical Journal, 4,* 169-174.

Working Party on Chinese Medicine (October 1991). *Working Party on Chinese Medicine Interim Report,* Hong Kong Government Printer.

PART IV

NEW STRATEGIES, NEW POSSIBILITIES

MARY ANN G. CUTTER

LOCAL BIOETHICAL DISCOURSE:
IMPLICATIONS FOR UNDERSTANDING DISEASE

I. THE MESSAGE

Together the essays in this volume explore how moral discourse, moral narratives, and moral commitments take different shape within particular cultures and traditions as we enter the Third Millennium. More specifically, the essays consider the challenges these different shapes pose to our moral relations between and among ourselves and to the character of medical knowledge, therapeutics, and bioethical reflections in our society.

Among the authors in this volume, we see varying accounts of bioethical discourse ranging from global to local accounts, from abstract to particularized perspectives (see Tonnies, 1957 [1887]). For some (e.g., Cheng, 2002; Holliday, 2002; Becker, 2002), the values that frame bioethical discourse are the same as those found in and between communities. On this view, there is available a set of fundamental or commonly-shared values that frame such discussions and an expectation that problems may be resolved given a recognition of such values. For others (e.g., Fan, 2002; Lee, 2002; Schmidt, 2002), the values that frame bioethical discourse vary in terms of content and procedure, as well as degree and kind. On this view, resolution of conflict takes place within particular communities and differences among value commitments are to be expected. Given the lack of consensus among authors, a global perspective that can frame discussions regarding what binds all in matters of epistemological and value commitments remains at best one among many perspectives. This is to say that a global bioethical perspective is not available.

One response to the disagreement noted above is to say that we may in the future have a method to resolve value differences. All one can do is wait. But this is a rather uninteresting intellectual response, for it precludes further discussion, a move not supported by the authors in this volume. A second response is that we do not have, and are not likely ever to have, a single method for resolving disagreement. In a pluralist world, there may be many sources of meaning and value and consequently a

Julia Tao Lai Po-wah (ed.), Cross-Cultural Perspectives on the (Im)Possibility of Global Bioethics, 319–334.
© 2002 *Kluwer Academic Publishers. Printed in Great Britain.*

multiplicity of points of view on particular issues (MacIntyre, 1981; Nagel, 1979, pp. 128-137). What this means is that fundamental debates about the possibility of a global or general bioethical response cannot be settled by bioethics or social philosophy, for these are only begging the question because a premise of the argument assumes the conclusion. One must assume some particular premises in order to obtain substantial moral conclusions (Engelhardt, 1995, Ch. 2) and such premises do not have global consensus. Different societies establish different rankings of priorities. Accordingly, a distinction between a global or international bioethics and a particular or local bioethics is in order (Fan, 2002). One can share a context-rich, localized bioethics only with those who are committed to the same fundamental moral premises and value rankings. Alternatively, a minimal global common morality can only be derived by default in terms of the right of forbearance: every society and community have the right not to do that which they take to be inappropriate. This is not because moral pluralism and multiculturalism are by themselves intrinsically valuable. This is because moral pluralism is the only option for a common procedural morality in the face of the philosophical inability to justify one particular morality without begging the question (see, Engelhardt, 2002). Either way, we are left at this point with an inability to resolve fundamental disagreements about meaning and value, thereby resulting in at best local bioethical discourses about particular subject matters.

II. ON DISEASE

An implication of the foregoing regarding the implausibility of a global bioethics is that the major concepts that frame medicine, such as health and disease, are local as well. This is the case because concepts of health and disease involve value commitments, which are vulnerable to the same difficulties in commanding global agreement noted above. What follows is an analysis of this implication, with particular attention to how new knowledge in genetics additionally contributes to a local account of disease in the new millennium. A consideration of prominent bioethical issues raised by a local account of disease concludes this essay. In this way, the axiological and epistemological challenges in medicine are shown to be inextricably bounded.

1. Why Study Disease?

Decisions about the meaning of disease have direct and important consequences for daily life and the allocation of significant portions of social resources. As Arthur L. Caplan (1993) notes, the emergent concern with disease in this century is a function of many forces. To begin with, disease has served as a major classification in medicine since its beginning and it is important to be clear about what medicine is talking about when it uses the term "disease." The efficacy of 20th century medicine, public health, and its attendant technology in preventing and reversing many forms of infection, dysfunction, and nutritional deficiency constitute reasons for concerns about disease. Reproductive technologies, genetic engineering, cosmetic surgery, and physician-assisted suicide, for instance, challenge our views about what constitutes proper domains of medical attention and why. Concerns regarding the role medicine plays in assessing the worth and value of human beings through its clinical classifications are additional reasons for interest in disease. One is reminded of the abuse in medicine against women (Smith-Rosenberg and Rosenberg, 1981), negroes (Cartwright, 1981 [1851]), Jews (Proctor, 1988), and Russian dissidents (Pope, 1996). Finally, and despite legal restrictions (e.g., Colorado Statute 10-3-1104.7, Equal Employment Opportunity Commission, 1995), institutions such as government, the insurance industry, and business increasingly play roles in determining what clinical conditions are worthy of recognition as disease by deciding which are covered and which are not. Depending on how disease is understood, the investment of resources (e.g., funds, personnel, power, legal protection) will be seen to be indicated or unnecessary, or ethically required or discouraged.

Put another way, our understanding of disease involves bioethical considerations. Decisions we make about what will be considered disease and which will be treated and at what level involve judgments concerning good and bad, and right and wrong, as these judgments concern the actions and character of human beings. Alternatively, bioethical reflection involves assumptions about what conditions are to be considered clinical classifications and which are worthy of response by health care professionals. The tie between epistemological and axiological matters in medicine is important for bioethicists to recognize because the conclusions they draw have impact on discussions in medical epistemology. Alternatively, reflections in medical epistemology give guidance to discussions in bioethics.

2. Procedural Character of Disease

We begin with a preliminary analysis of the concept of disease. The term 'disease' comes from the Latin roots *dis* (or priv-) plus *esse* (or to be), or privation of being. Procedurally speaking, disease refers to descriptions of pathological psychosomatic processes that occur in the human species and bring patients to the attention of health care professionals. In this way, disease may be seen to involve three components: descriptive, prescriptive, and social (Cutter, 1988). In terms of description (Albert et al., 1988), disease results from careful observation and well-reasoned arguments involving the formation and testing of hypotheses (or proposed explanations). The clinician starts with observations of the disease, moves to propose an explanation, tests the explanation against evidence from the presenting disease, and, with each presentation of the disease, tests new evidence against existing explanations. Through this method, medicine has demonstrated remarkable power to describe the world of disease in terms of symptoms, the underlying causes of disease phenomena, and relations that are open to a range of interpretations.

In this process of explanation, disease labels or classifications give rise to the development of norms. Such norms function prescriptively: they serve as the basis for judgments about how individuals with certain conditions ought (not) to be, (not) to act, and so on. Individuals ought to be able to function according to ideals of activity that are proper to an organism and those of form and grace that are worthy of achievement. Furthermore, we decide how to act, what to strive for, and what to resist in light of such norms. We ought, for example, to provide treatment to those who are suffering. In this way, disease functions as a treatment warrant for those qualified to offer a response. Disease-norms are clusters of characteristics and abilities that function as standards by which individuals and their conditions are judged to be "good" or "bad" instances of particular human conditions.

As framed here, the descriptive level of analysis in discussions of disease is tied to a prescriptive level. Facts and values, as Engelhardt (1995, Ch. 5) and others (e.g., Fleck, 1979 [1935]) teach us, interweave in complex ways. Observations in medicine are always ordered around theoretical commitments, including judgments concerning how to select and organize evidence into explanations. Further, observations are always ordered around evaluative commitments including those concerning what phenomena are assigned significance in terms of functional, instrumental, and aesthetic values and what actions are appropriate in order to achieve certain ethical goals (e.g., minimizing suffering, respecting patient autonomy).

Moreover, the prescriptive force is backed by social sanctions, fashioned in light of what goals are seen to be worthy of achievement in our collectives lives. If one conforms to a disease-norm, one increases the probability of being a recipient of social support (Parsons, 1951). If one diverges from such norms, one lessens the chance of being a recipient of social support. In short, disease involves descriptive, prescriptive, and social commitments. Yet, disease involves more. Even after the success of this account of disease, one will still be faced with a choice among particular accounts of disease. One will have to learn the rules of evidence and inference accepted by those working in the particular community one wishes to join. One will have to reflect on how to weigh certain judgments vis-a-vis particular outcomes, which involves listening to and understanding those involved in select clinical discourses. One will in addition have to accept that our understanding of disease evolves and changes with new knowledge and recognition of previously unrecognized values and disvalued states of affairs. Given the futility of discovering clinical reality and a simple sense of disease, we begin to appreciate that there will at times be differing, if not competing, interpretations of disease.

On this account, disease is a localized concept. To begin with, disease is localized in terms of what and how it describes. An example is coronary artery disease, which can be correlated with genetic, metabolic, anatomic, psychological, and social variables, depending on whether one is a geneticist, an internist, a surgeon, a psychiatrist, or a public health official. The construal will depend upon the appraisal of which etiological variables are epistemologically significant and are most amenable to manipulation, by whom, for what goal, and at what cost. For example, a surgeon may decide that the major factors in coronary artery disease include structural abnormalities, which may be altered through surgical techniques. A geneticist may find that the major factors in coronary artery disease include a specific set of alleles, which may be genetically engineered. The public health official may decide that the basic variables in coronary artery disease are elements of a lifestyle that include little exercise, overeating, and cigarette smoking. The clinician may then address these social variables and consider the disease to be, as Stewart Wolf (1961) puts it, a "way of life."

In addition, disease is localized in terms of what and how it values. Disease exists in order to cure, to care, to intervene. Since we understand disease in terms of responding to patient complaints, it follows that all disease takes place against a background of value presuppositions. Simply put, functional values tell us what ideals of activity are proper to an organism. Instrumental values tell us how to get from means to end, where the end is the maximization of benefit and minimization of harm. Aesthetic values tell us what ideals of form and grace are worthy of achievement. Since all may not share in the same views regarding what ideals are proper to be achieved, ethical values are at stake as well. To diagnosis, predict, and treat, while taking proper regard of patients, requires acknowledging the range and specific expressions of values embodied in how we understand disease.

Finally, disease is localized in terms of cultural forces. We frame disease not only in terms of the relation of the condition to the organism, but in terms of the relation of the organism to the environment (Reznek, 1987, p. 69). Dyslexia is not a disease in a pre-literate environment, but in a literate one. Insensitivity to growth hormone is not a disease among pygmies, but is one among Masai. Anorexia nervosa is a diagnosis in a resource-rich culture that places positive value on thin body figure, but not one in a culture where starvation is a result of limited resources. Even within certain clinical categories such as depression, seemingly universal afflictions can differ profoundly from culture to culture (Osborne, 2001, p. 100), thus leading the developers of the Diagnostic Statistical Manual of Behavioral Disorders IV (1994) to introduce a new category of psychiatric illnesses known as "culture-bound syndromes". In short, the descriptions, values, and socio-political influences that enter into how we understand and manipulate disease are localized, depending on the perspective taken by those providing the account.

III. LOCAL (LOCATING) DISEASE IN AN ERA OF GENETIC MEDICINE

In the Third Millennium, we can expect that molecular genetics will increasingly define how to understand and undertake disease. As Watson and Cook-Deegan put it, "The major impact of the genome project will be a slow but steady conceptual evolution – a change in the way that we think about disease and normal physiology" (Watson and Cook-Deegan, 1990, p. 3322). Let us consider the effect new knowledge in genetics will have on our understanding of disease in the new millennium and how it encourages a local account of disease.

1. Genetic Primer

Current work in genetics is devoted to developing detailed maps of the human genome and the genome of several other well-studied organisms (e.g., bacteria, yeast, fruit fly, and mouse), and to determining the complete nucleotide sequences of these genomes. A geographical analogy may be helpful for understanding how gene mapping and sequencing will be done (Walters, 1997, pp. 223-224). If you wanted to make a physical map of a country, you might first use a satellite photograph of the whole area. This satellite photo corresponds to locating the 23 pairs of human chromosomes. You might divide the country into 1,000 regions, each with a certain number of square miles. For this more detailed map, you might obtain photographs taken from airplanes. Similarly, scientists have begun to divide the human genome into major regions, each of which consists of perhaps 40,000 units or base pairs. In our geography project, there might be certain regions that require special attention even at this stage of mapping, for example, major metropolitan areas or areas with potentially dangerous geographical faults. In a parallel way, scientists have discovered that certain health- and disease-related genes are located in particular regions of particular chromosomes. They have begun to investigate these regions in greater detail at this stage of mapping. The final stage in the geography project might be a highly detailed map that indicates streets or even individual buildings on those streets. These fine details correspond to knowing the sequence in which some or all of the 3 to 3.5 billion base pairs are arranged in human chromosomes in the human genome.

But what exactly is a human genome? Human beings usually have 23 pairs of chromosomes in the nuclei in every cell. Chromosomes contain genes that code for proteins and other products (e.g., polypeptides) that our bodies need to function, as well as intervening stretches of DNA whose function is not yet understood (and has in the past been referred to as "junk DNA"). The smallest coding units within the genes and intervening sequences are called bases (i.e., a purine [i.e., adenine or guanine] or a pyrimidine [i.e., thymidine or cytosine] on one strand of the DNA that forms hydrogen bonds with a complementary base on the other strand). It is estimated that our 23 pairs of chromosomes contain approximately 30,000-50,000 genes (Weiss, 2001, A1) and 3 to 3.5 billion base pairs. Scientists seek to learn more about the structure and function of human genes, chromosomes, and extra-nuclear DNA, which, taken together is called the human genome.

Up until this point, the human genome sounds fixed and unchanging. It is this popular interpretation that leads the media worldwide to suggest that genes have a unilateral, single-causal relation to phenotype (i.e., observable characteristics of an organism produced by the organism's genotype interacting with the environment). Upon closer inspection, however, new knowledge in genetics challenges this simplistic or reductionistic view (Office of Science Education, 2000). Genetics is the study of inherited variation. Human genetics is the study of inherited human variation. Homo sapiens is a relatively young species and has not had as much time to accumulate genetic variation as have the vast majority of species on earth, most of which predate humans by enormous expanses of time. Nonetheless, there is considerable genetic variation in the human species. The human genome comprises about 3×10^9 base pairs of DNA, and the extent of human genetic variation is such that no two humans, save identical twins, ever have been or will be genetically identical. Between any two humans, the amount of genetic variation, or biochemical individuality, is about 0.1 percent. This means that about one base pair out of every 1,000 will be different. Any two people, therefore, will have about 3×10^6 base pairs that are different.

An implication of this view is that the terms "gene" and "genome" must always be specified with regard to the particular biochemical processes that it produces. For example, it has been suggested that there may be differences in the genetic etiology of type II diabetes between Mexican Americans and Scandinavians, with somewhat higher frequency of early insulin resistance in the former and an early pancreatic beta cell defect in the latter (Lander and Schork, 1994, p. 2039). Further, among those within a particular population, differences in genetic expression is expected to emerge, thus leading to differences in phenotypic expression. In short, all is not equal when it comes to genes.

In addition, genes never operate alone. Genes are always located in an environment, which further contributes to variation in their expression.

For example, sickle-cell trait is not pathological in malarial areas, but is at high altitudes. Assuming a genetic predisposition to alcoholism, one is not considered an alcoholic until one is in an environment in which alcohol is consumed in a way that lacks limits. This new way of thinking reflects a false dichotomy between nature and nurture and embraces the view that any explanation in biology and medicine must appeal to both genes and environment. As a consequence, complete predictive power in the case of genetic expression is rarely attainable.

2. Genetic Variation and Disease

As previously stated, new knowledge in genetics will "change the way we think about disease" (Watson and Cook-Deegan, 2000, p. 3322). Let us consider the implication of genetic variation on how we understand disease.

To begin with, much of human genetic variation is insignificant biologically. That is, it has no adaptive significance. Some variation (for example, a neutral mutation) alters the amino acid sequence of the resulting protein but produces no detectable change in its function. Other variation (for example, a silent mutation) does not even change the amino acid sequence. Furthermore, only a small percentage of the DNA sequence in the human genome are coding sequences (sequences that are ultimately translated into protein) or regulatory sequences (sequences that can influence the amount of gene product that results from a given gene). Differences that occur elsewhere in the DNA – in the vast majority of the DNA that has no known function – have no impact.

Some genetic variation, however, is positive, helping the species adapt to changing environments. A classic example is the mutation for sickle hemoglobin, which in the heterozygous state provides a selective advantage in areas where malaria is endemic. More recent examples include the CCR5 mutation, which appears to provide protection against AIDS. The CCR5 mutation is a deletion of 32 base pairs of DNA that alters an important receptor on the surface of macrophages. HIV, the virus that causes AIDS, cannot bind to the altered receptor and therefore does not enter the cell. Early research on this genetic variant indicates that it may have arisen in Northern Europe about 700 years ago, during the European epidemic of Northern Europe, an indication that it may have provided protection against infection by Yersinia pestis, the bacterium that causes plague (Office of Science Education, 2000, p. 9). The sickle cell and AIDS stories remind us that the biological significance of genetic variation depends on the environment in which genes are expressed. It also reminds us that differential selection and evolution would not proceed in the absence of genetic variation within a species.

Some genetic variation, of course, is associated with disease, with accounts of pathological psychosomatic processes that occur in the human species and bring patients to the attention of health care providers. Classic single-gene disorders such as cystic fibrosis and Huntington disease are examples. Here the genetic component is very large. Increasingly, research is also uncovering genetic variations associated with the more common diseases that are among the major causes of sickness and death in developed countries, e.g., heart disease, cancer, diabetes, and psychiatric disorders such as bi-polar and schizophrenia. Here the genetic component is smaller. Whereas disorders such as cystic fibrosis result from the effects of mutation in a single gene and are evident in virtually all environments, the more common diseases result from the interaction of multiple genes and environmental variables. These more common diseases, such as cancer, heart disease, and diabetes, are termed "complex".

The genetic distinctions between relatively rare single-gene disorders and the more common complex diseases are significant (Juengst, 1995; Office of Science Education, 2000, p. 9ff). Genetic variations that underlie single-gene disorders generally are relatively recent, and they often have a major detrimental impact, disrupting homeostasis in significant ways. Such disorders also generally exact their toll early in life, often before the end of childhood. In contrast, the genetic variations that underlie common, complex diseases generally are of older origin and have a smaller, more gradual effect on homeostasis. They also generally have their onset in adulthood. The last two characteristics make the ability to detect genetic variations associated with common diseases especially valuable because people have time to modify their behavior in ways that can reduce the likelihood that the disease will develop, even against a background of genetic predisposition, thus emphasizing the important role environment plays in genetic expression.

The lesson here is that new knowledge in genetics requires that we understand disease in terms of genes plus environment. Clearly, the excitement lies in continuing the work begun by nineteenth-century germ theorists (Hudson, 1987) in the search for specific causes of disease. While this may be the case, the work of contemporary geneticists differs from nineteenth-century pathologists in supporting the view of gene as inherited variation and requiring the coupling of nature and nurture in any analysis of disease. In supporting this, work by contemporary geneticists contributes to a localized account of disease in that any account of disease must be located specifically and concretely in terms of an individual's genes and their environments.

Put another way, explanations such as "gene for" and "in the gene" are confusing and lack support. One must specify what molecular biochemical processes are associated with what kind of phenotypical expression set within what particular environment. The task of specifying all the variables is daunting, but required in order to understand appropriately the role played by genes in disease expression. In addition, the task is important insofar as disease labels function as treatment warrants. In order for patients to be treated in an optimal way, careful specification of the condition under study and the recommended response are necessary.

IV. LOCAL (LOCATING) DISEASE: BIOETHICAL IMPLICATIONS

The essays in this volume concern the bioethical implications of new knowledge in medicine in the Third Millennium. In keeping with this focus, consider the bioethical implications of a local account of disease.

To begin with, questions arise concerning the legitimacy of any single set of standards or classificatory scheme of disease in medicine. As this essay argues, differences in how genes are expressed, in how genes are influenced by their environment, and in how cultures interpret such expressions must be taken into consideration in framing disease categories (Spector, 1996; Osborne, 2001). Differences turn in part on how normality and abnormality are understood. Traditionally speaking, health is considered normal and disease abnormal. Here normal can mean (Murphy, 1976, pp. 117-33): (1) statistical, (2) average or mean, (3) typical or expectable, (4) conducive to the survival of the species, (5) innocuous or harmless, (6) commonly aspired to, and (7) most perfect or excellent of its class. Genetics shows us that pathological processes are normal (in terms of 1, 2, 3, and especially 4). In this sense, it is normal to be diseased. Alternatively, genetics cannot convince many of us that it is good to be diseased. In this way, disease stands in contrast to that which is normal (in terms of 5, 6, and 7) and thus designated abnormal. The point here is that the relation between normal and abnormal (Canguilhem 1978 [1966]), or health and disease, becomes even more complex in an era of genetic medicine.

This is not to suggest that trans-cultural accounts of disease (e.g., sickle-cell anemia, heart disease) are impossible. Clinical communities can and do agree on a great many explanatory accounts because they are bound by constraints of scientific methodology and influenced by similar notions of what makes biological and psychological life normal versus abnormal, pleasant versus unpleasant, good versus bad, and right versus wrong. In other words, humans world-wide share genetic endowment and many share environments. Given that disease depends on what sort of organism has the condition and on the relation of that organism to the environment, it follows that those organisms that share biological and environmental contexts can and will be seen to have a similar disease. In fact, science and medicine require that we classify in a general way in order to provide a basis for professional knowledge, investigation, and organized response within institutional structures. Beyond this, individual differences are important to understand the disease expressed by a particular organism.

A localized account of disease necessitates a localized approach to treatment, which raises a new set of bioethical challenges. At one end of the spectrum are genetic tests intended to identify people at increased risk for the disease and to recognize genotype differences that have implications for effective treatment. Pharmacogenomics focuses on crucial differences that cause drugs to work well in some people and not at all, or with dangerous adverse reactions in others. For example, researchers investigating Alzheimer disease have found that the way patients respond to drug treatment can depend on which of three genetic variants of the ApoE (Apolipoprotein E) gene a person carries (Lander and Schork, 1994, p. 2039). Increasingly in the future, we can expect pharmaceutical companies to develop highly specific drugs for individuals with particular conditions (Office of Science Education, 2000). How they come to know such differences will likely raise a host of bioethical challenges in the research setting.

At the other end of the spectrum are new drugs and therapies that specifically target the biochemical mechanisms that underlie the disease symptoms or even replace, manipulate, or supplement nonfunctional genes with functional ones. Collectively referred to as gene therapy, these strategies typically involve adding a copy of the normal variant of a disease-related gene to a patient's cells. The most familiar examples of this type of gene therapy are cases in which researchers use a vector to introduce the normal variant of a disease-related gene into a patient's cells and then return those cells to the patient's body to provide the function that was missing. This strategy was used in the early 1990s to introduce the normal allele of the adenosine deanimase (ADA) gene into the body of a little girl who has been born with ADA deficiency (Lander and Schork, 1994, p. 2039). In this disease, an abnormal variant of the ADA gene fails to make adenosine deanimase, a protein that is required for the correct functioning of T-lymphocytes. Other approaches to gene therapy include correcting a patient's genetic defects that involve only a single base change (chimeraplasty) and introducing new genetic information into a patient's cells leading to a reversal of genetic defects.

The bioethical challenges raised by genetic treatment are numerous. Depending on how distinctions between normality and abnormality are drawn, individuals will be regarded as diseased or healthy and the investment of health care resources will be seen to be necessary or unnecessary. Treating everyone in a world of limited resources is simply not possible. Treating everyone is, according to population geneticists, contraindicated given the role mutations play in the adaptation of a species to environment. Treating anyone with germ-line, as opposed to somatic, therapy is considered by some a violation of natural law. Treating a select group of individuals raises issues concerning who will be treated and why. Then there are those who will choose not to be treated even in the face of recommendations to be treated. Limits on treatment will turn, then, on the availability of resources, on what constitutes known and appropriate foci and means of intervention, on the criteria employed, and on the wishes of patients. The bioethical issues of justice, utility (i.e., benefit/burden calculations), sanctity of life, informed consent, and patient autonomy become central in discussions of how we understand and undertake disease in the Third Millennium (www.nhgri.nih.gov).

V. CONCLUSION

Taken together, the essays in this volume illustrate the difficulties in grounding any single account of bioethics that transcends societies and communities. The inability to ground a global bioethics has implications for how we understand disease in the new millennium. This is the case because disease is seen to involve descriptive, values, and social commitments, all of which are vulnerable to the difficulties in commanding global agreement. Genetics offers additional support for a local, as opposed to global, account of disease. One of the goals of this book is to reflect upon the challenges raised by the implausibility of a global bioethics, and for that matter our understanding of major concepts in medicine, and offer viable responses to engender further discussions. This account of the implication of our discussions of the possibility of a global bioethics on our understanding of disease is offered as an encouragement to continue thoughtful consideration of the epistemological and axiological challenges faced by medicine in the Third Millennium.

Department of Philosophy
University of Colorado at Colorado Springs
Colorado Springs, Colorado, USA

REFERENCES

Albert, D. A. et al. (1988). *Reasoning in Medicine: An Introduction to Clinical Inference*. Baltimore: Johns Hopkins University Press.

American Psychiatric Association. (1994). *Diagnostic and Statistical Manual of Mental Disorders IV*. Washington, D.C.: American Psychiatric Association.

Becker, G. K. (2002). The ethics of prenatal screening and the search for global bioethics. In this volume.

Billings, P. (1992). Discrimination as a consequence of genetic testing. *American Journal of Human Genetics*, 50, 476-482.

Biological Sciences Curriculum Study. (1997). *The Puzzle of Inheritance: Genetics and the Methods of Science*. Colorado: Biological Sciences Curriculum Study.

Canguilhem, G. (1978) (1966). *On the Normal and the Pathological*. C.R. Fawcett (trans.), Dordrecht: D. Reidel Publishing Company.

Caplan, A. L. (1993). The concept of health, illness, and disease. In: W.F. Bynum and R. Porter (Eds.), *Companion Encyclopedia of the History of Medicine* (pp. 233-248), Vol. I, London: Routledge.

Cartwright, S. (1981) (1851). Report on the diseases and physical peculiarities of the Negro race. In: A.L. Caplan et al. (Eds.), *Concepts of Health and Disease: Interdisciplinary Perspectives* (pp. 305-325). Massachusetts: Addison Wesley.

Cheng, C. (2002). Bioethics and philosophy of bioethics: A new orientation. In this volume.

Cutter, M. A. G. (1988). Explaining AIDS: A case study. In E.T. Juengst and B. Koenig (Eds.), *The Meaning of AIDS: Perspectives From the Humanities* (pp. 21-29). New York: Praeger Scientific.

Cutter, M. A. G. (1998), Negotiating diverse values in a pluralist society: Limiting access to genetic information. In R.F. Weir (Ed.), *Ethical and Legal Implications of Stored Tissue Samples* (pp. 143-159). Iowa: University of Iowa Press.

Engelhardt, H. T. Jr. (1995). *The Foundations of Bioethics.* 2nd ed., New York: Oxford University Press.

Engelhardt, H. T. Jr. (2002). Morality, universality, and particularity: Rethinking the role of community and in the foundations of bioethics. In this volume.

Fan, R. (2002). Moral theories v. moral perspectives: Need for a new strategy for bioethical exploration. In this volume.

Fleck, L. (1979) (1935). *Genesis and Development of a Scientific Fact.* T.J. Trenn and R.K. Merton (Eds.), F. Bradley and T.J. Trenn (Trans.), Chicago: University of Chicago Press.

Holliday, I. (2002). Genetic engineering and social justice: Towards a global bioethics? In this volume.

Hudson, K. et al. (1995). Genetic discrimination and health insurance: An urgent need for reform. *Science,* 27 (October 20), 391-393.

Hudson, R. P. (1983). *Disease and Its Control: The Shaping of Modern Thought.* Westport: Greenwood Press.

Juengst, E. T. (1995). Prevention and the goals of genetic medicine. *Human Gene Therapy,* 6 (December), 1595-1605.

Lander, E. S. and Schork, N. T. (1994). Genetic dissection of complex traits. *Science,* 265 (September 30), 2037-2048.

Lee, S. (2002). Reappraisal of the foundation of bioethics: A Confucian perspective. In this volume.

MacIntyre, A. (1981). *After Virtue.* Notre Dame: University of Notre Dame Press.

Murphy, E. (1976). *The Logic of Medicine.* Baltimore: Johns Hopkins University Press.

Nagel, T. (1979). *Mortal Questions.* England: Cambridge University Press.

Office of Science Education, U.S. National Institutes of Health. (2000). *Human Genetic Variation.* Washington, D.C.: U.S. National Institutes of Health.

Osborne, L. (2001). Regional disturbances. *The New York Times Magazine,* (May 6), 98-102.

Parsons, T. (1951). *The Social System.* Glencoe: The Free Press.

Pope, V. (1996). Mad Russians. *U.S. News and World Report,* (December 16), 38-43.

Popper, K. (1963). Prediction and prophecy. In K.R. Popper (ed.), *Conjectures and Refutations,* London: Routledge & K. Paul.

Proctor, R. (1988). *Racial Hygiene: Medicine Under the Nazis.* Massachusetts: Harvard University Press.

Reznek, L. (1987). *The Nature of Disease.* London: Routledge and Kegan Paul.

Schmidt, K, (2002). Stabilizing or changing identity? - The ethical problem of sex reassignment surgery as a conflict among the individual, community and society. In this volume.

Smith-Rosenberg, C. and Rosenberg, C. (1981). "The female animal": Medical and biological views of women and her role in nineteenth-century America. In A.L. Caplan et al. (Eds.). *Concepts of Health and Disease: An Interdisciplinary Perspective,* Massachusetts: Addison-Wesley.

Spector, R. E. (1996). *Cultural Diversity in Health and Disease.* 4th ed. Connecticut: Appleton and Lange.

Tonnies, F. (1957) (1887). *Community and Society.* Michigan: Michigan State University.

Walters, L. (1997). Reproductive technologies and genetics. In: R.M. Veatch (Ed.), *Medical Ethics* (pp. 209-238). Massachusetts: Jones and Bartlett Publishers.

Watson, J. and Cook-Deegan, R. (1990). The human genome project and international health. *Journal of the American Medical Association*, 263, 3322-3324.

Weiss, R. (2001). Genome puzzle yields big picture. *The Sunday Denver Post*, February 11, 2001, A1, 18A.

Wolf, S. (1961). Disease as a way of life: Neural integration in systematic pathology. *Perspectives in Biology and Medicine*, 4, 288-305.

www.nhgri.nih.gov/98plan/elsi (Ethical, Legal and Social Implications [ELSI] Research: Goals and Related Research Questions and Education Activities for the Next Five Years of the U.S. Human Genome Project).

CHUNG-YING CHENG

BIOETHICS AND PHILOSOPHY OF BIOETHICS:
A NEW ORIENTATION

I. INTRODUCTORY REMARKS:
A DEEPER UNDERSTANDING OF HUMANITY

Bioethics must emerge from biological findings. Without biological findings or without a basic biological understanding of the human person there can be no bioethics. Here I take the view that even though biological findings are scientific discoveries, the reason why they can become the basis for biomedical technology or simply bio-genetic engineering skills is that, relative to a deeper understanding of humanity, they generate values for human pursuit, and values generate norms of human action. The present state of bio-medical technological development and bio-genetic engineering research is so full of miraculous achievements and results that we are under pressure to ascertain their values, and also to formulate or decide what norms and rules of action can be formulated, adopted and justified.

Take, for example, the case of cloning. Granted that we now know how to clone a human person and all living things including animals and vegetables, where values of cloning really lie and what conditions define or confine its values remain basic questions. We know arbitrary cloning in grains may create a destructive dependency, which may lead to farming disasters and agricultural monopolies. Further, as eaters of grain, the human species even may be endangered. But there are obvious gains or values in cloning special grains. Our question is about the boundary conditions of these values as well as the boundary conditions of our actions and policy-making. When we come to the cloning of animals, we are equally confronted with the problem of quality of the results of cloning and the consequent problem of power domination and price monopoly, to say the least. The impact on humanity would be tremendous and irrevocable. This means that we need a strong and deep ethical reflection on the biogenetic technology and biogenetic skills and their power of genetic, ethnic, social and economic impact and transformation. For this ethical reflection we need also to delve into a deeper knowledge

Julia Tao Lai Po-wah (ed.), Cross-Cultural Perspectives on the (Im)Possibility of Global Bioethics, 335–357.

of such skills and watch and wait for clinical results which may take decades to yield significant information.

Or take another case for assessment: contract pregnancy and surrogate motherhood. Artificial insemination and ova replacement are no doubt technically feasible and in fact have been practiced on a fairly large scale. But relatively speaking, the time during which this has been practiced is still too short to tell its social impact. But if these techniques are widely adopted (when people see their convenience and profitability), we shall face unfathomable long-term societal consequences. It has been pointed out that this practice would lead to a strong commodification of human life, which would demean human life. Furthermore, it would slowly disintegrate the existing system of family, for it would result in a loss of the inherent power of mutual attraction between sexes, and the life-rooted commitment of the married couple to care for children. We simply do not know (or know only very little of) the extent to which these suggestions might be true. Hence, even though we can see values of such technology, we may not recognize the limitations and boundary conditions of these values. Without a clear understanding of such values, how do we formulate and justify our norms of action? Hence, again we need to examine ethically the biomedical technology of contract pregnancy and surrogate motherhood and, for this purpose, dig more into the case studies of such a practice.

All in all, it seems clear that bioethics needs a new orientation which would oblige us to systematically examine the values and norms of our given knowledge of biomedical and biogenetic skills and theories on the one hand, and to carry on more research on and study of the empirical and practical impact of such technologies on society on the other. We need to pose questions to the scientific researchers regarding the status of present findings, possible outcomes of current research projects or programs, assessment reports and clinical data so that ethicists can see what values would proximately emerge and what norms would shape up in light of this certain and probable information. If there is no critical understanding of the research work, there can be no critical evaluation of its ethical implications.

Even with regard to the ethical assessment of the research activity itself, we need an understanding of what poses as an issue or problem. Take for example the recent report (in *The New York Times*, April 20, 2000) on the research findings on fetal cells' ability to cure 30% of Parkinson's patients under sixty. Does this justify fetal cell research?

Does this contribute to the redeeming value of abortion? Without more trials there is no way to reach a conclusive judgment.

For this dual purpose of re-examination of bio-ethical principles and their foundations, we have to examine our understanding of the human person or the human individual so that we know what general values we have to uphold and preserve. We should then reflect on the technology of bio-medical engineering and research to see how they are related and placed with our existing values and norms, and to what extent and in what direction they would benefit humanity and therefore lead to the recognition of new values and adoption of new norms of action.

II. LEVELS OF HUMAN EXISTENCE AND HUMAN PERSONALITY

Before introducing my understanding of what a human person is, it is important to make a distinction between a human individual and a human person. For this purpose we need to explain what a human individual is first, for it is on the basis of this conception that we can build our concept of a human person. An individual is an entity which has a distinguishable substantial identity which endures in time. Hence material objects such as the sun and moon are individuals, animals and insects such as dogs and beetles are individuals, and human beings are individuals.

Individuals furthermore can be named and identified by designating terms, but this does not mean that they are actually named. Normally there do not exist social rules for behaving toward material and animal individuals. There nevertheless exist some general well-understood rules governing our behavior toward material things, plants and animals. Do not destroy things. Do not mistreat animals. Do not destroy flowers and grasses. Unless we have a specific purpose to serve, there is no reason and no cause to destroy things, destroy plants or mistreat animals or insects. We may regard these general principles as belonging to the Daoist principle of protecting the environment and respecting life. This is the general principle we may call the No Harm Principle. We may also call it the Follow Nature Principle, in the sense that we should not disturb or destroy natural ways of things, plants and animals without good reasons or simply for selfish interests and greed.

What then is a human individual? A human individual is an individual that has the biological characteristics of a human being. It is first of all a human organism, and has the individual and overall general functions and abilities of a human being, both biologically and psychologically identifiable. Biologically, a human being is a featherless biped with a round head and square feet. To qualify as a member of the human species, a human individual has to have a certain cerebral volume, and he is also required to have the potential ability to think and speak. Psychologically, a human individual must have consciousness and self-consciousness so that he is aware of his own uniqueness and is even capable of casting doubt on his own existence. Furthermore, a human individual is assumed to have inner feelings and desires and to be capable of pleasures and pains, sometimes consciously and sometimes unconsciously. He may also harbor an existential anxiety over his finite existence and mortality as Heidegger has suggested.

We see that, in the above specification, a human individual is both externally (behaviorally) definable and internally (self-reflectively) definable. In a broad sense we attribute to a biological human body with organic functions, a psychological configuration of inner abilities and capacities. Thus a human individual is always regarded as capable of seeking pleasure and avoiding pain. By reflection and observation a human individual is capable of observing the principle of reciprocity of human feelings, as suggested in the Confucian and Neo-Confucian philosophy of the human person. In fact, the Confucian and Neo-Confucian philosophers always conceive human persons as capable of self-cultivation and self-transformation.

What then is a human person? Normally we can regard all human individuals as human persons. But it is certainly possible to regard a human person as demanding more than just a human individual does: A human person is an agent who is capable of acting on other individuals to cause pain and pleasure, and hence who can be held accountable or responsible for actions originating from him or her. We have seen many controversies surrounding the idea of a person, but we cannot have a notion of a person without having a notion of the human person, which is founded on a notion of the human individual.

When a human individual does not have the ability to act on others in a direct or calculated way (with intention directed to another individual or to a non-individual thing), we may not regard the individual human as a full person or a person. In this sense we can in fact speak of degrees of personhood or 'personal powers'. In this sense we do not have to assume that a person must act morally, for a human person may have powers exercised morally, non-morally or immorally insofar as we can describe those powers according to the ways of their exercise or the consequences of their impact or influence. Hence we may have a human individual who is a moral or an immoral person. We may even separate being a person from being a human individual or even from being an individual. We may speak of God as a person who is not human or even may not be an individual. The Christian God in some historical Christian theology is an individual person whereas the Brahman in the Indian Veda literature is a person which is sometimes an individual and sometimes not. The Dao in Daoism is neither a person nor an individual but assumes a power equal to the most powerful person, namely God.

We may use the term 'person' in various ways; one way is to use it metaphorically in terms of the human person. After all, a person is a *persona* or a mask in its original Latin sense. As a human being sees his own image as a human person, his personality therefore is considered the character behind his personal image as presented to others. Here it may be pointed out that in my definition of the person as an individual having active power to change things or create things, I may appear to deviate from the original meaning of person as *persona*. It is clear that the idea of a person as *persona* focuses on the experiences of an observer toward an object presented to the observer, whereas the idea of a person as an agent focuses on the experiences of the change and transformation as induced by an agent, whether the agent is visibly presented or not.

This agency-based personal individual is actually conceived in all descriptions of creator-gods or first causes. Without such powers of change, transformation and creation, the creator-god or first cause would not be so labeled. Hence it seems that the notion of agency or power is essential to our conception of God as a divine person or a spiritual person. One reason why we call God a divine person seems to be that we need to conceive of God in one way or another and that to conceive of God as having a *persona* (a mask or human face) is both convenient and useful. After all, as a creator, God is like a father and as a father he can be conceived to lead, guide, chide, punish, protect, patronize, encourage and award. Hence God is conceived to be a personal God or a Divine Person, with all His creative powers of change and transformation and yet having the human *persona*. In reality, God need not be conceived of as having a human *persona*, even if God is conceived of as a person in my sense of the person as agent.[1] One may wonder why a human person has the human *persona* at all, if not made in the image of God. This could be useful hypothetical reasoning, but in light of our imaginative faculty, it might be suggested that it is in the image of the male human man that God is conceived and that this is needed for endowing dignity and uniqueness to the human existence. On the other hand, it is possible also to think that there is a creative power in the universe which is infinitely intelligent and possesses a subtle way of self-sustainable development and growth in terms of which the human individual with his human face evolves into existence. In fact, we see in the Indian philosophy of 'spanda-karika', the Upanishads belief regarding Kashmir Saivism to the effect that God is pure energy and pure action and yet has no persona or personality at all. This would avoid the question of the gender of a personal God. But again, the idea of an intelligent and creative universe is no doubt compatible with the existence of the human person as well as with the idea that the human person endows such a power with a human face. This is because human persons have the ability and desire to anthropomorphize all higher forms of being and even give a human character to lower forms of beings, which have their own characteristic forms, such as horses, birds, dogs and cats.

Apart from a divine person such as God, we can also speak of a legal person that is in fact a matter of legal incorporation resulting in the creation of a business company or an organization. The reason that we call it a legal person is obviously that it has the ability to act and conduct itself according to rules of its charter, formulated according to its purpose and objective of incorporation. In other words, a legal person is a legal agent who can induce changes and initiate transactions. When the corporation becomes insolvent and dissolves, the legal person of the corporation also dissolves, for it can no longer assume the function of an active power. Similarly, when a human person loses her ability to take care of herself, her function as a person also declines and, therefore, she cannot be trusted or relied upon as a person, but instead could be loved or cared for as a human individual or human being.

It is to be noticed that when we redefine the human person as an active power, we do not specify whether the active power vested in the human person is good or bad (good in the sense of to do good things, and bad in the sense of to do bad things), as it is ordinarily understood. Without going into a detailed examination of good and bad, in the sense we have experienced them as human beings, it is nevertheless clear that good indicates an action or a process which produces benefit to oneself, others, or one's community, whereas bad is an action or a process which produces harm to oneself, others, or one's community.[2]

One may indeed query how such a conception of good and bad is related to right and wrong. Again, without examining our common use of the notions of right and wrong, we suggest that what is right is what conforms to a norm and what is wrong is what violates a norm. We must admit that there are degrees of conformity to and degrees of violation of a given norm. We must also admit that not all types of action are normalized in the sense that we have unquestionable norms for those actions. In fact, what we have defined as norms of action are results of practice and informal legislation which have become incorporated in the community life or accepted as guides and/or principles of our action. They may or may not be further rationally justified or legitimated in a second-order norm or principle or theory. They may not necessarily form a system, which could cover all new cases of action.

Given such an understanding of norms in a given community, what is regarded as right action need not be right in this higher sense of right, i.e., conform to a different principle or a principle of a higher order, and what is regarded as wrong action need not be wrong in this higher sense of wrong, i.e., violate a different principle or a principle of a higher order. Hence there is always a need for argumentation and justification for innovative actions, which appear to violate or comply with a given norm. Nevertheless, given this understanding of norms, we may still argue and point out that norms of action are accepted or tolerated because they are benefit-productive or harm-destructive. To say the least, these norms must not be harm-productive nor benefit-destructive. It is in this sense that we reconcile the good and bad of an action with the right and wrong of an action in a framework of both act-utilitarianism and rule- or code-utilitarianism.

III. PROCESS-AGENCY THEORY OF THE HUMAN PERSON

Now to go back to our notion of the human individual or human being, a human being is a human person if he or she has the power to act on others in a community or the world. In order for such action to happen and to bear a result, it is no doubt necessary that human beings have wills and motives to act, as well as have intelligence and reason to make actions effective and efficacious. He must, in other words, have both a sense of end (teleological rationality) and a sense of means (instrumental rationality) to accompany his will to act. As his action is to bear on others in the community, he is judged good or bad and his action judged right or wrong depending upon how he bears out his action and his general conduct. He is required to do certain things and/or not to do certain things in order to be accepted into a community. His doing and/or not doing these certain things become his obligations; he is guaranteed liberties to do certain things and/or not to do certain things and his doing and/or not doing these certain things become his rights. It is clear that I am here taking obligations and rights as community-endowed in their origins, directly or indirectly linked to the good and bad, right and wrong of the actions of the human person.

What is projected here is an understanding of the human person as always community-oriented and community-involved, even though as a human being he has the power to feel pain and pleasure and to possess other emotions, and to be capable of desiring satisfaction for both the material and spiritual needs of his life. But it is in his ability (or on behalf of his ability) to act and in his actual actions that he becomes a human person. It is obvious that a human being in his or her normal life cannot but be a human person. The point of my distinguishing the human being from the human person is to recognize the human status of human babies and even human zygotes and human fetuses insofar as these entities are confirmed as being human or being capable of definitely becoming human, i.e., possessing human genes which internally sustain, one way or the other, the form and function of the human person. As human beings or potential human beings (human beings-in becoming) they deserve human care and human love as much as we are able to extend. We, as full human persons, should extend to them care and protection regarding their potential well-being or even rights to no harm insofar as these do not conflict with a common or larger good. On the other hand, when we come to see the presence of the active powers in human beings, we have to treat them in accordance with demands of doing good and avoiding harm. We would require them to have virtues, (as abilities to do good) which are bases for developing or configuring obligations and rights, duties and freedoms.

It is also important to see that human beings are to become human persons just as potential human beings are to become actual human beings who are then potential human persons. (One might see that a human fetus is a potential human person.) There are different philosophical theories which propose that human beings have a nature which is either good, bad, or a mixture of good and bad. Now we may ask whether and how these theories of human nature have to do with understanding the human being and the human person. Insofar as Chinese philosophy is concerned, it is clear that these theories of human nature are geared toward explaining how and why human good can be achieved.

According to the theory of Mencius, human nature contains the seeds of goodness in the form of feelings and initial desires which are benefit-productive. Hence it is our obligation as human beings to firm up, cultivate and develop these feelings into actual moral actions and practices so that a society and a state can be founded on the basis of these moral actions and practices. On the other hand, the theory of human nature, as conceived by Xunzi, projects human nature as sources of human evil and immorality, because it inclines human beings to excesses of selfishness, licentiousness and self-abandonment. Thus, in order to achieve a harmonious and productive society, we need to curb these inner tendencies of human nature and train or educate human individuals with reason, rules of order and propriety which come into existence as a result of rational self-reflection on the part of enlightened human persons, i.e., the sages and morally superior persons (*junzi*).

Even though these two theories of human nature are opposite in identifying essential sources of human good and human bad, they nevertheless agree in affirming the importance of the transformation and development of human individuals. Although there is a difference between the transformation that lies in affirming the autonomy of the individual human being in the Mencian theory, and the transformation that lies in affirming the importance of external behavior modification in the Xunzian theory, these two theories further agree on the ideal of transforming the human being, not only into the human person but into the human person toward perfection, namely the supreme moral person described as the sage (*shengren*).

There is a reason for bringing up the subject of the *shengren*: we need to see that our reflections on bioethics can not simply or justly stop at the level of the human person as given. This is because unless we understand all the potentialities of the human being or the human individual, we cannot know how biological and biomedical discoveries and inventions will help our becoming, not just a full person but a supreme moral person or the *shengren*. One thing is quite clear: the idea of a *shengren* requires an understanding of the moral developmental abilities of human nature. It requires that the human being be actively involved in developing himself and that he relates to others through his actions for good. All this of course presupposes that the human being is a free and creative agent with autonomy of his own will, and that he is capable of self-discipline and self-control. Confucius says in the *Analects* that at seventy he can do what he desires to do without transgressing the moral rules. This reveals how morality requires an inner source of development and creativity, even though morality has to be evaluated in terms of rules and consequences.

It is clear that my conception of a human person is different than that of Professor H. Tristram Engelhardt, Jr. Whereas Professor Engelhardt distinguishes between persons in a strict sense and persons in a social sense,[3] I see only a distinction between a human individual and a human person. A person in the strict sense is a full-fledged human individual endowed with self-consciousness, moral concern and rationality. He is no doubt an agent and therefore qualifies as a person under my definition. But a person in a social sense could be an infant or a senile patient. Persons of this sort are only derivatively defined, rather than defined in terms of the characteristics of the persons themselves. For the strict definition of the human person would exclude them from being human persons. But in my definition, an infant is a human being, as is a senile invalid. Even a human fetus could be recognized as a human individual, albeit a potential one, but certainly not as a person in the social sense.

The point is that a human being is a being developed in a sequence of certain specifiable stages: there is an early stage of being a biological individual or living individual, followed by the stage of consciousness and self-consciousness in terms of perceptions, desires and feelings. Lastly comes the stage of moral will and action, including moral practice and creative productivity. We can see that becoming a human person usually entails going through these stages in sequence, for there cannot be a moral will to act before one has moral feelings and experiences of life (in terms of pain and pleasure, strife and harmony). But what a person has acquired in a later stage can be lost in a still later stage, thus causing a lapse into an earlier stage of development with fewer abilities and more restricted visions and wills. We can see this lapse in a normal life cycle when one becomes old and senile, or a collapse when one encounters an illness such as Parkinson's disease or Alzheimer's disease, in which a person can become disoriented and lose a great quantity of memory. Hence a person will lose his ability to act and conduct useful and creative work.[4] In these cases we seem more willing to say that these persons are less social persons or persons in a social sense but more human individuals who lose certain powers of agency and capacities to make an impact. They are nevertheless human beings who are capable of suffering and even love and who deserve human care and human love.

Given our description of the human person from a process-agency point of view, we may now summarize our description of the human being/becoming (to be called a 'human process-entity') in terms of the five levels of existence or five stages of its development:

1. Human process-entity as a physical object;
2. Human process-entity as a biological individual;
3. Human process-entity as a self-conscious moral being;
4. Human process-entity as a moral agent and creative power;
5. Human process-entity as a full realization of moral perfection.

We call a human process-entity a human person when he or she has reached the 4th stage, whereby social relations and social well-being are possible. But we recognize that even though a human being normally develops into the 4th stage and even comes to embrace a desire to become fully realized as a sage, represented by stage 5, he or she could lapse back into stage 3, 2 or even 1, at his or her death. The importance of this understanding is that there could be different ethical principles corresponding to each and every stage above. We may indeed suggest the following ethical principles or considerations in dealing with individuals on each stage or level:

1. Ethics of aesthetic appreciation;
2. Ethics of vitality and strength;
3. Ethics of love and support;
4. Ethics of respect and responsibility;
5. Ethics of emulation and fulfillment.

IV. PRINCIPLES AND THEORIES OF ETHICS IN STAGES

We may now ask how and why we have the above different forms of ethics in line. The basic justification of any ethics is a recognition of reality as reality which is capable of emerging or embodying a value and hence presenting a norm. If we simply follow the distinction between value and fact, in the tradition of David Hume and G. E. Moore,[5] we would not have any reality-rooted ethics at all. All values and norms would become a matter of mental invention and subjective projection, which would then either require objective justification or lead to relativistic controversies. On the other hand, if we grant ethics a realistic-ontological basis, there is a common ground and shared base for more universal agreement and commensurable understanding.

As a matter of fact, there is no reason why the subject and object of knowledge and action may not form a unity based on the underlying unity between the subject and object in an ontological reality. It is only by appeal to an ontological understanding that ethics emerges from values, and values issue from facts. Of course, even with this understanding we still can maintain a distinction between fact and value for they could be regarded as belonging to two different discourses serving different human purposes. Yet with the mediating reality, fact can become evaluated and transformed into value, and value can become objectively analyzed and therefore transformed into fact or knowledge of fact. In both processes the human person has performed or acted in light of his deeper understanding of reality and his deeper understanding also becomes effective because of his action and practice.

With this being said, we can see that on the level of the human process-entity as a physical object, a human process-entity commands our attention as a physically individuated object, which has presented its physical forms and physical organization as any other physical object. Like with regard to any physical object, there is no reason for us to do anything except to relate to the human process-entity as an object of aesthetic appreciation, for any other action would call for a justification. To appreciate an object aesthetically is to see or observe the object without any exterior motive. Kant has argued that there is an intrinsic value (hence no external interest) in appreciating an aesthetic object. Or, to put it a different way, for Kant, when we pass a judgment of beauty in recognizing something as aesthetically attractive or beautiful, we realize an intrinsic value in ourselves in viewing the aesthetic object.

Here I take a much broader view than Kant. I postulate that any physical object is potentially an object of aesthetic attraction and in fact that every physical object is realizable as an object of aesthetic appreciation whether positively beautiful (aesthetically attractive) or negatively ugly (aesthetically repulsive). That a physical object is capable of evoking such an aesthetic response resides in the nature of the physical object in relation to us as conscious human beings. If we become merely physical objects, our relationship to other physical objects could still be thought of as aesthetically attractive or aesthetically repulsive and this relation of course could be analyzed as a relation of order or disorder, harmony or disharmony, in our imaginative reflection. We may also adopt a Whiteheadian language of mutual prehension or a Yijing language of complementary *yin-yang* interaction in describing the relation between two physical objects, which would give rise to an understanding of value. As value is realized, the minimum norm for us as agents is to strive to preserve the value if it is positive and to discard or transform the disvalue if it is negative.

Another observation needs also to be made explicit. We have to consider humanity as an organically related whole because there is a time process of development from potentiality to actuality, and from an early stage to a later stage, with the possibility of lapsing into a quasi-earlier stage. But from the organic holistic view, we have to see any stage as an integral part of the whole process and we have to consider the importance of the value in an earlier stage in its contribution to the later development and the whole process as a structure. In this sense, in any later stage of development of the human process-entity, we must appreciate the contribution made by the human process-entity at an earlier stage. Not only does the earlier stage process-entity form a base and starting point for the development of the later stage process-entity, but the earlier stage process-entity can be regarded as carrying the very motive power for such development.

It is in this regard and by this understanding that we can derive the principle stating that a process-entity has value not just in itself, but in terms of the place and function it possesses and plays in the development of the process-entity in later stages. Hence the norm which emerges from such a value must be one which would encourage us to preserve and appreciate as much as we can the given process-entity as a phase of the later development or as a part of the whole process-reality. It is on the basis of this understanding that it is easy to justify the Confucian demand that we pay filial piety to our parents, especially when they become old. When our parents become old, they lose their powers of action and may come to exist as dependents with reduced being and energy. But do we therefore abandon them? No, we have every reason to support them, love them and appreciate them because they have done their utmost to love and support us as children so that we can grow to become a later day process-entity.

Because of this holistic and organic consideration, it appears that not only should we aesthetically appreciate a human being as an aesthetic object, but we also have a special obligation to pay respect to a human being as a physical object when that human being is intimately and organically related to us. Even when our parents become deceased, we do not discard them as waste, but still treasure their bodies as something dear to us. This is the ethics of aesthetic appreciation realized as an ethic of filial piety when applied to the parent-child relationship. This ethic of aesthetic appreciation no doubt could also extend to our aesthetic appreciation of history, culture, and the whole of nature.

V. ETHICS OF VITALITY AND STRENGTH AND THE JUSTIFICATION OF BIOEMEDICAL TECHNOLOGY

What then is the 'ethics of vitality and strength'? When we come to confront the human individual in the biological sense, we see the human individual as an organism coming into existence at the time of conception or birth and ending its existence at the time of death. In this sense an organism is a finite being occupying a limited period of time for life and having a restricted region of activity. During this period of time the human organism can suffer from attacks of disease, become handicapped, or be destroyed by diseases. Even if this organism does not contract fatal diseases, it will still slowly age and become old in the sense that all its organs will become worn out and become subject to attacks of fatal diseases. Even if we have wiped out all causes for disease, the organism will finally grind itself to a stop because its organs will run out of energy and power. This means that the organism will die a real natural death.

Given this understanding of the human biological entity, it seems natural for a human individual to wish that his life span be prolonged and his body be rejuvenated when it is old, and purified or cured if he suffers from disease. What makes it natural to wish these two things is that one would consider living as an intrinsically good and desirable state of existence, and living in whole health as an even better and more desirable state of living (more desirable than being diseased or dead). What make living and wholesome living a desirable or more desirable state of existence is not only that we enjoy the state of living, but that we see that the state of living enables us to accomplish what we hope to do with our heart, mind and spirit, which it is the purpose of life to achieve and the end of life to reach. In other words, we wish to prolong our life and increase our abilities of action because we wish to achieve our values and reach our ends with our heart and mind. In this sense, what science and technology can do to enhance our living, cure our diseases, and prolong our lives is desirable and valuable by our measure of power and energy.

The question we face today is whether science, technology, and, in this case, biomedical technology, which serves a useful purpose for life, may presuppose or require an immoral or inappropriate use of resources or devices and come to benefit only a few selected people. Here the term *immoral* is intended to suggest that these researches may prejudice or shortcut the same kinds of values we are supposed to bring about or increase through our biomedical research. This enables us to see that the issue of biomedical technology is not just an issue for bioethics, but also an issue for the valuation and evaluation of biomedical technology— namely, an issue for whether the development of biomedical technology serves the whole purpose of the society of human persons. In fact, this enables us to see that the value of increasing the power and length of human life does not lie with such an action, but lies with what such an action requires and what it brings about, for the human organism is not just a human organism, after all, and there is not just a single human organism to take into consideration.

The point to drive home from the above is two fold: If we were to consider the human organism as simply an organism or biological individual, independent of its being more than just a biological individual and independent of there being other biological individuals, merely desiring such a power increase is a reason for doing it. In this sense biomedical technology, which cures disease and prolongs life, is desirable. But as the human organism is also a human individual and a human individual could also be a human person, and a human person could reach an ideal of moral perfection, we have to judge biomedical technology in light of and in terms of requirements of value on these higher levels of being. This implies that there are two basic standards of valuation of biomedical technology which supervene on the ethics of vitality and strength, namely:

1.   The Holistic Principle: Biotechnology has to conform to the potential and inherent ends of a whole person, not just the ends of a biological individual;

2.   The Universalistic Principle: Biotechnology has to conform to the potential and inherent ends of the whole community of human persons, not just the ends of a single biological individual or human person, or a select community of biological individuals or human persons.

In general, in using these two principles, there does not seem exist any serious difficulty or problem for justifying efforts and results dealing with medical cure, healthcare and anti-aging research. But if we look closely into the ethics of medical cures, which involves organ transplants, there are questions of personal identity in brain cell transplants apart from the problem of the commodification of human organs. Hence there is always the issue of the control and legitimacy of the acquisition of organs. There is also the intrinsic question of personal and familial-social adjustment which needs special consultation and arrangement and hence a reckoning with personal and social cost factors.[6]

Or take the anti-aging ethics of using Viagra® and similar drugs. One may question whether, being basically effective with male human organisms rather than with female human organisms, the use of Viagra® by the male human organism would lead to a male dominance and a consequent discord between the male and the female organisms. Perhaps we need more empirical data to warrant a stable judgment of value and thus more study and observation to even formulate a norm of conduct.[7] This no doubt brings out the question of the so-called 'ethical risk' and likewise question of 'ethical luck'. We have to venture an ethical evaluation though even we do not command the full data of the conditions for the development of a biomedical technology and the results of its use. Like buying a stock, one has to venture according to its long-term benefits and short-term performance, for the good or the bad will not usually justify its investment or divestment. But curiously enough, as people buy stocks for different purposes, one may suggest that the ethical risk or the ethical luck of the use of a biomedical technology could vary with various kinds of people. In this sense, we need not have only one type of biomedical technology, but as many kinds as are possible, insofar as they are compatible and consistent with our two principles of ethical valuation and evaluation.

When we move to the area of manufacturing birth by test tubes, contract pregnancies, and cloning, we enter into an area with an even higher factor of ethical risk. This is because we do not know the right ethics of birth in this area of birth technology. We see enough litigation over contract pregnancy and surrogate motherhood cases. But the real problem is the consideration of the biological individual involved in contract pregnancy with regard to his or her becoming a moral human being. How would this individual feel in relation to three women who could claim to be his mother? Or how would the individual human being feel in relation to his genetically cloning parent or the doctor who did the cloning? Will this feeling radically change the traditional human relationship and hence the nature of the traditional community and society? Is this change for the better or for the worse? Again there are hard questions to answer and only in time can we answer them.[8] But again, there is a high ethical risk involved.

It will be a far more serious problem when man learns the secret of how to manufacture life, not just to manage or arrange the birth of life. We shall confront the fundamental question of power of creation and the correlative question of domination or supreme domination of power. If we combine such power with the power of dictating death, these few selected or privileged human individuals would become Lords of Life and Death, and the whole crowd of human beings would have to obey the rules of life and death as commanded by these few Lords or just one Lord. Perhaps we would have to start our culture and civilization over again, superseding the present one as obsolete and antiquated.

This unlikely vision (which may be so only from the present point of view of human beings) brings out the moral message implicit in adopting the ethics of vitality and strength, namely, that science and technology as practiced by humankind today are basically geared toward and premised on the ethics of vitality and strength alone, which are infinitely desirable independent of considerations of the Holistic and Universalistic principles for moral valuation and evaluation. This ethic has a strong appeal to human individuals because the human individual feels that in light of this ethic, he would be a benefactor as well as a beneficiary, even though he would, in all probability, be a villain as well as a victim. Hence we see the charm of the Nietzschean argument for a morality of masters over a morality of slaves. Even slaves wish that someday they will become masters and therefore be able to dominate other would-be slaves who are not now slaves themselves. Because of this Nietzschean charm, it is possible that science and technology will revolutionize our society and hence slowly supplant a western traditional society. Similarly, biomedical science and technology will revolutionize our life forms and our images of the human individual and society on a fundamental level. I can envision such a process of revolutionary change which may well become inevitable, but I wish to argue for a face-to-face or back-to-back transformation in full view of the reality of the human individual as a moral being and as a responsible person, and hence in full view of the ethics of love and care and the ethics of respect and rights (and hence of responsibility).

There is in fact more to say about the ethics of vitality and strength which can be also called the 'ethics of the will to power' in light of the Nietzschean philosophy. Even though we need to deal with the power of biomedical technology to manufacture birth and dictate death through an ethics of birth and an ethics of death, we have not covered the full range of issues to do with the ethics of vitality and strength. In the present day we are familiar with the issue of abortion, which can be regarded as a matter belonging to the ethics of no birth, and the issue of dying with dignity (the issue of euthanasia), which can be regarded as a matter belonging to the ethics of death. There is of course the issue of artificial life or the issue of bringing life into a machine or merging a machine with life, which is no doubt linked to the issue of creation of organic life. This issue may be called the paradigm issue of an ethics of abolishing death.

All these issues deserve separate discussions which I will not be able to give on this occasion. But to mention them in connection with the ethics of vitality and strength is highly significant, for they are the issues which directly challenge the life of the human individual and hence invite solutions by biomedical science and technology. These issues belong to the very foundational level of human existence as a biological organism. They are the issues which are derived from the human experiences of birth, aging, illness and death, which require understanding, explanation, solution and overcoming. It is on the basis of these fundamental experiences that Satya Sukyamuni has come to a deep understanding through his deep meditation. His understanding and solution are transcendental in a sense, and not technological. He asks us to see these four phases of life (life, old age, disease and death) as mere deep-seated illusions of human nature, which have no substance at all. Hence one must overcome these experiences by a deep understanding called 'enlightenment' and transcendence into nothingness. With this proposed understanding, a practice of the Buddhist ethics of life is dictated and evolved. Similarly, we see how Christianity settles these issues of life and death and proposes its own way of conducting life in a Christian theology of God of perfect wisdom.

Given these historical examples of solutions by way of a fundamental understanding, we might suggest that a biomedical science and technology which would or could revolutionize our understanding of life and death needs to produce and develop a way of life and an ethics based on its own fundamental insights into life and death. For the time being, we do not yet have these insights. Therefore we have to still appeal to the ethics of love and care on the level of human individuals and to the ethics of respect and responsibility on the level of human persons and even to the ethics of emulation and comprehensive benefit on the level of the moral perfection of the human person.

VI. FORMS OF ETHICS OF A HIGHER LEVEL

In dealing with the ethics of a higher level, it is clear that we enter into philosophical reflection on the nature of human being and a subsequent understanding of human existence. It is on this basis that we first observe that as human individuals, we have inner feelings which constitute our responses to things happening to us from the outside world and which also constitute our inner world of aspirations and visions. It is clear that on this level of feeling we come to see the central importance and value of love and care. It is by enduring love and self-giving care that we become human individuals in a community of human individuals: We come into being together with the coming into being of our communities. We are simply being-in-community, not just biological organisms, whether alone or aggregate.

It is then clear that based on this requirement of love and care, vitality and strength become only relatively important, and not absolutely important. From this point of view, a biomedical technology is to serve the purpose of such an ethics, not vice versa. Any biomedical technology which would overturn this ethics of love and care (such as is possibly presented by Confucianism or Christianity) would not be pursued or would be discarded. A not altogether perfect example, perhaps, is that of the practice of Christian Scientologists in the U.S., who reject modern medicine by sticking to a traditional but systematic form of life of their own. As an ethics of love and care forms the central core of present day communities, it is important to see that any change of human life and human practice resulting from the biomedical technological revolution must not overturn but instead should preserve the value and practice of love and care for the human community. In light of this requirement, we cannot really substitute a biomedical system of love and care (via machines and computers) for the real human touch. Instead, we need to develop a human-centered, humane and friendly implementation of biomedical technological care in hospitals and elsewhere.

When human beings are regarded as full human persons who are responsible agents in the community and the world, we must require a high sense of responsibility from the development of biomedical technological skills. These skills not only must demonstrate a high degree of self-responsibility in providing love and meeting needs of care, they must also be consistent with a just distribution of obligations and rights. In other words, these biomedical technological inventions should not be used as instruments of domination and political power, but must instead contribute to the cause of justice as held up by the ethics of respect and responsibility. It is clear that new rights, new obligations, and new rules for distributing or attributing such rights and obligations, require wide discussion and long term working out, similar to the case of working out the rights to intellectual inventions which improve our abilities and speed of hearing, viewing and writing or publishing. Whether any one biomedical skill is acceptable or not depends on a rigorous ethical examination of each case. It is clear that all cases of ethics of birth, ethics of no birth, ethics of disease, ethics of cure and healthcare, ethics of aging, ethics of anti-aging, ethics of death and ethics of abolishing death fall into this examination for the future development of humankind.

Finally, we come to the ethics of emulation and comprehensive benefit. Even though human beings are not ethically or morally perfect, they have the right and natural ability to aspire to something better than themselves. It is in this sense that we can speak of a morally perfect person, to whom we may not only aspire, but whom we may emulate in our lives. In such emulation we may become more human in the sense that we may become able to develop and bring about more love and care for humanity and create conditions provide more love and care. In this sense an actual person could act ideally as an ideal person. In view of this understanding, it is clear that any biomedical technology must not prohibit a human person from becoming idealistic in this moral-perfect sense, to say the least. It would be even more desirable that biomedical technology enhance and promote such a possibility of moral perfection. However, for the moment, it seems unclear how any biomedical technology could have any such value in sight. Apparently, the present biomedical technology has basically only the ethics of vitality and strength in view and has not ventured beyond considerations of human beings as biological organisms.

It is interesting to note that Heidegger urged the unfolding of authenticity in human beings by way of reflection on the finitude of life. It is by such reflections that one could be awakened to make decisions to do the right thing in one's life, in order to make it significant and meaningful. It is equally interesting to speculate on what a human individual would do when he or she is given immortality by bio-medical technology. Would he act to make his life more meaningful and more valuable or would he be lost in a platitude of prolonged life routines and delayed moral decisions and reflections?[9]

VII. CONCLUSION: TOWARD THE FUTURE WITH THE PAST IN RETROSPECT

We have examined and explored five levels of existence for human individuals and distinguished five forms of ethics of human existence based on these four levels of human existence. It is important to see the human individual as an object of aesthetic appreciation, as a biological entity to be perfected or strengthened, as a moral being able to establish human communities of love and care, and as a rational agent who acts with a sense of self-respect (dignity) toward the fulfillment of his responsibilities and assertion of his rights. Finally, a human individual must be seen as being capable of aspiring to and emulating an ideal of moral perfection so that his life would be endowed with an ultimate sense of significance and meaning.

In view of these basic facts about human persons and human life, we further suggested two fundamental principles for moral valuation and evaluation: the Holistic Principle of the Human Person and the Universalistic Principle of the Human Community.

Given the framework of these moral considerations, we have discussed how biomedical technology is to be valued and evaluated on different levels of human existence and relative to the different levels of requirement of the different forms of ethics. It was made clear that the human development of biomedical technology is itself a matter of the ethics of vitality and strength, and would naturally serve the purpose of the ethics of vitality and strength (of the ethics of the will to power). But at the present time we have to require any biomedical technology to strive to conform to the standards of ethics of a higher order, namely the three forms of ethics above and beyond the ethics of vitality and strength. In this connection, I pointed out that development of the present-day biomedical technology together with the present-day biomedical science may actually revolutionize human society, and therefore cause an upheaval of the present-day family and community structures, and even eventually change our understanding of human nature and the human person.

However, even though such a revolution is unavoidable, I do wish to see the transformation of society and human nature monitored and critiqued by what we have in our existing understanding of the human person and our existing forms of ethics of the higher order. I pointed out that there is always a factor of ethical risk involved in any use of biomedical technology, and certainly that the ethical risk is greater and higher, as and when it is pregnant with a fundamental challenge to our knowledge of good and evil, as when we speak of a biomedical technological revolution.

It is important to keep in mind this ethical risk on the one hand, and, on the other, to open our minds to a brave new world of social and human change under the impact of biomedical revolution and thus look forward to a positive (rather than negative) revamping of our way of life and ethics. As human persons at this juncture of our civilization, we need this double-faced, Janus-like wisdom in the understanding, valuation and evaluation of biomedical technological development. This is why I would like to call for a new orientation in thinking on bioethics, as well as a new orientation in thinking in the philosophy of bioethics.

Department of Philosophy
University of Hawaii at Manoa
Honolulu, USA

NOTES

[1] In other words, God can be purely spiritual, functioning in the world or some super-world in some fashion, without assuming any personal face.
[2] I have defined four types of good in the following order of magnitude: benefiting others at the expense of oneself; benefiting both oneself and others; benefiting others without doing harm to oneself; benefiting oneself without doing harm to others. Similarly I have defined four types of bad in the following order of magnitude: harming others for the purpose of self-benefit; harming others without self-benefit; harming both others and oneself; harming oneself without harming others.
[3] See Engelhardt, 1996, pp. 136.

[4] We have to admit that there are incapacitated persons like Steve Hawking, the British physicist, who have successfully exercised great will power in thinking and acting, and thus made an impact on the world through both bio-technical instrumentation and a moral will to live and think.

[5] David Hume initiated the distinction between fact and value in his classic work *A Treatise of Human Nature*. G. E. Moore criticized what he called the naturalistic fallacy in various definitions of good made in terms of natural qualities such as pleasure and survival abilities in his classic work *Principia Ethica*. While I agree that values cannot be simply defined in terms of natural qualities, I also hold that the naturalistic fallacy bespeaks some unresolved riddle on how values are to be conceived in relation to objective facts. My position is that we need to see values as a unity of the objective fact and the subjective affirmation of the desirability of a state of experience to be embodied in the objective entity or state of fact.

[6] See Murray, 1999.

[7] In a comment on this point, Professor Engelhardt related that women did enjoy using Viagra® in enhancing their sexual vitality. In this regard, depending on degrees of intensity and distribution, my worry on the partiality of Viagra® use could become groundless or imaginative.

[8] See ibid.

[9] At present human beings do not yet have immortality. The question is how we as human beings confront death. See an interesting article by Hardwig, 1999.

REFERENCES

Engelhardt, H. T., Jr. (1996). *The Foundations of Bioethics*, 2nd edition. New York: Oxford University Press.

Hardwig, J. (1999). Is there a duty to die? In: J. D. Arras & B. Steinbock, *Ethical Issues in Modern Medicine*, 5th edition (pp. 292-302). Mountain View: Mayfield Publishing.

Murray, T. H. (1999). Families, the marketplace, and values. In: J. D. Arras & B. Steinbock, *Ethical Issues in Modern Medicine*, 5th edition (pp. 460-488). Mountain View: Mayfield Publishing.

HYAKUDAI SAKAMOTO

A NEW POSSIBILITY OF GLOBAL BIOETHICS AS AN INTERCULTURAL SOCIAL TUNING TECHNOLOGY

I. INTRODUCTION

In the last decade I have worked to develop and publicize bioethics in Japan and in Asia, especially East Asia. I first founded the Japanese Association of Bioethics in Tokyo in 1989, and then in 1995, I organized the East Asian Association of Bioethics, in Beijing, China. After ten years of activity with these two associations, I have come to believe that bioethics is, at the end of the 20th century, confronting a big turning point, i.e., it is turning from traditional western bioethics to a new global bioethics for the 21st century, and it now requires a new background philosophy as well as a new methodology and policy for its globalization. What then should we hope for in the new global bioethics?

II. WESTERN BIOETHICS AS A TECHNOLOGY ASSESSMENT OF MODERN SCIENTIFIC INNOVATION

To begin with, I will examine the essential nature of Western bioethics. In the Western world, bioethics, distinguished from medical ethics, emerged only in the late 1960s, when, at the culmination of the 'innovation' of science and technology, the movement of 'technology assessment' (TA) was promoted in the advanced nations, especially in the U.S.A., with fear of the unexpected harmful effects of the development of modern science and technology including, especially, bio-science and bio-technology. I assume the very original idea of bioethics was brewed as one of the byproducts of this movement of technology assessment, as is implicitly shown in V. R. Potter's book *Bioethics, Bridge to the Future* (1971).

Now, what was the criterion to assess science and technology at that time, i.e., the 1960s and 1970s? I assume the criterion in the Western world at that time was clearly the moral standard of 'modern humanism'; thus some sciences and technologies were rejected because they were anti-humanistic. However, what is the nature of humanism here?

Julia Tao Lai Po-wah (ed.), Cross-Cultural Perspectives on the (Im)Possibility of Global Bioethics, 359–367.
© 2002 *Kluwer Academic Publishers. Printed in Great Britain.*

Historically speaking, 'humanism' in the Western world was originated in the age of renaissance and then developed and theorized through the age of, especially, the 17th and 18th centuries of enlightenment movement, in which the focus was put on the establishment of virtue of human individuals as citizens. Therefore, this type of European humanism has been rather 'human-centric', or 'anthropocentric', which was, as its natural consequence, backed up by the human "frontier mentality" according to Daniel Chiras's naming (1985). Also, this humanism was fortified by the modern idea of 'person' and 'human dignity' as found in the 17-18th centuries in philosophy, especially of John Locke (1690, Book 2, Ch.27, 8-29) and Immanuel Kant (1785). Thus, a 'person' is identified as a *rational* being, and therefore, a human being, as a person, is free and given a wide-ranging freedom including, ironically enough, heroic freedom to conquer Nature in the extreme and also, later on, the promise of a variety of human rights, eventually the fundamental human rights in the course of the French revolution.

At the first stage of the technology assessment movement, the criterion of the assessment was clearly to protect 'human beings' from technological disasters, and this aim was easily identified with the protection of human rights. This general mood reflected in the bioethics of this first stage described "bioethics" as "the way to protect human rights and human dignity from the invasion of bio-technologies" and through the course of debates on this issue, the traditional paternalism, such as political paternalism and scientific paternalism, which would include medical paternalism, was severely rejected, and instead 'self determination', 'informed consent', 'patients' rights,' etc. were recommended. In this first stage, almost all issues of bioethics were treated by this principle, i.e., the 'protection of human rights'. For instance, in the U.S.A., 'bioethics' often meant the "establishment of a legal system about bioethical issues" from the viewpoint of human rights.

III. A TURNING POINT OF BIOETHICS IN THE 1980s AND 1990s

However, bioethics came to a big turning point in the 1980s and 1990s. This was brought by, first, the extremely rapid development of genetics; second, by the rise of the environmental approach to bioethics; and third, by the introduction of the Asian (or non-Euro-American) paradigm.

First, at the end of the 20th century, we almost obtained the ability to manipulate human genes or genomes, by the way of recombinant DNA, i.e., the ability to alter the genetic character of a human body artificially. This technology made gene therapy possible. And also very recently, we have almost established the possibility of human cloning, showing the reproductive omnipotence of the bodily cells of all animals. These innovations imply the possibility of 'artificial evolution', altering humankind to another kind, or, on the contrary, not altering it (by using the cloning technology) and rejecting the possible natural evolution according to the man-made objectives of our own value system. Now, I propose to introduce the new concept of 'artificial evolution', in contrast to the ordinary concept of 'natural evolution'. Here, a new philosophy is badly needed to back up these new technologies of 'artificial evolution' in relation to the concept of fundamental human rights.

Now, does this technology violate fundamental human rights? The Council of Europe in this regard made the first apparent bioethical attack against biotechnologies in 1982 by its Recommendation 934 on genetic engineering from the standpoint of human rights. It says, "human rights imply the right to inherit a genetic pattern which has not been *artificially* changed" (1983, p. 19). However, we are now going to admit 'gene therapy', which necessarily changes the human genetic patterns artificially by the name of medical treatment, which might promise the future improvement of human utility and therefore, 'human happiness'. The distinction between somatic cell genetic therapy and germ-line therapy is only provisional. Here, we have to notice that the concept of human happiness has, in a sense, become a concept contradictory to the concept of the protection of human rights.

Second, through urgent environmental debates, there occurred conflicts between two different types of idea concerning the 'protection of the environment'. One idea is to protect the environment in order to preserve the best living condition for human beings of present and future generations. The other is to protect Nature for its own sake. The former is typically human-centric, and it has been gradually replaced by the latter under the influence of recent developments of ecological knowledge together with the severe regret and criticism of the 'frontier mentality' of modern humanism. People are now tacitly confirming the value of Nature (or Earth) itself, instead of the value of human being. Human beings are not the owner of Nature or the Earth.

Third, the range of vision to look at bioethics has been expanded to the regions outside Europe and U.S.A., especially to Asia. Many bioethical incidents happened in Asia, which were quite strange to European mentality. For instance, the Japanese rejection against heart transplantation from a brain dead body (we restarted heart transplantation only in 1999 after 30 years' interval) was quite odd for Euro-American minds. Also, many incidents suggested an Asian hostility to the sovereign idea of 'fundamental human rights', such as the Tian An Men incident in China, and other events in Singapore and Malaysia. People began to notice the peculiarity of Asian minds in considering bioethical issues. Something is fundamentally different. First of all, in many countries in East and Southeast Asia, the sense of 'human rights' is very weak and foreign, and they have no (or a weak) theoretical background for the concept of human rights. Rather they are more concerned with overcoming starvation and poverty *not* by human rights *but* by national or regional wealth and mutual aid. The recent introduction of the European idea of human rights rather caused moral, ethical and political conflicts among Asian societies.

The Asian view of Nature is also historically heterogeneous for the European. Nature is something not to be conquered but something with which to live. Generally speaking, an Asian has a holistic way of thinking instead of the European individualistic way. Therefore, Asian people put a higher value on the holistic happiness and welfare of the total group or community to which they belong, rather than on their individual interests. Here we can find the biggest significance of the Asian participation in the field of bioethics.

Now, in the present post-modern age, it is quite necessary for our human society to globalize bioethics for its future development. But it is almost impossible to do this by insisting on the universality of human rights, hence, the universality of Euro-American bioethics. Here is the reason why the new global bioethics is needed.

IV. THE CHARACTERISTICS OF THE ASIAN ETHOS

Before proceeding on to the global bioethics, we have to examine the characteristics of the Asian ethos on which the possible Asian bioethics or bioethical way of thinking stands, and which is supposed to be essentially different from the European one.

Generally speaking, the essence of the Asian ethos is said to be 'a holistic harmony' in contrast to the modern European inclination to dualistic individualism. The Asian worldviews and general ways of thinking have the following remarkable characteristics.

1. They put a higher estimation on total and social 'well-orderedness' than on individual interests or individual rights and dignity, and this 'well-orderedness' is accomplished by the sound assignment of social roles to people, and also by the fulfillment of the corresponding responsibilities of those people (individuals, groups and communities). Standards of this 'well-orderedness' depend on the social system of each respective period of time. In the tradition of Confucianism, it was interpreted in a feudalistic way as a matter of course. However, people could equally enjoy the peace of their society and their ordinary individual life. Here, peace means not only the state of the non-existence of war, but also mental peace.

2. Ethics, as well as social justice are interpreted in very realistic ways, as, for instance, social tuning techniques or the like. There is no unique or absolute God, categorical imperative, free will, or autonomy from which to deduce the concepts of goodness and justice, or precepts to control people's behaviors in order to pursue social peace. Every ethical and moral code is essentially relative to its period and region. Eventually, there has been only small room for the idea of 'fundamental human rights', if any.

3. Fundamental naturalism is pervasive in every Asian system of thought. According to Asian naturalism, our prima facie, non-natural and artificial human activities are ultimately included in Nature as its small parts. Before Nature, human beings are quite incompetent, and therefore, the distinction between natural and artificial is always blurred. Thus, to be 'natural' and to be 'artificial' are not contradictory at all. In short, there is no antagonism or clear-cut dichotomy between Nature and human being in the depth of the Asian way of thinking, and way of living. Also, it is well known that the Asian way of thinking has always been resistant to the mind-body dichotomy.

4. Asians are inclined *not* to believe or pursue any 'invariance' or 'eternity'. Buddhist precepts always tell us "all things flow and nothing is permanent." On the contrary, Western culture has always sought 'invariance' or 'eternity' which remains identical through every change. Thus various 'conservation laws' have been established in the history of the sciences, such as the 'law of energy conservation' and the 'parity

conservation', etc. In the same fashion, Western philosophy introduced 'personal identity' which remains invariant through all possible changes as a human being. This idea of invariance is somewhat foreign to the traditional Asian ethos. This is one of the most significant differences between Asian and Western ways of understanding nature and human being.

V. THE NATURE OF THE POSSIBLE GLOBAL BIOETHICS

Then what is the nature of the possible global bioethics?

1. *New humanism*: First, the possible global bioethics should stand on the new philosophy concerning the relation between nature and the human being. At least, the 'frontier mentality' and human-centrism of 18th-century-type humanism must be abandoned. Also the simple-minded naturalism of 'laissez faire' is impossible, for we have already acquired the ability and technology to control the evolution and future of humanity. We should now establish a new humanism without human-centrism, a new methodology to complement this new humanism, and modern sciences and technologies to control human evolution, adopting the Asian ethos and wisdom to avoid the European excessive inclination to the manifold natural-artificial dualism.

2. *Minimization of Human Rights*: Second, we should reconsider the nature of the human being apart from the 18th century philosophical anthropology of Kant and other idealists, which gave ground for the idea of the universality of human rights. Why is only humankind bestowed with such 'rights', and why are those rights universal? Kant might refer to the notion of 'person' or 'personhood' (1785) which is essentially universal, rational and free. But the concept of 'person' is, as its etymology (*persona*) shows, only a fictional masque socially given to humankind, and therefore, it can be neither universal nor *a priori*. On the contrary, the human genetic character is, in a sense, inborn and *a priori* and universal in a pattern common to all human species. Therefore, 'personal identity' is conceptually different from 'genetic identity'. Also, why can we discriminate with reference to some sorts of natural rights, such as the 'non-human rights' or 'rights of non-human beings' belonging to animals and trees? At the same time, the idea of the 'dignity of human being' should be reconsidered. Why is human being exclusively dignified? At least in some traditional ways of thinking in Asia, there is

no idea of human dignity distinguished from animals and others. We must take the standpoint of 'value relativism', which is a challenge to Western bioethics.

At the same time, we have to appreciate the fact that in Western societies, most people, even professional ethicists, are still inclined to believe in the absolute universality of 'human rights' and, therefore, the absolute university of bioethics, for this idea properly functions in leading and controlling their social systems, especially their legal systems. Our urgent task here is to find a way to make both positions concerning 'human rights' and 'bioethics' compatible in order to find a new refined methodology for global bioethics.

3. *Holistic harmony*: Third, we have to investigate a new philosophy for the foundations of global bioethics. I believe it must be grounded in the traditional 'ethos' of each region, which might be fundamentally different from the European 'ethos' in many respects. It is not easy work to unite these differences. However, this work is the most crucial part of global bioethics, which is expected to harmonize and to bridge over all kinds of global ethos, East and West, South and North. In this sense the new Global Bioethics should be 'holistic' in contrast to European 'individualistic'. Taoism, Confucianism, and Buddhism are all overwhelming in the Asian ethos. Their doctrines and precepts are all holistic in general. They tend to put a higher value on nature, society, community, neighborhood and mutual aid than individual ego. It is sometimes a sort of severe anti-egoism. But it is not necessarily altruism either. It always seeks some sort of holistic harmony of the antagonists. In this respect, Asian people have sometimes shown extraordinary, dexterous ability to harmonize social and moral conflicts and disorders in their respective and ad hoc ways without referring to any universal or common principle. This might be called 'social fine tuning technology'. The possible global bioethics of the new century should provide this sort of social technique to control and harmonize no-principle-oriented and value-mixed, chaotic societies. Here the society is not the well-ordered deductive system, which is ruled by some set of invariant principles.

4. *The Policy of Global Bioethics as a Social Tuning Technology*: Now, what is the best suitable policy for the 'harmonization' of the near future global bioethics as a 'social tuning technology'? Here, I would like to suggest that we replace the ambiguous word *harmonize* with the more realistic word *compromise* or *bargain*[1], because the best policy or strategy of the future global bioethics seems not to be reasonable, well organized

and 'principle-oriented'. In contrast to the principle-ism of most ongoing bioethics, it might be unreasonable, chaotic, ambiguous, and anarchistic.

When one is bargaining with someone, it is not always a good strategy for one to make one's principles explicit to one's bargaining counterpart and persuade her/him reasonably. If one shows one's principle explicitly from the beginning, the other party will try to knock that principle down in order to obtain profit from the bargain. "Don't tell a lie" cannot be an absolute principle any more. You can save the life of some people by telling a lie to vicious pursuers. The honesty-maxim is no longer an example of the categorical imperative. Also, the 'no-harm principle' could not be an absolute and universal principle any longer, for to harm some people or even to kill a person might be a good strategy to protect a nation, society or community. In the reality of many present day societies, which claim to guarantee fundamental human rights in their constitutions, to kill a person in the name of capital punishment is a proper legal procedure. The notion of 'justice' is also dubious. If you insist on your justice whatever it is, then another party will invent another kind of antagonistic justice, and war will inevitably ensue.

I suggest that the global bioethics of the new century should not refer to any kind of 'universal principle', 'justice', 'categorical imperative', or the like for its policy. The only policy possible here will be continuing dialogue without reference to any rigid principle, or, on the contrary, with reference to all antagonistic principles as impartial bargain alternatives, ultimately soothing the opposition and antagonism among the principles to reach a 'consensus of any kind', even though it might be quite unreasonable or absurd. I call this policy 'dialogue bargain policy', or 'bargain consensus'. This would be the only possible way to realize the Asian ideal of 'harmony' in the third millennium.

In the 'bargain dialogues', people of both parties would bring in all sorts of reasonable as well as absurd claims which are fundamentally and eventually derived not from reason or justice, but from hedonistic desires, and through the sometimes long, and sometimes patient bargaining dialogue we can hope for a possibly very absurd but harmoniously acceptable 'bargain consensus'.

We can imagine here that, through the process of reaching a bargain consensus, some sort of common feeling or compassion between both parties would be effectively born. I tentatively assume the existence of such a common feeling in all people as the 'Feeling of Ache and Pity'. In order to strengthen my assumption in the depth of traditional Asian thought, I will only quote here the lines of Mencius (Meng Zi): "Every person, as a human being, is innately bestowed the Heart of Ache and Pity," and "The Heart of Ache and Pity is *Jen* (perfect virtue of humanity) itself."[2] Now, I assume that we could best reconstruct a new, post-modern humanism based on this 'feeling of ache and pity' for the third millennium.

In this concept of 'bargain', I would include not only bargains between people, but also between people and communities and even between human beings and Nature. In reality, we have lost many valuable things because of the failure to bargain with Nature including other kinds of lives on the earth, through these centuries. This is the real status of the recent environmental crisis, and thus, harmony between human beings and Nature is looked for very urgently.

Professor Emeritus
Aoyama-gakuin University
Tokyo, Japan

NOTES

[1] The word 'bargain' has recently often been used in the texts of legal-sociological discourse as a term to illustrate a new pragmatic and feasible way to reach a social agreement and consensus between mutually opposing parties. Cf. David Gauthier (1987).

[2] See *Mencius* (11A: 6). My own translation.

REFERENCES

Annuaire Europeen – European Yearbook Vol. 30. (1983). Hague: Martinus Nijhoff Publishers.

Chiras, D. D. (1985). Environmental Ethics; Foundations of Sustainable Society. In: *Environmental Science, A Framework for Decision Making*. San Francisco: The Benjamin/Cummings Publishing Company, Inc.

Gauthier, D. (1987). *Morals by Agreement*, Oxford: Clarendon Press.

Kant, I. (1785). *Grundlegung zur Metaphysik der Sitten*.

Locke, J. (1690). *An Essay concerning Human Understanding*.

Mencius (Meng Zi).

Potter, V. R. (1971). *Bioethics, Bridge to The Future*. Upper Saddle River: Prentice-Hall.

RUIPING FAN

MORAL THEORIES VS. MORAL PERSPECTIVES:
THE NEED FOR A NEW STRATEGY FOR
BIOETHICAL EXPLORATION

I. INTRODUCTION

There is a puzzle in contemporary bioethical approaches. The same
normative moral theory may be used to generate different bioethical
accounts, whereas different moral theories can be used to produce the
same bioethical account. For example, David Friedman and Peter Singer
are both committed to the utilitarian theory, but they have developed
quite different bioethical accounts. Friedman supports a free market
distribution of health care (1991), while Singer argues for an entirely
state-controlled health care system (1976). Both Robert Veatch and
Norman Daniels use contractarianism to build their bioethical approaches,
but Daniels requires justice to be realized as the first moral priority in
society (1985), while Veatch holds that the requirement of justice is only
one of a set of equally binding moral principles that are often in
competition (1981). Finally, although Tom Beauchamp is a utilitarian and
James Childress a deontologist, they have come to the same set of
bioethical principles and together argue for the same moral solutions to
varieties of bioethical issues (1994).

How is this possible? How can the same normative moral theory be
used to produce different bioethical accounts (as in the cases of Veatch
and Daniels, Friedman and Singer)? And how can different normative
moral theories be used to generate the same bioethical account (as in the
case of Beauchamp and Childress)? These circumstances suggest that
differences among various bioethical accounts cannot be entirely
accountable through the differences among different moral theories. If a
normative moral theory could determine all the moral substance of a
bioethical account, it would not have been possible for one normative
moral theory to generate different bioethical accounts, while several
moral theories lead to the same bioethical account. Although many seem
to believe that bioethical accounts, as well-organized moral approaches to
provide reasons, arguments, conclusions, and solutions to bioethical
issues, are derived from normative moral theories, these circumstances

*Julia Tao Lai Po-wah (ed.), Cross-Cultural Perspectives on the (Im)Possibility of Global
Bioethics*, 369–390.
© 2002 *Kluwer Academic Publishers. Printed in Great Britain.*

indicate that the role of moral theories in bioethical accounts has been exaggerated. The general character of a normative moral theory does not exhaust all the moral content involved in the use of such a theory.

In order to understand bioethical accounts more clearly, this paper argues that we need to examine a tripartite interplay among moral theories, moral accounts, and moral perspectives. A moral theory identifies general moral statements or principles formulateable within a moral perspective. A moral perspective, on the other hand, is most comprehensive of a morality. It includes the full content of a moral tradition lived by members of a moral community. Finally, a moral account reconstructs a moral perspective regarding a domain of moral issues by following a formal structure provided by a moral theory. Hence, these three concepts (moral perspective, account, and theory) should be distinguished in order to understand bioethical exploration and justification in a clear way. Again, a moral perspective is the image of a real moral life lived by a group of people. It is concrete, canonical and content-full; it cannot be entirely discursive. On the other hand, a moral theory is both general and systematic. It is composed of abstract statements that are inevitably vague, ambiguous, and underdetermined without further interpretations. Finally, a bioethical account cannot be substantiated by a moral theory. It is rather a conceptual restructuring of a moral perspective along a formal framework provided by a moral theory. In other words, what a moral theory contributes to a bioethical account is only a thin structure in which concrete moral substance is arranged. The content-full moral substance of a bioethical account can only be offered by full-bodied moral perspectives.

A popular strategy in bioethical explorations has been exploring moral theories in the hope of reaching universally justified moral conclusions and solutions to bioethical issues. This paper calls for a new strategy: disclosing moral perspectives. In order to understand or establish a bioethical account, this new strategy recommends, energies ought to be invested in closely investigating a particular moral perspective shared by a moral community and in comprehending its specific moral content. Only in this way, I contend, can bioethical explorations provide clear and profound moral instructions for the perplexing bioethical issues we face today as well as contribute to a better understanding of the appropriate relationship among individual, community and society in the contemporary world.

II. MORAL THEORIES: TOO GENERAL TO OFFER SPECIFIC
MORAL GUIDANCE

There are several logical possibilities when two (or more) conflicting moral conclusions are drawn from the same normative moral theory. First, the theory may be incoherent in itself. Hence one part of the theory leads to one conclusion while the other part leads to another. Second, although the theory is coherent, some mistakes may be made in drawing inferences from the theory. Third, two or more different types of theories may be hidden behind an apparently single theory; hence, different accounts draw on different theories. Finally, the theory may by itself be insufficient to generate any specific conclusion unless further interpretations of the theory are introduced. The first possibility can be put aside because we are considering the authentic theories such as utilitarianism and deontology in bioethical discussion. The second possibility can also be put aside because no such logical mistake has ever been found in the prestigious bioethicists' works we are considering. The third possibility is an *ad hoc* explanation. One can always identify different theories if one finds different conclusions arising from a theory. But that would lead to a very loose definition of normative moral theories and result in the erosion of the distinction between moral theories and moral accounts. The only plausible explanation is the last possibility: a normative moral theory is by itself insufficient to draw any concrete conclusion unless further interpretations of the theory are introduced. Under this explanation, it is easy to understand why different conclusions have been derived from the same normative moral theory: because different further interpretations[1] of the theory have been provided. Consequently, introducing different interpretations to a moral theory leads to different concrete moral conclusions in different moral accounts.

Similarly, when two different types of theories lead to the same bioethical account (as in the case of Beauchamp's utilitarianism and Childress' deontology), there can also be several possibilities. First, some misunderstandings are involved concerning one or both of the theories. Second, each of the two theories is sufficient to generate a bioethical account, and they happen to generate the same account. Finally, neither of the two theories is by itself sufficient to generate a bioethical account without introducing further interpretation of the theory, and a similar further interpretation is offered for each of the two theories so that they lead to the same bioethical account. The first possibility can be put aside, at least in the case of Beauchamp and Childress. It is unlikely that they have misunderstood the nature of the utilitarian or deontological theory. The second possibility can also be ruled out because other utilitarians and deontologists have worked out other bioethical accounts differing from theirs. I hold that the third possibility is the best explanation of the Beauchamp and Childress case. That is, they each offer a similar interpretation of the moral theories to which they each respectively appeal.

These analyses disclose three important facts: (1) we are not able to use a normative moral theory to draw concrete moral conclusions without introducing an interpretation of the theory, (2) the theory allows for different interpretations, and (3) different interpretations of the theory lead to different conclusions on particular bioethical issues. This is to say, a normative moral theory is general in the sense that it is underdeterminate for particular moral issues.[2] Indeed, in the process of establishing a moral account, a normative moral theory can simply be identified as an abstract moral principle or doctrine. For instance, the utilitarian moral theory can be identified as the utilitarian principle of utility: as long as one follows the principle of utility that "the right act is the one that produces at least as much social utilities as any other alternative" to build one's moral account, one is following the utilitarian theory. The same goes for Kant's deontological moral theory. As long as one employs the categorical imperative that "one should act in the way as if the maxim of one's act were to become by one's will a universal law of nature," one is following Kant's deontological theory. It becomes clear that neither of such principles can produce concrete moral conclusions regarding particular moral or bioethical issues without adding detailed interpretations of the principle in the very context of its application.

Accordingly, it is a normative moral theory together with some specific interpretations of the theory that are able to provide concrete moral substance in a moral account. Such interpretations must have been added to the theory from outside of the theory. They cannot be obtained from the theory itself because if they could, there would not have been the possibility that conflicting moral accounts can be developed from the same theory as the bioethical puzzle shows. This can be explained more clearly in terms of the formal logic of moral arguments. When used in moral arguments regarding particular issues, the general principles or doctrines of moral theories serve as major premises of the arguments. In order to draw conclusions, further specific minor premises must be added to the major premises to complete the arguments. Such minor premises are not contained in the major premises. Instead, they are brought in from sources independent of the major premises. Hence, although moral theories determine major premises in our moral arguments, they leave open minor premises. Whereas major premises determine general directions or frameworks of the arguments, different minor premises lead to different moral conclusions. This is why we can produce various moral accounts based on one moral theory.[3]

People often expect to obtain moral standards from moral theories so as to justify their moral conclusions or solutions regarding particular moral issues, especially when they encounter difficult moral issues. However, moral standards are in fact at home in our non-theorized everyday moral lives. Our inherited moral traditions, values, ideas, and rules as well as our moral intuitions and judgments accompany us in our ordinary moral practices and offer specific moral instructions. We do not have to hold a moral theory in order to obtain moral instructions. On the other hand, since these ordinary moral instructions are usually non-discursive, non-theorized, and not well-structured, people expect to use a normative moral theory to systemize them as a deductive moral account as if they had thus been justified in providing moral guidance to practical issues. This explains why, in our understanding or establishing a bioethical account, it is easy for us to concentrate on a moral theory and overlook the real sources of our moral standards. We fail to recognize that what a moral theory offers to a bioethical approach is not so much moral substance, but a general moral structure along which we display our concrete moral instructions from everyday life. What is worse, focus on moral theories tends to generate the false consciousness of moral consensus and ignore the complexities and nuances of real moral practices, explorations, and elaborations. Since principles or doctrines identified by moral theories are abstract formal statements, they can easily obtain the support of the intuitions of individuals, overlooking a number of conflicting principles and doctrines that are similarly supported by contrary intuitions. In short, instead of offering justification for particular moral instructions, focusing on normative moral theories has led to the ways in which the diversity and plurality of actual moral commitments and convictions are ignored. It obscures the picture of what a bioethical account is.

Thus, in order to understand bioethical accounts better, it is necessary to recognize something larger than the scope of content covered by a normative moral theory. We need a clear notion to indicate the source of possible specific interpretations of a moral theory in order to reach a moral account. I use the notion of *a moral perspective* to play such a role and thus lay out a tripartite relation among a bioethical account, a normative moral theory, and a moral perspective in bioethical exploration. In short, a bioethical account is a systematic moral approach structured by a moral theory and substantiated by a moral perspective. It cannot be built up simply on the basis of a moral theory without concrete moral resources offered by a moral perspective. A moral perspective is the root of all possible interpretations of a moral theory and thus provides specific moral substance for a bioethical account.

III. MORAL PERSPECTIVES: WAYS OF LIFE

A moral perspective is best understood as a way of life. It is what the members of a moral community share, including the full content of a morality. Given that bioethical accounts cannot be exhausted by moral theories, the notion of "moral perspectives" can be used to illuminate the substance of bioethical accounts and characterize the details of our moral activities. The prominent examples of moral perspectives include the Confucian moral perspective, the Buddhist Moral perspective, the Christian moral perspective, the Aristotelian moral perspective, the modern Western liberal moral perspective, the contemporary Western feminist moral perspective, etc. Since a moral perspective is the constellation of a moral community's moral understandings and commitments, it is not entirely systematic, discursive, or deductive.

I would like to follow Thomas Kuhn's notion of a paradigm for scientific activities to explain the notion of a moral perspective for moral explorations.[4] Instead of moral principles or theories, what is crucial for a moral perspective is its exemplars, namely, concrete rules, examples, and stories. For the sake of clarity, I would like to identify three types of components for a moral perspective, although we should bear in mind that any real moral perspective must be inclusive, not as analytic as displayed in this explanation.

The first type of component of a moral perspective is specific moral exemplars, such as classical moral narratives, examples, paradigm case-analyses, specific problem-solutions, stories, rituals, rules, and so on. These ingredients are much more concrete than general statements. They are shared by members of a moral community, being the most intimate elements in their moral lives and guiding their moral experiences in concrete ways. They convey a particular conception of the good life, informing the individual regarding what examples he should follow in making a good life, which virtues he should nurture in forming a good character, and what strategies he should use in resolving moral problems. Confucianism and Roman Catholicism offer good examples of these kinds of moral exemplars.

The second type of component of a moral perspective is prominent moral commitments. Such commitments restrict individuals to particular ways of life which manifest their moral perspectives with moral significance. To be sure, such commitments are already implicit in the concrete exemplars of a moral perspective, but they are worth picking out as a type of component for their symbolic significance for a moral perspective. Indeed, they are probably the most well-known part of a moral perspective in general society. For instance, the virtue of filial piety (*xiao*) for Confucians, the virtue of compassion for Buddhists, the prohibition of abortion for Christians, circumcision for Judaists, individual liberty and independence for liberals, and women's liberation for feminists are all symbolically important for their respective moral perspectives. One cannot be a Confucian without being committed to the virtue of filial piety, just as one cannot be a feminist without being committed to the cause of women's liberation.

The third type of component of a moral perspective is formal statements provided by the perspective. Typical examples of such formal statements include the principle of *ren* for the Confucian moral perspective, the doctrine of *karma* for the Buddhist moral perspective, the Thomistic principle of love for the Roman Catholic moral perspective, the doctrine of individual rights for the modern Western liberal moral perspective, and a call for care for the contemporary Western feminist moral perspective. Compared to prominent moral commitments, these statements are more abstract and general, like normative moral theories. Extra explanations must be offered in order to make full sense of them and apply them in practice. Indeed, detailed interpretations of a formal statement are already contained in the concrete exemplars of a moral perspective, so that people living the perspective know how to use the formal statement in practice and how to explain it to other people in concrete terms.

Some formal statements may be formulateable within more than one moral perspective. For instance, a principle of love, the Kantian principle of respecting persons as ends, a principle of justice, and a principle of beneficence can easily be integrated into a number of different moral perspectives. It may not be difficult for many moral perspectives to use them as formal statements for their perspectives in structuring a particular moral account. However, the specific interpretations of these principles may be dramatically different from one perspective to another. People have good reason to doubt whether they remain the same principles across different moral perspectives. Although it may be convenient and useful for a moral community to draw on a new formal statement to restructure its moral substance regarding a practical issue, the detailed interpretation of any new statement is already implicit in the full content of a moral perspective, namely, in its exemplars.

Indeed, what is cardinally important to a moral perspective is moral exemplars.[5] Different moral perspectives may use the same set of formal statements (e.g., they may use the same normative moral theory), but they remain different moral perspectives if they hold different exemplars. On the other hand, even if two moral accounts use different formal statements, they may still reflect the same moral perspective if they offer the similar interpretations to their respective formal statements and share the same specific moral commitments as well as appeal to the similar concrete solutions to problems and/or paradigm case analyses in dealing with issues. Of course, the bioethicists of these accounts may hold different metaethical, metaphysical, and religious views. As long as these views do not make a real difference in their moral accounts, it is reasonable to understand that their moral accounts are representative of the same moral perspective.[6]

With this notion of moral perspectives at hand, we can obtain a clearer picture of popular bioethical accounts. When an ethicist argues for his moral theory through a well-organized moral account, he is best understood as displaying a particular moral perspective regarding a domain of moral issues. The moral theory is best understood as a tool for organizing detailed moral content. For instance, when Kant explores the issues of suicide, promise-keeping, cultivation of talents, and assisting others in need in his *Grounding for the Metaphysics of Morals* (1981, pp. 30-32), he is actually establishing moral accounts by laying out a moral perspective according to an organized framework offered by his categorical imperative (as the core of his deontological moral theory). But the categorical imperative offers him only a line of structure along which he arranges his moral substance. The moral substance of his accounts is from the full-bodied moral perspective that he holds.

This significant role of moral perspectives played in our moral accounts invites us to consider a new bioethical strategy that differs from the current strategy of exploring a moral theory. The strategy of exploring a moral theory pushes theorists to focus on a moral theory or principle to generate moral conclusions and solutions regarding bioethical matters. This strategy is mistaken because moral theories, without further interpretations, are underdeterminate of specific moral conclusions or solutions. The better strategy we should adopt is that of disclosing a moral perspective. Under this new strategy, we are encouraged to look closely at particular moral exemplars (as well as specific moral commitments implicit in them) shared by a moral community. Instead of exploring a moral theory, we understand that moral theories or principles can only be used as means to organize moral substance in moral accounts. In order to understand or establish a bioethical account, we must get access to the rich and detailed moral content of moral communities and experiences. Compared to the old strategy, this new strategy has two eminent merits: (1) it is complete and coherent in the sense that it can offer specific conclusions and solutions drawing on all plausible resources from a moral perspective (unlike the old strategy which cannot lead to specific conclusions from a moral theory without offering further interpretations to the theory from outside of the theory); and (2) it is sincere and dynamic in the sense that it encourages individuals to convey their living moral standards honestly and to engage in serious moral communications with other moral perspectives concerning specific moral commitments as well as concrete narratives or examples (unlike the old strategy which concentrates on generalized formal statements and abstract elaborations without touching on the deep levels of moral experiences).

IV. AN EXAMPLE OF THE SAME MORAL THEORY BUT DIFFERENT MORAL PERSPECTIVES: PETER SINGER VS. DAVID FRIEDMAN

This section uses the case of Peter Singer vs. David Friedman to explain how to employ the strategy of disclosing a moral perspective in understanding bioethical accounts in a clear way. Both Singer and

Friedman follow the utilitarian theory to construct their bioethical accounts. However, their accounts are incommensurable in their basic philosophical arguments as well as their particular solutions to bioethical issues. The best explanation for this puzzle is that their accounts reflect different moral perspectives, although they adopt the same moral theory to reconstruct their respective moral perspectives. This can be seen more clearly by examining what concrete solutions and case analyses they have offered, what specific moral commitments they hold, and what detailed interpretations of the utilitarian principle that they have provided. In short, in order to understand their bioethical accounts clearly, we must go beyond their utilitarian principle and closely look at the specific exemplars that they have offered.

1. Concrete problem-solutions

One of the major issues studied by both Singer and Friedman is that of justice in health care allocation, namely, what is the just way for society to distribute health care resources. Both Singer and Friedman, as utilitarians, contend that the just way of allocating health care resources lies in maximizing people's interests. However, Singer argues that only a single-tier, all-encompassing, and state-run health care allocation can maximize the interests of people (1976), while Friedman argues that only through the free market mechanism without government intervention can health care allocation obtain the best consequences (1991). For Friedman, medicine must be treated as a commodity like food or housing in the market in order to assert its best function. But for Singer, medicine must be controlled entirely by the state so that people's health care interests can be secured and promoted. As a result, Singer and Friedman offered sharply contrasting solutions to the problem of justice in health care allocation. This has heuristically shown that stopping at the utilitarian theory is stopping too soon to be able to understand their bioethical accounts, although both have employed the utilitarian theory to establish their accounts.

The issue of just health care explored by Singer and Friedman in each of their accounts includes a series of problems and items that are too multiple to be examined completely in the scope of this paper. Instead, to illuminate the crucial role played by moral exemplars in their bioethical approaches, it is sufficient to touch on only a couple of their solutions to problems: the problem of blood supply by Singer (1985) and the problem of helping the poor by Friedman (1991).

The issue of blood supply is a heuristic issue for Singer. He uses it as an example to show why the free market does not fit health care allocation. The issue is: should blood for medical purposes be obtained from voluntary donors and commerce in blood be prohibited? Or should both voluntary donations and commercial transactions be allowed? In countries like Britain, commercial transactions are prohibited and only voluntary donations are encouraged, while in other countries like the United States, both voluntary donations and commercial sales are available. Many argue, Singer concedes, that even if commercial dealing in blood is allowed, anyone who wants to give can still do so. Prohibiting commerce in blood is a denial of the freedom of those who wish to sell; permitting commerce, on the other hand, does nothing to limit the freedom of those who wish to give. Since freedom is a goal that most of us value, says Singer, this seems to be a strong argument against prohibition of commerce. However, Singer argues that this is not the case.

> This notion of freedom is superficial in the most literal sense of the term; it refuses to probe beneath the surface. In the particular case we are considering, this notion of freedom is satisfied in the American situation because a person can give blood voluntarily *if* he chooses to do so. It is, in this notion of freedom, irrelevant to consider that, ... the existence of a commercial system may discourage voluntary donors. It appears to discourage them, but not because those who would otherwise have donated their blood voluntarily choose to sell it instead if this alternative is available. In fact, donors and sellers are, in the main, different sections of the population. Rather, voluntary donors are discouraged because the blood's availability as a commodity, to be bought and sold, affects the nature of the gift that is made when blood is donated. If blood is a commodity with a price, to give blood seems merely to save someone money; it has a cash value of a certain number of dollars. As such, the importance of the gift will vary with the wealth of the recipient. If blood cannot be bought, however, then its value as a gift depends on the recipient's need. Often, it will be worth life itself. Under these circumstances, giving blood becomes very special, an act of providing for strangers, without hope of reward, something they cannot buy and without which they may die. The gift relates strangers in the community in a manner that is not possible when blood is a commodity (1985, p. 9, italics original).

In short, from Singer's view, allowing commerce in blood makes altruism unnecessary, and "so loosens the bonds that can otherwise exist between strangers in the community" (p. 10). If a community puts a money value on everything, it will restrict ways in which strangers can help each other and set the interests of one person against those of another. This will make a "materialistic community in which each looks out only for his own interests" (p. 11). Hence, Singer's major concern is promoting altruism, not protecting freedom, because he takes altruism, not freedom, to be the main contribution to the best consequence. For him, a community may legitimately restrict the freedom of individuals to sell blood in order to encourage the development of altruistic attitudes that he takes are important to the best interests of the community. In other words, the utilitarian principle sets down the general framework for his bioethical account in which he is bound to search for moral justification in terms of the best consequence for society. But what produces the best consequence is determined by some particular exemplars of a moral perspective that he holds. (In this case, it is promoting altruistic attitudes and restricting individual freedom that produce the best consequence for society.)

On the other hand, Friedman's solutions manifest different exemplars. He also considers some cases that seem to suggest that the government should regulate or redistribute health care, but his analyses lead to quite different solutions from Singer's. A prominent case he considers is the problem of the poor. As is well known, the traditional standard utilitarian argument for redistribution to help the poor begins with the belief of marginal utility: for a given individual, the marginal utility of income declines as income increases. From this belief it follows that, on average, the poor have a higher marginal utility for income than the rich. Hence, if government provides medical care (or housing, or food, or money) to the poor at the expense of the rich, total utilities in society increase. Thus, government should mandate such redistribution in order to increase people's total interests.

Friedman provides three reasons to refute this argument. First, it is not necessarily true that the marginal utility of income is correlated inversely with income. If income is got by inheritance, lottery, or some other chance mechanism, that belief is plausible. But it is implausible if different income is the result of different effort. "An individual who greatly values the things that money buys will be more willing than others to give up other goods, such as leisure, in order to get income, so he will, on average, end up with a higher income" (1991, p. 280). In such a case, Friedman contends, it is more reasonable to assume that income and the marginal utility of income are positively correlated.

Second, the argument of marginal utility does not have special relevance to medical costs or allocation, unless different medical bills are an important cause of inequalities in the marginal utility of income among individuals. From Friedman's view, this is not always the case (pp. 280-281).

Third, even if we should prefer outcomes that are biased towards the poor, everything else being equal, we have yet to decide whether we should favor political or private market to achieve the outcomes. According to Friedman, both available theory and actual evidence are not very clear about this issue. For one thing, public choice theory does not offer any clear answer as to whether government intervention is likely to distribute to or from the poor, to increase or decrease inequality in society:

> On the one hand, votes are more evenly distributed than income, which should tend to make the political market more egalitarian than the private market. On the other hand, many of the characteristics that give groups and individuals political influence are closely related to income. Education reduces information costs, labor skills differentiate their possessors into (relatively) concentrated interest groups, stockholders have their interest represented by well organized firms and skilled (hence highly paid) workers by well organized unions, and so forth. So far as theory is concerned, it is difficult to predict whether the political system is more likely to transfer money down the income ladder or up (pp. 281-282).

In addition, Friedman argues, the actual evidence is also unclear. Some government programs, such as food stamps, welfare, and the like, are meant to help the poor. Some programs, such as state subsidies to higher education, are actually to help the not-poor at the expense of the poor, because the poor, rather than the not-poor, are most probably not going to benefit from higher education. Still other programs, such as Social Security, have ambiguous effects. "Poorer individuals, on average, start work earlier and die earlier, hence pay more years and collect for fewer" (p. 282). Hence, for Friedman, the program may have harmed rather than benefited the poor. Unlike Singer who believes the cultivation of altruism is the most important thing for the best consequence, Friedman's concern is with an unhampered free market for maximizing the interests of individuals. As in the case of Singer, the utilitarian principle similarly sets down the general framework for Friedman's bioethical account in which he must search for moral justification in terms of the best consequence for society. But what produces the best consequence is again determined by some particular exemplars of a moral perspective that he holds (in his case, it is promoting individual free choices and avoiding government intervention with the free market that produce the best consequence for society).

2. Specific moral commitments

Some specific moral commitments are implicit in the solutions offered by Singer and Friedman as well as in their other discussions throughout their moral accounts. Evidently, although they hold the same utilitarian formal statement (namely, we are morally obligated to seek the best consequences for members of our society – and ideally, for humans in the whole world), they are committed to quite different concrete moral values and convictions. In the case of Singer, one of his adamant moral commitments is equality. For him, total interests cannot be maximized without being equalized among individuals. Second, he believes that the notion of freedom is superficial and individuals' free choices must be restricted in order to achieve the best consequences for a community. Third, he believes developing altruistic attitudes in society is enormously important for promoting general happiness. Fourth, he emphasizes the avoidance and elimination of suffering in the consideration of interests. That is why he has a strong concern for getting rid of suffering in both humans and animals. Finally, he has grave doubts about the effect of the free market mechanism. For him, government must intervene to equalize the distribution of vital things (such as food, shelter, medical care) among people in order to achieve the maximum social interests.

On the other hand, Friedman worries about the risks and side-effects of government intervention in the free market. For him, any government intervention is suspicious at best and disastrous at worst. He strongly believes that individuals' unbounded free choices will lead to the maximum total interests in society. So he is strongly opposed to any government redistribution project.

3. The formal statement and its specific interpretations

As utilitarians, both Singer and Friedman take the fundamental utilitarian principle as a formal statement to construct their bioethical accounts: one ought to adopt the course of action or a public policy that most likely maximizes the interests of all that are affected. But they hold different understandings of this principle and offer different interpretations of it. Obviously, the crucial thing for an interpretation of the principle lies in which course of action or public policy most likely maximizes interests. Singer's detailed interpretations are manifested in his particular arguments in relation to a wide range of ethical issues, including animal rights, world famine, defective newborns, health care allocation, and so on. Basically, Singer interprets the utilitarian principle in terms of equality. He has a strong commitment to egalitarian consideration of interests. According to him, the utilitarian principle is equivalent to a more general principle of equality, namely, the principle of equal consideration of interests. This latter principle requires one to take into account the interests of all who are affected by one's decision, no matter whose interests they are. For him, the best way to realize the goal of this principle is through government intervention in the free market. And total interests cannot be maximized without equally realizing everyone's interests. This is why he argues for radical government intervention, regulation, and redistribution in order to maintain equal interests for everyone in society. In short, for Singer, equality maximizes interests.

Friedman's explanations of the utilitarian principle are found in his doctrines of natural rights and economic efficiency, as well as his arguments about private ownership and operation in general and health care allocation in particular. For Friedman, the utilitarian requirement can only be realized through the free market mechanism in light of a doctrine of natural rights, namely, negative rights that "can be described in terms of entitlements" (1991, p. 261). Although everybody has such rights, Friedman argues, they are rights only in the negative sense, namely, they imply that others have obligations not to intervene in one's own affairs. Accordingly, although I have rights, they are "not a complete statement of what I ought to do, but they are a complete statement of my claims against others and theirs against me" (p. 261). For instance, for Friedman, that I have a right to health means that no other individuals should act so as to harm my health. But it does not mean that others should have a moral obligation to provide medical care to maintain my health. Hence,

Friedman's care for total utilities and efficiencies do not have any egalitarian concern. Instead, he argues that only a free market can bring to individuals the best consequences. Basically, he interprets the utilitarian principle in terms of freedom, which for him maximizes interests.[7]

V. CONCLUDING REMARKS

This paper argues that a normative moral theory, being general and abstract, is insufficient to generate concrete conclusions regarding bioethical issues. Only a moral perspective that compasses both a normative moral theory formulateable within the perspective and concrete moral exemplars (which implicitly contain the detailed interpretations of the theory) is able to disclose the entire picture of moral deliberation reflected in a bioethical account. It is a moral perspective (in particular, its exemplars), not a moral theory, that accounts for the character of concrete bioethical commitments and approaches. Only moral perspectives are close to real life moralities. Moreover, the strategy of disclosing moral perspectives emphasizes the investigation of particular exemplars rather than formal moral statements or symbolic moral commitments. This is because formal statements and moral commitments are already implicit in concrete moral exemplars (such as specific solutions to problems and paradigm case analyses), and they are far from being exhaustive of the richness and nuances of moral exemplars. As the forgoing example of Singer and Friedman shows, we need to check the particular exemplars that a bioethicist adopts in order to understand adequately what is involved in providing a bioethical account as well as in offering a moral justification. Only through this augmented strategy can we hope adequately to understand the nature, complexity, and diversity of contemporary bioethical practice and exploration.

This strategy also differs from the strategy of casuistry proposed by Albert Jonsen and Stephen Toulmin (1988).[8] Jonsen and Toulmin argue that real moral deliberations take place in referring to paradigm cases rather than following general principles or rules. Individuals, they observe, often reach agreements on particular cases or problems, but not on general principles or theories. Hence, they recommend, we should focus on cases, rather than on principles, in conducting bioethical studies. In particular, their strategy of casuistry suggests, in facing a bioethical problem, one should not try to locate a moral principle to offer an answer.

Rather, one should search for a previous case as the paradigm case to direct one's solution to the current case. This strategy, however, is inadequate. It is correct to say that moral explorations and deliberations involve more than the following of moral principles or rules, because one has at first to understand the principles or rules in order to follow them. But such understanding involves one's inarticulate or tacit knowledge which goes beyond the scope of the principles or rules, as Michael Polanyi persuasively illustrated (1962). It is such tacit knowledge that offers personal appraisal and final approval of the principles or rules that one will be able to apply or follow. Hence, moral explorations are not simply principle- or rule-following. On the other hand, however, it is incorrect to contend that moral explorations are primarily referring to precedent (paradigm) cases. Referring to paradigm cases is also a complicated moral activity. For one thing, there can be innumerable previous cases that are relevant among which one has to decide which case is the paradigm case for illuminating the current case. Hence, one has to appraisal, classify and generalize the relevant aspects of the current case as well as previous cases in order to determine a paradigm case. Moreover, in order to make such appraisals, classifications and generalizations, one has to appeal to certain concepts, ideas and principles as guiding frameworks or clues. This is to say, one has to use and follow some rules or principles in order to decide a paradigm case. Consequently, while the strategy of exploring moral theories exaggerates the role of principles in moral explorations, the strategy of casuistry overlooks the role of principles.

The strategy of disclosing moral perspectives requires a complete geography of moral explorations. Although it recognizes a crucial role played by the detailed interpretations of formal moral principles in bioethical accounts, it does not downplay the function of formal moral principles. Simply put, formal principles provide the structure, while detailed explanations provide the content, for moral accounts. Each element is irreplaceable.

Moreover, while Jonsen and Toulmin notice that individuals more easily agree on the morality of particular cases than on principles, this paper demonstrates the other side of the coin: scholars may also easily agree on some basic principles (e.g., Singer and Friedman on the utilitarian principle) while disagreeing on specific moral commitments and concrete problem-solutions or case-analyses. The geography of moral explorations is more complicated than the pictures painted either by "principlists" (such as Beauchamp and Childress) or by casuists (such as Jonsen and Toulmin). It involves both general statements and particular exemplars. From the strategy of disclosing moral perspectives, "principlists" overlook the huge iceberg of particular exemplars below the water of general statements, while casuists downplay the role of general principles implicit in particular moral exemplars.

Finally, from the strategy of disclosing moral perspectives, the prospect of global bioethics (i.e., a promise of providing substantive bioethical accounts for universal application) looms dim. Looking at the formal principles that people accept, the ways in which they specify, rank, and balance such principles, their particular moral commitments, and their concrete problem-solutions and case-analyses, it is clear that disagreements in morality are extensive and profound. This is simply because humans live with different moral perspectives. They may hold the same general principles, but giving different interpretations to them. They may happen to offer the same solution to a problem, but holding different specific moral commitments. All of this leads them to embracing different classical moral narratives and paradigm case-analyses. They do not even hold the same general moral principles. Certainly, as Jonsen and Toulmin observe, people sometimes arrive at the same solutions to particular moral cases even when they do not hold the same moral principles for regulating the cases. However, it is also easy to find cases in which the hope of "different-principles-but-the-same-solution" is totally disappointed. Because people's moral disagreements are extensive regarding all elements of moralities, it is highly improbable that all will convert to the same moral approaches or solutions to bioethical issues. Because their moral disagreements reach the very starting points of morality in their moral perspectives, it is impossible to justify one particular set of moral standards for universal application. Indeed, each substantive bioethical account reflects a particular moral perspective. By illustrating a tripartite interplay among moral perspectives, moral theories and moral accounts, the strategy of disclosing a moral perspective offers us an effective way of understanding the appropriate relationship among individual, community, and society.

Department of Public and Social Administration
City University of Hong Kong, Hong Kong

NOTES

[1] As will be shown in the subsequent sections, I use "interpretation" in a very broad sense. On the one hand, interpretations offer clear explanations to key concepts involved in a moral theory. On the other hand, they may also provide specification, ranking, and balancing of different principles or values covered in a theory.

[2] Indeed, a theory must be general in the sense that it needs to be applicable to different contexts. If it is not general, it cannot guide broadly.

[3] Beauchamp and Childress observe that a normative moral theory commonly refers to "(1) abstract reflection and argument, (2) systematic reflection and argument, and (3) an integrated body of principles that are coherent and well-developed" (1994, pp. 44-45). They mention that they do not claim that their principle-based theory has reached a level suggested by (3). This essay does not have to disagree with their observation. Instead, it assumes that a normative moral theory stays at an abstract level. Even if a moral theory does better than what Beauchamp and Childress' theory and reaches a level suggested by their condition (3), it remains abstract. A moral theory can be as systematic and well-developed as possible, but it will still be content-thin in the sense that it is not able to draw specific conclusions without further interpretations of the theory. For example, Kant's deontology, consisting of the categorical imperative, is well-argued and systematic, but it is not content-rich until Kant offers specific interpretations (by adding particular premises) to the categorical imperative and develops concrete moral standards. However, after such specific interpretations are added to the categorical imperative, what we find is already a moral account substantiated by a moral perspective. The same goes for Beauchamp and Childress' "principlism" as a moral theory.

[4] Just as Thomas Kuhn uses "paradigm" to characterize normal scientific activities and communities (1970, especially "Postcript"), "perspective" is used in this paper to characterize ordinary moral activities and communities. Of course, following Kuhn's example of "paradigm" does not suggest that science and morality are similar practices or disciplines. My notion of "a moral perspective" is only a formal, not substantive, "borrowing" from Kuhn's notion of "a paradigm." But the similarity between the two notions is a recognition that in either science or morality there is something bigger than abstract theories that plays an essential role in guiding activities.

Thomas Sowell proposes the notion of "a vision" to refer to our "pre-analytic cognitive act" (1987, p. 14). According to him, a vision is what we sense or feel before we have constructed any systematic reasoning that could be called a theory. The primary examples of visions he offers are the constrained and unconstrained visions of human nature presented respectively by Adam Smith and William Godwin (pp. 19-25). A vision is an almost instinctive sense of what things are and how they work, while a "paradigm" is much more intellectually developed entity, including scientific law, theory, and application together (p. 204). Hence, my notion of "perspective" is much more like Kuhn's notion of "perspective" rather than Sowell's notion of "vision." Most probably, each moral perspective contains and is developed from (a) vision(s). But a moral perspective is a totality of moral views (shared by a moral community) in which moral theories and general statements are all formulateable.

[5] Moral exemplars provide concrete solutions to problems and/or paradigm case analyses which are worth special attention. Applying moral principles and rules to our moral practice relies on our understanding of these principles and rules. However, our understanding involves more than verbal means. As Michael Polanyi observes, "all knowledge by which

man surpasses the animals is acquired by the use of language," but "the operations of language rely ultimately on our tacit intellectual powers" (1962, p. 95). In other words, our understanding of moral principles and rules is not derived from our ability to give them verbal expression (articulate knowledge). Rather, it is acquired through our tacit intelligence of how to use them (inarticulate knowledge). Since the verbal statements of principles and rules are virtually impotent, our tacit knowledge of how to use them is not learned by reciting such statements. Instead, it is acquired by concrete examples of how they function in our moral practice. Such examples are provided in concrete solutions and/or paradigm case analyses within a moral perspective.

6   Indeed, metaethical beliefs are often implicit in the moral exemplars of a moral perspective. Such beliefs regard the nature of morality, the methods of moral justification, and appropriate moral epistemology. The nature of morality concerns whether morality is objective or subjective, rational or emotional, and relative or absolute. The methods of moral justification concern how to justify moral rules, ideas, judgments, theories, and so on. The often-mentioned methods include self-evident axioms, reflective equilibrium, transcendental approaches, and so on. Moral epistemology cares about what constitutes appropriate moral evidence as well as what are appropriate rules of inference.

7   It is also interesting to take a look at Singer's and Friedman's metaethical beliefs (especially the method of moral justification) involved in their moral exemplars. Singer is strongly critical of popular appeals to a reflective equilibrium as a method of testing and/or establishing moral theories. On his view, the method of reflective equilibrium has analogies with the scientific method. It gives no sense to the idea of a correct theory other than the theory that best fits the data, after the data have been subject to possible revision in the light of plausible theories (1974, p. 494). Singer points out that John Rawls and all his followers are subject to the charge of moral subjectivism because of their appeal to the method of reflective equilibrium.

> If I am right in attributing this version of the reflective equilibrium idea to Rawls, then Rawls is a subjectivist about morality in the most important sense of this often-misused term. That is, it follows from his views that the validity of a moral theory will vary according to whose considered moral judgments the theory is tested against. There is no sense in which we can speak of a theory being objectively valid, no matter what considered moral judgments people happen to hold. If I live in one society and accept one set of considered moral judgments, while you live in another society and hold a quite different set, very different moral theories may be "valid" for each of us. There will then be no sense in which one of us is wrong and the other right (p. 494).

Moreover, Singer argues that Rawls misunderstands Henry Sidgwick as appealing to reflective equilibrium in establishing his utilitarianism. Sidgwick, Singer contends, believes that morality is objective. The validity of a moral theory should depend on the self-evidence of the "primary intuition" and the soundness of the reasoning used in its application, not on whether its results match our considered moral judgments (p. 503). The ultimate appeal is to the carefully considered intuitive judgment of the reader, not one which aims at matching a moral theory with the considered moral judgments either of the reader, or of some widely accepted moral consensus (p. 514).

> [A]ll the particular moral judgments we intuitively make are likely to derive from discarded religious systems, from warped views of sex and bodily functions, or from customs necessary for the survival of the group in social and economic circumstances that

now lie in the distant past. In which case, it would be best to forget all about our particular moral judgments, and start again from as near as we can get to self-evident moral axioms (p. 517).

Accordingly, Singer distinguishes *considered moral (intuitive) judgments* from *self-evident moral axioms*. Our ordinary intuitions or judgments are inevitably distorted by our religious or other prejudices. On the other hand, Singer contends, Sidgwick teaches us that there are "real ethical axioms - intuitive propositions of real clearness and certainty" (Sidgwick, 1962, p. 373). Sidgwick finds three, and only three, genuine axioms: impartiality (whatever action any of us judges to be right for himself, he implicitly judges to be right for all similar persons in similar circumstances), prudence (the idea of impartial concern for all parts of our conscious life), and objectivity ("as a rational being I am bound to aim at good generally, …not merely at a particular part of it") (pp. 510-511). Accordingly, Singer concludes that a better ethical method than that of reflective equilibrium is:

> search for undeniable fundamental axioms; build up a moral theory from them; and use particular moral judgments as supporting evidence, or as a basis for *ad hominem* arguments, but never so as to suggest that the validity of the theory is determined by the extent to which it matches them (p. 517).

Indeed, in Singer's well-known article 'Famine, affluence, and morality' (1972), he is evidently using the strategy of fundamental axioms to build his arguments for our moral obligation to prevent others from starving.

On the other hand, it is not quite clear how Friedman views the method of ethics. He is a utilitarian. It might be misleading for him to claim, as he did, that his view of natural rights is close to Robert Nozick's. Nozick is a deontological libertarian. Natural rights exist because they are right-making conditions, independent of outcome considerations. However, as a utilitarian, Friedman has to base his arguments on utility calculation or consequence consideration to argue for a conception of natural rights. Indeed, he does argue that natural rights correspond fairly closely to the laws of a pure free market society (1991, p. 261).

From this paper's view, Singer's strategy of fundamental axioms is simply illusory. Even if we can all agree on the three Sidgwick-Singerian "genuine axioms" (impartiality, prudence, and objectivity), these axioms cannot produce substantive moral accounts without further specific interpretations. Again, people will secure different interpretations for these axioms from different moral perspectives which in turn lead to different moral accounts.

[8] In working on the National Commission for the Protection of Human Subjects of Biomedical and Behavioral Research in the early 1970s, Jonsen and Toulmin found that although the commissioners argued interminably on matters of ethical principles, they agreed quickly about the morality of cases. This state of affairs inspired them to study the traditional moral strategy of casuistry such as the casuistry manifested in the Raman Catholic tradition. As a result, they proposed the strategy of casuistry for bioethical explorations in competition with the strategy of "principlism" (Jonsen, 1998, p. 82).

REFERENCES

Beauchamp, T. L. and Childress, J. (1994). *The Principles of Biomedical Ethics*, 4th edition. New York: Oxford University Press.

Daniels, N. (1985). *Just Health Care*. Cambridge: Cambridge University Press.

Friedman, D. (1989). *The Machinery of Freedom: Guide to a Radical Capitalism*. La Salle: Open Court.

Friedman, D. (1991). Should medicine be a commodity? An economist's perspective. In: T. J. Bole, III and W. B. Bonderson (Eds.), *Rights to Health Care* (pp. 259-305). Dordrecht: Kluwer Academic Publishers.

Friedman, D. (1996). *Hidden Order: The Economics of Everyday Life*. New York: Harper Business.

Jonsen, A. and Toulmin, S. (1988). *The Abuse of Casuistry: A History of Moral Reasoning*. Berkeley: University of California Press.

Jonsen, A. (1998). *The Birth of Bioethics*. New York: Oxford University Press.

Kant, I. (1981). *Grounding for the Metaphysics of Morals*. J. W. Ellington (trans.). Indianapolis: Hackett Publishing Company.

Kuhn, T. (1970). *The Structure of Scientific Revolutions*, 2nd edition. Chicago: University of Chicago Press.

Nozick, R. (1974). *Anarchy, State, and Utopia*. New York: Basic Books, Inc.

Polanyi, M. (1962). *Personal Knowledge*, Chicago: University of Chicago Press.

Rawls, J. (1971). *A Theory of Justice*. Cambridge: Harvard University Press.

Sidgwick, H. (1962). *The Methods of Ethics*, 7th edition. Chicago: University of Chicago Press.

Singer, P. (1972). Famine, affluence, and morality. *Philosophy & Public Affairs* 1, 229-243.

Singer, P. (1974). Sidgwick and reflective equilibrium. *Monist* 58, 490-517.

Singer, P. (1976). Freedom and utilities in the distribution of health care. In: R. M. Veatch and R. Branson (Eds.), *Ethics and Health Policy* (pp. 175-193). Cambridge: Ballinga Publishing Company.

Singer, P. (1985). Arguments against markets: Two cases from the health field. In: C.L. Buchanan and E. W. Prior (Eds.), *Medical Care and Markets: Conflicts Between Efficiency and Justice* (pp. 2-19). Sydney: George Allen & Unwin.

Sowell, T. (1987). *A Conflict of Visions*. New York: William Morrow and Company, Inc.

Veatch, R. (1981). *A Theory of Medical Ethics*. New York: Basic Books, Inc.

NOTES ON EDITOR AND CONTRIBUTORS

Julia Tao Lai Po-wah, Ph.D., Director, Center for Comparative Public Management and Social Policy; also, Associate Professor, Department of Public and Social Administration, City University of Hong Kong, Hong Kong.

Derrick K. S. Au, M.D., FRCP(Glas.), FHKAM, Chief of Service, Department of Rehabilitation, Kowloon Hospital, Hong Kong.

Gerhold K. Becker, Ph.D., Professor, Department of Religion and Philosophy; also, Research Fellow, Center for Applied Ethics, Hong Kong Baptist University, Hong Kong.

Chan Ho Mun, Ph.D., Associate Professor, Department of Public and Social Administration, City University of Hong Kong, Hong Kong.

Chung-Ying Cheng, Ph.D., Professor, Department of Philosophy, University of Hawaii, USA.

Mary Ann Cutter, Ph.D., Associate Professor, Department of Philosophy, University of Colorado at Colorado Springs, Colorado Springs, Colorado, USA.

H. Tristram Engelhardt, Jr., M.D., Ph.D., Professor of Philosophy, Rice University, Houston, Texas, USA.

Ruiping Fan, Ph.D., Assistant Professor, Department of Public and Social Administration, City University of Hong Kong, Hong Kong.

Anthony Fung, Ph.D., Associate Professor, School of Journalism and Communication, The Chinese University of Hong Kong, Hong Kong.

Corinna Delkeskamp-Hayes, Ph.D., Director of European Programs, International Studies in Philosophy and Medicine, Buchbergstrasse, Freigericht, Germany.

Ian Holliday, Ph.D., Chair Professor, Department of Public and Social Administration, City University of Hong Kong, Hong Kong.

Rev. Fr. Thomas Joseph, Director of Pastoral Research, International Studies in Philosophy and Medicine, North St. Petersberg, Florida, USA.

George Khushf, Ph.D., Humanities Director of the Center for Bioethics and Medical Humanities; also, Associate Professor, Department of Philosophy, University of South Carolina, Columbia, SC 29208, USA.

Shui Chuen Lee, Ph.D., Professor, Director, Institute of Philosophy, National Central University, Chungli, Taiwan.

Sara Marchand, M.A., Consultant, World Health Organization, Geneva, Switzerland; also, Ph.D. candidate, Department of Philosophy, University of Wisconsin, Madison, USA.

Ren-Zong Qiu, Professor, Director, Program in Bioethics, Centre for Applied Ethics, Chinese Academy of Social Sciences, Beijing, PRC.

Hyakudai Sakamoto, Professor Emeritus, Aoyamagakuin University, Tokyo, Japan.

Kurt W. Schmidt, Ph.D., Doctor of Theology, Zentrum fur Ethik in der Medizin am St. Marksu-Krankenhaus (Center for Medical Ethics at the St. Marcus Hospital), Frankfurt, Germany.

Stephen Man-Hung Sze, Ph.D., Principal Lecturer, General Education Center, Hong Kong Polytechnic University, Hong Kong.

Daniel Wikler, Ph.D., Senior Staff Ethicist, World Health Organization, Geneva, Switzerland; also, Professor, Department of Philosophy and Department of History of Medicine, University of Wisconsin, Madison, USA.

Yu Kam Por, Ph.D., Senior Lecturer, General Education Center, Hong Kong Polytechnic University, Hong Kong.

Hu Xinhe, Ph.D., Professor, Director, Department of Philosophy of Science and Technology, Institute of Philosophy, Chinese Academy of Social Sciences, Beijing; also, Professor, Research Center for Philosophy of Science and Technology, Shanxi University, Taiyuan, PRC.

INDEX

Philosophy and Medicine

Philosophy and Medicine

Philosophy and Medicine

39. M.A.G. Cutter and E.E. Shelp (eds.): *Competency. A Study of Informal Competency Determinations in Primary Care.* 1991   ISBN 0-7923-1304-6
40. J.L. Peset and D. Gracia (eds.): *The Ethics of Diagnosis.* 1992
          ISBN 0-7923-1544-8
41. K.W. Wildes, S.J., F. Abel, S.J. and J.C. Harvey (eds.): *Birth, Suffering, and Death.* Catholic Perspectives at the Edges of Life. 1992 [CSiB-1]
          ISBN 0-7923-1547-2; Pb 0-7923-2545-1
42. S.K. Toombs: *The Meaning of Illness.* A Phenomenological Account of the Different Perspectives of Physician and Patient. 1992
          ISBN 0-7923-1570-7; Pb 0-7923-2443-9
43. D. Leder (ed.): *The Body in Medical Thought and Practice.* 1992
          ISBN 0-7923-1657-6
44. C. Delkeskamp-Hayes and M.A.G. Cutter (eds.): *Science, Technology, and the Art of Medicine.* European-American Dialogues. 1993   ISBN 0-7923-1869-2
45. R. Baker, D. Porter and R. Porter (eds.): *The Codification of Medical Morality.* Historical and Philosophical Studies of the Formalization of Western Medical Morality in the 18th and 19th Centuries, Volume One: Medical Ethics and Etiquette in the 18th Century. 1993   ISBN 0-7923-1921-4
46. K. Bayertz (ed.): *The Concept of Moral Consensus.* The Case of Technological Interventions in Human Reproduction. 1994   ISBN 0-7923-2615-6
47. L. Nordenfelt (ed.): *Concepts and Measurement of Quality of Life in Health Care.* 1994 [ESiP-1]   ISBN 0-7923-2824-8
48. R. Baker and M.A. Strosberg (eds.) with the assistance of J. Bynum: *Legislating Medical Ethics.* A Study of the New York State Do-Not-Resuscitate Law. 1995
          ISBN 0-7923-2995-3
49. R. Baker (ed.): *The Codification of Medical Morality.* Historical and Philosophical Studies of the Formalization of Western Morality in the 18th and 19th Centuries, Volume Two: Anglo-American Medical Ethics and Medical Jurisprudence in the 19th Century. 1995   ISBN 0-7923-3528-7; Pb 0-7923-3529-5
50. R.A. Carson and C.R. Burns (eds.): *Philosophy of Medicine and Bioethics.* A Twenty-Year Retrospective and Critical Appraisal. 1997 ISBN 0-7923-3545-7
51. K.W. Wildes, S.J. (ed.): *Critical Choices and Critical Care.* Catholic Perspectives on Allocating Resources in Intensive Care Medicine. 1995 [CSiB-2]
          ISBN 0-7923-3382-9
52. K. Bayertz (ed.): *Sanctity of Life and Human Dignity.* 1996
          ISBN 0-7923-3739-5
53. Kevin Wm. Wildes, S.J. (ed.): *Infertility: A Crossroad of Faith, Medicine, and Technology.* 1996   ISBN 0-7923-4061-2
54. Kazumasa Hoshino (ed.): *Japanese and Western Bioethics.* Studies in Moral Diversity. 1996   ISBN 0-7923-4112-0

Philosophy and Medicine

55. E. Agius and S. Busuttil (eds.): *Germ-Line Intervention and our Responsibilities to Future Generations.* 1998 ISBN 0-7923-4828-1
56. L.B. McCullough: *John Gregory and the Invention of Professional Medical Ethics and the Professional Medical Ethics and the Profession of Medicine.* 1998 ISBN 0-7923-4917-2
57. L.B. McCullough: *John Gregory's Writing on Medical Ethics and Philosophy of Medicine.* 1998 [CiME-1] ISBN 0-7923-5000-6
58. H.A.M.J. ten Have and H.-M. Sass (eds.): *Consensus Formation in Healthcare Ethics.* 1998 [ESiP-2] ISBN 0-7923-4944-X
59. H.A.M.J. ten Have and J.V.M. Welie (eds.): *Ownership of the Human Body.* Philosophical Considerations on the Use of the Human Body and its Parts in Healthcare. 1998 [ESiP-3] ISBN 0-7923-5150-9
60. M.J. Cherry (ed.): *Persons and Their Bodies.* Rights, Responsibilities, Relationships. 1999 ISBN 0-7923-5701-9
61. R. Fan (ed.): *Confucian Bioethics.* 1999 [APSiB-1] ISBN 0-7923-5853-8
62. L.M. Kopelman (ed.): *Building Bioethics.* Conversations with Clouser and Friends on Medical Ethics. 1999 ISBN 0-7923-5853-8
63. W.E. Stempsey: *Disease and Diagnosis.* 2000 PB ISBN 0-7923-6322-1
64. H.T. Engelhardt (ed.): *The Philosophy of Medicine.* Framing the Field. 2000 ISBN 0-7923-6223-3
65. S. Wear, J.J. Bono, G. Logue and A. McEvoy (eds.): *Ethical Issues in Health Care on the Frontiers of the Twenty-First Century.* 2000 ISBN 0-7923-6277-2
66. M. Potts, P.A. Byrne and R.G. Nilges (eds.): *Beyond Brain Death.* The Case Against Brain Based Criteria for Human Death. 2000 ISBN 0-7923-6578-X
67. L.M. Kopelman and K.A. De Ville (eds.): *Physician-Assisted Suicide.* What are the Issues? 2001 ISBN 0-7923-7142-9
68. S.K. Toombs (ed.): *Handbook of Phenomenology and Medicine.* 2001 ISBN 1-4020-0151-7; Pb 1-4020-0200-9
69. R. ter Meulen, W. Arts and R. Muffels (eds.): *Solidarity in Health and Social Care in Europe.* 2001 ISBN 1-4020-0164-9
70. A. Nordgren: *Responsible Genetics.* The Moral Responsibility of Geneticists for the Consequences of Human Genetics Research. 2001 ISBN 1-4020-0201-7
71. J. Tao Lai Po-wah (ed.): *Cross-Cultural Perspectives on the (Im)Possibility of Global Bioethics.* 2002 ISBN 1-4020-0498-2
72. P. Taboada, K. Fedoryka Cuddeback and P. Donohue-White (eds.): *Person, Society and Value.* Towards a Personalist Concept of Health. 2002 ISBN 1-4020-0503-2
73. J. Li: *Can Death Be a Harm to the Person Who Dies?* 2002 ISBN 1-4020-0505-9

KLUWER ACADEMIC PUBLISHERS – DORDRECHT / BOSTON / LONDON